Quantum Theory and the Flight from Realism

This book is a critical introduction to the long history of debate concerning the conceptual foundations of quantum mechanics and the problems it has posed for physicists and philosophers. Quantum theory is widely held to resist any realist interpretation and to mark the advent of a 'postmodern' science characterised by paradox, uncertainty, and the limits of precise measurement. Keeping his own realist position in check, Christopher Norris provides a remarkably detailed and incisive account of the positions adopted by parties on both sides of this complex debate.

In a sequence of closely argued chapters, Norris examines the premises of orthodox quantum theory, as formulated most influentially by Bohr and Heisenberg, and its impact on various later philosophical developments. These include various proposals advanced by W V Quine, Thomas Kuhn, Michael Dummett, Bas van Fraassen, and Hilary Putnam. In each case, Norris argues, these thinkers have been influenced by the orthodox construal of quantum mechanics as requiring a drastic revision of principles which had hitherto defined the very nature of scientific method, causal explanation, and rational enquiry.

Putting the case for a realist approach which adheres to well-tried scientific principles of causal reasoning and inference to the best explanation, Norris clarifies these debates for a non-specialist readership and for students of philosophy, the history of science, and related disciplines. *Quantum Theory and the Flight from Realism* shows very strikingly how work of this kind can contribute to a better understanding of issues in the scientific domain.

Christopher Norris is Distinguished Research Professor in Philosophy at the University of Cardiff. He is the author of many books on aspects of philosophy and critical theory, among them, most recently, *Reclaiming Truth: Contribution to a Critique of Cultural Relativism*, *Resources of Realism: Prospects for 'Post-Analytic' Philosophy*, *New Idols of the Cave: On the Limits of Anti-Realism*, and *Against Relativism: Philosophy of Science, Deconstruction and Critical Theory*. His current research interests are in epistemology and philosophy of science.

Critical Realism: Interventions
Edited by Roy Bhaskar, Margaret Archer, Andrew Collier,
Tony Lawson and Alan Norrie

Titles in the series include:

The Possibility of Naturalism
A Philosophical Critique of the Contemporary Human Science, 3rd edition
Roy Bhaskar

Critical Realism
Essential Readings
Edited by Roy Bhaskar, Margaret Archer, Andrew Collier, Tony Lawson and Alan Norrie

Being and Worth
Andrew Collier

From East to West
Odyssey of a Soul
Roy Bhaskar

Quantum Theory and the Flight from Realism

Philosophical Responses to Quantum Mechanics

Christopher Norris

Routledge
Taylor & Francis Group

LONDON AND NEW YORK

First published 2000
by Routledge
4 Park Square, Milton Park, Abingdon, Oxon OX14 4RN

Simultaneously published in the USA and Canada
by Routledge
605 Third Avenue, New York, NY 10017

Routledge is an imprint of the Taylor & Francis Group, an informa business

Typeset in Baskerville by
Prepress Projects, Perth, Scotland

British Library Cataloguing in Publication Data
A catalogue record for this book is available
from the British Library

Library of Congress Cataloging in Publication Data
Norris, Christopher.
    Quantum theory and the flight from realism : philosophical
    responses to quantum mechanics / Christopher Norris
    Includes bibliographical references and index.
    1. Quantum theory – Philosophy.   2. Realism.   I. Title.
    II. Series.
    QC174. 13.N67   1999
    530.12'01–dc21   99-38146
    CIP

ISBN 13: 978-0-415-22321-8 (hbk)
ISBN 13: 978-0-415-22322-5 (pbk)

For Alison and Geraldine

Contents

Acknowledgements

Once again I should like to thank my colleagues and postgraduate students in the Philosophy Section at Cardiff for their interest in this project over the past two years and for the many occasions when they have offered help, advice, and intellectual stimulus. I was fortunate to have various opportunities, at home and away, for developing these arguments in the form of seminar presentations and guest lectures. Mike Greenhough in the Physics Department at Cardiff provided one such welcome forum while Harry Collins (Sociology) tried vainly to convince me in open debate that there is really no need for philosophy of science – let alone all this talk of 'reality' and 'truth' – once the sociologists have done their stuff. Visits to Canada, Finland, and India came at just the right time since they gave me an opportunity to discuss these issues in a number of exceptionally fruitful exchanges with philosophers, physicists, and cultural historians. I should also like to thank Manuel Barbeito, Robert Barsky, Mike Beaney, Roy Bhaskar, Norman Geras, Alison Hall, Peter Holland, Kathy Kerr, Juha Koivisto, Mikko Lahtinen, Paul Norcross, Pauline Phemister, Chris Philippidis, Garry Potter, Ato Quayson, Dhruv Raina, Duncan Salkeld, Srikant Sarangi, Jon Simon, Shiva Srinivasan, and David Sullivan, all of whom helped to create the occasions for some lively and much-needed dialogue. Michael Coleman invited me to Jyväskylä University in Finland, where I was offered a memorable guided tour of the particle accelerator and assured that the people working with it – designers, technicians, physicists – took a sturdily realist view of these matters. No doubt this is a quick and dirty version of the argument but one that should at least count for something in the context of quantum-theoretical debate.

My work on the book was interrupted for a while by the fresh outbreak of science-wars controversy provoked by the physicist Alan Sokal and his famous (not to say infamous) spoof article 'Transgressing the Boundaries: toward a transformative hermeneutics of quantum gravity'. From our various debates on air and in print I would guess that we are now pretty much agreed in thinking that Sokal was right to speak out against current anti-realist and cultural-relativist fashions, but also that his spoof was picked up and exploited to tactical advantage (without much grasp of the issues) by a good many dubious allies. Anyway, I am grateful to Heidi Rettmaier of the Institute of

Contemporary Arts for giving us a whole evening in which to discuss those issues; also to Boyd Tonkin – literary editor of the *Independent on Sunday* – for allowing me rather more than the allotted space in which to review the book (coauthored by Sokal and Jean Bricmont) that appeared in the wake of that curious affair. I should add, with the benefit of having since read some of Bricmont's essays on the topic, that we are agreed – though for different philosophical reasons – in rejecting the orthodox QM line and its presumed anti-realist consequences.

Otherwise the chief debts incurred during the writing of this book were to Alison, Clare, and Jenny, who as usual had a good deal to put up with. It is dedicated to Alison and Geraldine for reasons that are just about the best in the world but have nothing whatever to do with the philosophy of quantum mechanics.

Cardiff
October 1999

Note
Chapters 1, 4, 5 and 6 are revised versions of material that has previously appeared in the journals *Foundations of Science*, *Southern Humanities Review*, *Inquiry*, and *Philosophical Forum*. I am grateful to the editors and publishers concerned for their permission to reprint this material.

Introduction

In this book I examine various aspects of the near century-long debate concerning the conceptual foundations of quantum mechanics (QM) and the problems it has posed for physicists and philosophers from Einstein to the present. They include the issue of wave–particle dualism; the uncertainty attaching to measurements of particle location or momentum; the (supposedly) observer-induced 'collapse of the wave-packet'; and the evidence of remote superluminal (i.e. faster-than-light) interaction between widely separated particles. I also show in some detail how the orthodox 'Copenhagen' interpretation of QM has influenced current anti-realist or ontological-relativist approaches to philosophy of science, among them the arguments advanced by thinkers such as Michael Dummett, Thomas Kuhn and W.V. Quine. Moreover, there are clear signs that some philosophers – including Hilary Putnam – have retreated from a realist position very largely in response to just these problems with the interpretation of quantum mechanics. So it is important to grasp exactly how the problems arose and exactly why – on what scientific or philosophical grounds – any alternative (realist) construal should have been so often and routinely ruled out as a matter of orthodox QM wisdom.

Perhaps a few personal reminiscences would not be out of place at this point. Eight years ago I moved from the Department of English to the Department of Philosophy in Cardiff, having previously published several books on literary theory that might be construed – so it struck me now – as going along with the emergent trend towards anti-realism and cultural relativism in various quarters of 'advanced' theoretical debate. What brought this home with particular force was the advent of a new postmodernist fashion which seemed to count reality a world well lost for the sake of pursuing its own favoured kinds of hyperreal fantasy projection. The results were evident not only in literary studies – a fairly safe zone for such ideas – but also in other disciplines which had likewise taken the postmodern-textualist turn, among them history, sociology, political theory, and even philosophy of science. So it seemed important to challenge this burgeoning academic trend, especially with regard to its impact on sociology of knowledge and 'science-studies' where cultural relativism had by now established a strong disciplinary hold.

I offer the above brief remarks by way of explaining why an erstwhile literary theorist should have switched to the history and philosophy of science and then, yet more improbably, to conceptual problems in the foundations of quantum theory. For this has been among the most fertile sources for people in the (erstwhile) humanistic disciplines who wish to give 'scientific' credence to their claim that realism is a thoroughly outmoded doctrine which no self-respecting physicist would nowadays endorse. Then there is the range of often far-fetched speculative 'solutions' that QM theorists have produced in response to what they take as the resultant crisis now afflicting all forms of 'classical'-realist or causal-explanatory thought. The so-called 'many-worlds' and 'many-minds' interpretations are among the most widely known since no doubt the most appealing in their sheer ontological extravagance and range of suggestive science-fiction possibility. Elsewhere there is the vague notion that since quantum mechanics is deeply mysterious therefore it must be somehow connected with other such likewise mysterious matters as the nature of consciousness or the possibility of human freewill as against the claims of old-style scientific determinism. Thus one often finds it said that present-day science has abandoned any notion of an objective or mind-independent 'reality' and at last come around to an outlook of full-fledged postmodernist scepticism with regard to such values as truth, objectivity, and method. This thesis can be made to look all the more plausible by citing authorities like Bohr, whose statements often invite such a reading on account of their highly paradoxical quality and fondness for all sorts of far-reaching speculative claims. Indeed a good many fashionable forms of anti-realist and cultural-relativist doctrine take for granted this idea that their position finds support from the latest findings of theoretical physics. Typical of these is Jean-François Lyotard's strangely placid assurance that 'postmodern' science has nothing to do with truth – even truth at the end of enquiry – but everything to do with uncertainty, undecidability, chaos, paralogistic reasoning, the limits of precise measurement, and the observer-dependent nature of (so-called) physical 'reality'.

So it seemed worthwhile – even a matter of some urgency – to examine the source of these ideas and determine how far they had taken hold through a failure (or refusal) to acknowledge the existence of alternative accounts. More constructively, my book presents various arguments in favour of one such alternative, the 'hidden-variables' theory developed since the early 1950s by David Bohm and consistently neglected or marginalized by proponents of the Copenhagen doctrine. This is a version of the pilot-wave hypothesis, first put forward by Louis de Broglie, according to which the particle is 'guided' by a wave whose probability amplitudes are exactly in accordance with the well-supported QM predictions and measured results. Where it challenges the orthodox theory is in Bohm's realist premise that the particle *does* have precise simultaneous values of position and momentum, and furthermore that these pertain to its objective state at any given time, whatever the restrictions imposed upon our knowledge by the limits of achievable precision in measurement. On this basis, I suggest, one can begin to sort out the various deep-laid philosophic confusions – especially that between ontological and epistemological issues – which characterize Niels Bohr's writings on the topic, and which can still be seen in a great many present-day treatments of QM theory. Very often these involve paradoxical claims about the 'unreality' of time, not only within quantum physics and cosmology (e.g. John Wheeler's speculations about observer-induced retroactive causality over billions of light-years' distance), but also in the thinking of anti-realist philosophers – such as Michael Dummett – who deny the existence of verification-transcendent truths with respect to past events other than those (very few) for which we possess adequate documentary warrant. Anti-realism is nowadays a widespread trend among thinkers of various persuasions, from its sophisticated (Dummett-type) logico-semantic form to Putnam's more pragmatic 'internal realist' or framework-relativist version, and – at the farthest extreme – postmodernist ideas about the eclipse of reality and the obsolescence of truth. What these otherwise diverse approaches all have in common, I argue, is a notion of quantum mechanics as having destroyed the case for scientific realism or created such problems with it as to require a radical redefinition of what 'realism' entails, whether in the subatomic or the macrophysical domain. And this despite the well-known paradox of Schrödinger's Cat, which amounts to a *reductio ad absurdum* of that doctrine when extended to the realm of macrophysical objects and events.

These confusions took hold at an early stage in the history of quantum physics (more specifically, in the well-known series of debates between Einstein and Bohr), and cannot be resolved – only deepened or pushed to one side – by adopting the orthodox instrumentalist line. They emerge most clearly in subsequent discussions of the 1935 Einstein–Podolsky–Rosen (EPR) paper, which laid down criteria for a realist interpretation compatible with the known laws of physics, among them those of relativity theory. The EPR argument in turn gave rise to J.S. Bell's equally famous theorem to the effect that any such interpretation – one that entailed the existence of 'hidden variables' – would also entail some highly problematic consequences, including (what Einstein refused to accept) nonlocal effects of quantum 'entanglement' at arbitrary space–time distances. However, as I argue, Bohm's theory is able to accommodate this problem while also maintaining a realist ontology and producing results in accordance with the well-established QM observational results and predictions. Moreover, it avoids the kinds of extravagant conjecture – such as the 'many-worlds' interpretation currently championed by David Deutsch – which take orthodox QM as their basis for proposing a massive (scarcely thinkable) revision to our grasp of what constitutes a 'realist' worldview. In Chapters 4 and 5 I take issue with the premises and the logic of Deutsch's argument, while remarking on the way that it unwittingly repeats whole chapters from the history of pre-Kantian speculative metaphysics.

According to Deutsch, the many-worlds (or multiverse) theory is the sole plausible, i.e. physically and logically consistent solution to the various well-known QM paradoxes of wave–particle dualism, remote simultaneous interaction, the observer-induced 'collapse of the wave-packet', and so forth. According to this hypothesis we must assume that *all possible outcomes* are realized in every such momentary 'collapse' since the observer splits off into so many parallel, coexisting, but epistemically non-interaccessible 'worlds' whose subsequent branchings constitute the lifeline – or experiential world series – for each of those endlessly proliferating centres of consciousness. Deutsch concedes that his multiverse theory is highly counter-intuitive but none the less takes it to be borne out beyond question by the huge observational–predictive success of QM and the conceptual dilemmas that supposedly arise with alternative (single-universe) accounts. Moreover, he claims that this theory resolves a range of long-standing and hitherto intractable philosophic problems, among them the mind–body dualism, the various traditional paradoxes of time, and the freewill versus determinism issue.

I suggest, on the contrary, that Deutsch's argument involves a largely unwitting transposition of speculative themes from the history of rationalist metaphysics into the framework of present-day quantum debate, often with bizarre or philosophically dubious results. Moreover, it discounts at least one highly promising alternative, i.e. Bohm's 'hidden variables' theory, which offers a realist interpretation perfectly consistent with the full range of QM predictive–observational data. I then consider various possible reasons for the resistance to Bohm's theory among proponents of the 'orthodox' (Copenhagen) version and also for the strong anti-realist, at times irrationalist, bias that has characterized much of this debate since Bohr's well-known series of exchanges with Einstein. Chapter 5 concludes by pointing out some relevant contrasts between Deutsch's ontologically extravagant use of the many-worlds hypothesis (one that bears a close though unacknowledged kinship to the thought of speculative metaphysicians from Leibniz to David Lewis) and those realist modes of counterfactual reasoning – e.g. in Kripke and the early Putnam – which deploy similar resources to very different causal-explanatory ends.

In more general terms, my book makes the case for an alethic (objective, truth-based and verification-transcendent) conception of realism, as opposed to the epistemic conception which on principle denies the possibility of truths beyond reach of our present-best knowledge, evidence, or powers of observation. This latter viewpoint has dominated much of the debate about quantum mechanics, not only among orthodox theorists but also among those – including, arguably, the EPR authors – who have sought to defend a realist interpretation. Indeed, it was just this ambiguity in the EPR paper which gave a hold for the apparently decisive counter-arguments mounted by Bohr and his followers. I show how the orthodox (instrumentalist) stance gave way to a strain of dogmatic thinking which on the one hand refused to admit any question of the reality 'behind' or 'beyond' QM appearances, while on the other it effectively raised this refusal to the status of a full-scale metaphysical creed with distinct irrationalist leanings. In short, the philosophy of quantum mechanics has remained in a state of Kuhian 'crisis' for the past six decades and more compared with the theory's remarkable success in matters of applied technological progress and predictive–observational warrant. If anything, the situation is now more confused – as a result of Bell's theorem and its subsequent experimental proof – than when Planck and Einstein first proposed the quantum hypothesis in response to various anomalies encountered with phenomena such as black-body radiation and wave–particle dualism.

However, this gives all the more reason to think that the orthodox theory is indeed 'incomplete' in some crucial respect, and that Bell was justified – despite his own results – in holding out for a possible realist solution along the lines suggested by Einstein and Bohm. Such an argument will gain additional weight if one accepts the 'classically' well-established principles of causal reasoning and inference to the best (most adequate) explanation. To interpret QM – on the orthodox account – as having somehow undermined those principles can scarcely be warranted given its conspicuous failure to resolve the kinds of problem pointed out by physicists, like Einstein and Schrödinger, who had themselves made decisive contributions to the theory at an early stage, but who later became deeply dissatisfied with the Copenhagen version. Still less can philosophers be justified when they invoke these unresolved problems in support of a programmatic anti-realism extending far beyond the specialized domain of quantum-theoretical debate. Thus it is *preposterous* in the strict sense of that term – an inversion of the rational order of priorities – when thinkers claim to draw far-reaching ontological or epistemological lessons from a field of thought so rife with paradox and lacking (as yet) any adequate grasp of its own operative concepts. At any rate, there is something awry about a theory that has exerted such widespread influence while effectively raising incomprehension to a high point of orthodox principle. My book seeks to clarify these issues for the benefit of philosophers with a interest in theoretical physics and for physicists willing to consider philosophical questions that are often ignored or declared off-bounds in standard treatments of the topic.

1 Is it possible to be a realist about quantum mechanics?

I

Recent years have seen something of a growth industry in books on the topic of quantum mechanics, some of them unabashedly populist while others – often written by practising physicists – are pitched toward a 'serious' yet non-specialist readership.[1] Much of this writing is highly impressive in its ability to convey quantum-theoretical ideas in a language that somehow overcomes the resistance created by a range of distinctly problematic and counter-intuitive arguments. Not that intuition is by any means a reliable guide in these matters. After all, many crucial advances in mathematics, geometry and physics over the past two centuries and more – indeed right back to Copernicus and Galileo – have involved a decisive (often difficult) break with certain kinds of common-sense knowledge, intuitive self-evidence, or supposed a priori truth.[2] As Peter Holland remarks, '[t]he concept of "intuition" is like that of "human nature": it is a function of history and not eternally frozen. The notion that a body persists in a state of uniform motion unless acted upon by a resultant force would be counter-intuitive to Aristotle but natural for Galileo. Quantum phenomena require the creation of quantum intuition.'[3] Yet Holland himself writes from a realist standpoint and as one who firmly rejects the orthodox view – orthodox at least among many quantum physicists and philosophers of science – that whatever the notional reality 'behind' those phenomena it cannot be grasped, described, or represented in conceptual-intuitive terms. Such is the peculiar challenge of quantum mechanics, one that emerged during the early decades of this century and which continues to generate deep and widespread disagreement. My aim in this book is partly to clarify the various philosophic issues involved and to show how they have often been misunderstood by parties on both sides of the realism–anti-realism dispute. But it is also to argue – more constructively – that the case for realism with regard to quantum mechanics is a great deal stronger than is commonly thought by proponents of the received view and likewise by non-specialist readers whose grasp of those issues is very largely shaped by that same orthodox consensus.

What most interested lay persons will have gathered from the current literature can perhaps best be summarized as follows. (1) QM has given rise to a number of problems and paradoxes – among them the wave–particle dualism – as regards physical 'reality' and the kinds or degrees of exactitude in scientific knowledge that we can hope to gain concerning it. (2) Those problems have to do with certain limits that apply to the detection or measurement of quantum phenomena, such as the impossibility of assigning precise simultaneous values of location and momentum, or the fact – famously enshrined in Heisenberg's uncertainty principle – that any observation of subatomic particles, for instance through an electron microscope, will involve their exposure to a stream of other such energy-bearing particles and will thus affect or in some sense determine what is 'actually' there to be observed. As for the quantum paradoxes (3), these take rise from the necessity, as it seems, of abandoning local realism (i.e. Einstein's rule that no causal influence can propagate faster than the speed of light) in favour of remote superluminal interaction between particles at no matter how great a distance.[4] For there is now a large body of experimental evidence that such nonlocal effects can indeed be shown to exist and that any realist interpretation will consequently need to take them on board, thus creating additional (some would say insoluble) problems for its own case.

That is, one can take a singlet-state pair of particles whose combined angular momentum is zero and then project them on divergent paths towards two detectors or measuring devices (in this case Stern–Gerlach magnets) set up to determine their spin-value with respect to some given orientation. Thereafter, if a measurement is carried out on particle A and produces the value 'spin-up $= +\frac{1}{2}$' for a given parameter, then any measurement conducted simultaneously on particle B will produce the inverse value 'spin-up $= -\frac{1}{2}$'. (Of course they might yield any range of likewise anti-correlated 'up' or 'down' spin-values depending on the polarization component which the device was set to detect.) This follows from orthodox quantum theory but also from the classical law concerning the conservation of energy as applied to angular momentum. In other words, it is known in advance that the two particles will always yield a sum-zero value for any given parameter if measured at any point in their trajectory and whatever the extent of space–time separation between them.[5]

So far there is nothing in the least paradoxical about this situation. After all, it is analogous to the case in which one tears a playing-card in two and sends each piece to a geographically remote correspondent, one of them (say) in London and the other in Christchurch, New Zealand. If they are aware by pre-arrangement of what's going on, each will know with full certainty which half the other has received as soon as they examine the content of their own package. Where the paradox shows up is with the further requirement – again as specified by orthodox QM – that any results thus produced with respect to either particle will depend upon the kind of measurement carried out, i.e. the setting of the spin-detector and hence the particular outcome in this or that case. Moreover, that result will decide the outcome of any measurement which might be performed simultaneously on the *other* particle, since it follows – by the inverse-correlation rule – that this must always be the case for quantum-mechanical systems (or particle pairs) that have a common source or which have interacted at some previous stage.

But then, what precisely can be meant by the terms 'simultaneous' and 'previous', as used in the foregoing sentence or in any attempt to describe or explain what is happening here? For it also follows from orthodox QM that these events must transpire in a space–time framework that permits violations of special relativity, or which allows for superluminal (faster-than-light) interaction between particles at any distance from each other. In which case there can be no appeal to Einstein's principle for establishing simultaneity relative to the speed of light, the latter taken as an absolute limit on causal propagations of whatever sort.[6] Some commentators – Maudlin among them – have argued that this need not be the case since special relativity only requires that any space–time metric be Lorentz-invariant, which on a certain construal might allow for the existence of superluminal transmission.[7] All the same, there is clearly a marked tension (if not perhaps a downright inescapable conflict) between the orthodox interpretation of quantum mechanics and Einsteinian relativity theory. Moreover, any talk of 'previous' states or events – such as the particles' orientation when separated at source or the spin-values that might have been measured at some 'earlier' stage in their trajectory – is likewise rendered highly problematic. That is to say, it takes for granted the impossibility that those events could somehow be affected – or those measurements somehow retroactively determined – by whatever occurs at a 'later' stage in the system's space–time evolution.

Such were some of Einstein's chief objections in the famous series of debates with Niels Bohr, when he argued that the orthodox (Copenhagen) theory of quantum mechanics was necessarily 'incomplete' since it entailed the existence of unthinkable phenomena such as instantaneous remote correlation or 'spooky action-at-a-distance'.[8] Although he had been among the chief contributors to the early development of quantum mechanics, Einstein was by now deeply dissatisfied with what he saw as its failure to provide any adequate realist or causal-explanatory account of QM phenomena. This change of mind went along with his shift from a broadly positivist (or instrumentalist) approach according to which a scientific theory need achieve no more than empirical-observational and predictive accuracy to a realist position that entailed far more in the way of express ontological commitment. Hence the highly charged character of Einstein's debates with Bohr, addressed as they were to such fundamental issues such as the limits of precise measurement, the observer-independent status (or otherwise) of physical reality, and the extent to which quantum theory entailed a radical break with existing ideas of scientific method and truth.

Thus Einstein maintained that orthodox QM was demonstrably 'incomplete' in so far as it failed in the basic task of providing a description of quantum phenomena that was consistent with the full range of observational–predictive results while also explaining those results in terms of a credible realist ontology and an account of the underlying causal mechanisms that produced them. Since the doctrine as it stood offered no such account – since it refused on principle to venture beyond the empirical evidence so as to avoid certain highly paradoxical or counter-intuitive consequences with regard to the supposed reality 'behind' quantum-phenomenal appearances – therefore (he argued) it fell far short of the requirements for an adequate physical theory. To Bohr's way of thinking, conversely, orthodox QM was indeed 'complete' in all basic respects, and any problems had to do with the limits of our classical-realist concepts and categories when applied to quantum mechanics. Only by adopting an empiricist approach – one that sensibly acknowledged those limits and resisted the temptation to speculate on matters beyond its conceptual grasp – could thought be prevented from creating all manner of needless problems, dilemmas, or antinomies. Thus Bohr's philosophy of science can be seen as a mixture of Kantian and pragmatist themes, one that confines knowledge to the realm of phenomenal appearances while quantum 'reality' is taken as belonging to a noumenal realm that lies beyond reach of any concepts we can frame concerning it, and which thereby justifies the pragmatist equation of truth with what effectively counts as such for all practical (predictive-observational) purposes.

This is why Bohr disagreed so sharply with Einstein on the issue of whether the orthodox theory might yet turn out to be 'incomplete', or to leave room for some future advance that would reconcile quantum mechanics with the aims and methods of classical physics, including – most importantly in this context – the special and general theories of relativity. For one major problem with orthodox QM was that it seemed to entail the existence of nonlocal simultaneous (faster-than-light) 'communication' between particles that had once interacted and then moved apart to whatever distance of space–time separation. This problem arose – ironically enough – as a consequence of Einstein's last and most determined effort to refute Bohr on the measurement issue and to show that one could, at least in principle, obtain a full range of precise values for every component of the system. After all, it followed from orthodox QM (as well as from the classical conservation laws) that if two particles had once interacted and at that time possessed a sum-zero joint angular momentum then their combined angular momentum at every time thereafter – no matter how far from the point of interaction – would always necessarily be zero. In which case, Einstein reasoned, one could obtain a value for some given parameter (e.g. spin-component) on particle A of the separated pair and know for sure *without conducting any physical measurement on it* that particle B would possess an anti-correlated value for that same parameter. Meanwhile, one could carry out a physical measurement for the other parameter on particle B and thus establish – again by the conservation rule – a precise anti-correlated value for particle A. In other words, contrary to orthodox QM fiat, there was no reason in principle why one should not assign determinate (objective) values to every parameter despite Heisenberg's uncertainty principle and the limits it placed on our capacity for physically observing or measuring those same values.

The crux of these debates – to Einstein's way of thinking – was not so much the epistemological issue with regard to the problems of quantum observation–measurement but rather the *ontological* issue of whether such values could be thought to exist independently of the given experimental set-up or means of obtaining observational results. What he refused to accept in the orthodox (Bohr–Heisenberg) account was its idea that those results were actually produced – along with any notional quantum 'reality' beyond or behind appearances – by the very act of observation or the particular localized or momentary choice of measurement parameter. This seemed to Einstein a gross dereliction of basic scientific principles and one which effectively opened the way to all manner of pseudo-scientific speculation. Worst of all, it abandoned the belief in objective (observer-independent) truth and replaced it with the instrumentalist notion that truth *just was* whatever could be known from some partial perspective imposed upon us by the limits of our current observational means, technological resources, or powers of descriptive and conceptual-explanatory thought. Thus Einstein's final response to Bohr – written up jointly with his colleagues Podolsky and Rosen, and thereafter known as the 'EPR paper' – took the form of this classic thought experiment which claimed to establish the existence of objective values for all components of a quantum system, and hence the error of supposing that the empirical limits of observation–measurement were also the limits of quantum 'reality' so far as we could possibly conceive it. To confuse these issues, so Einstein believed, was a category mistake of the worst sort since it left one with the choice between a doctrinaire empiricism that blocked any adequate (causal-explanatory) grasp of quantum phenomena or, on the other hand, a philosophy of quantum physics that could easily fall prey to all kinds of paradoxical, speculative, or even irrationalist and quasi-mystical ideas.

Hence Einstein's series of attempts to prove that Bohr had ignored certain crucial factors which, if taken into account, would avoid the quantum paradoxes and deliver an alternative construal consistent with local realism and relativity theory. Yet at each stage Bohr produced yet more ingenious arguments showing – or purporting to show – that Einstein had himself overlooked some further, strictly unavoidable problem concerning (for instance) the limits of precise measurement, the impossibility of obtaining simultaneous independent values for both particles, or the lack of any shared (space–time invariant) coordinate system against which to determine their supposed trajectory from one measurement to the next. Einstein had failed to reckon with the nonlocal character of quantum interactive systems as required by orthodox ('Copenhagen') QM. Thus, he assumed that any causal influence – or any passage of mysterious 'forces' between particles – would have to occur within the framework of special relativity according to which nothing could propagate faster than the speed of light. However, it was just in order to accommodate the QM prediction of phenomena such as these that Bohr came up with his series of arguments against the possibility of a 'classical' (i.e. a local-realist and space–time invariant) interpretation of the evidence. In each case, he countered, Einstein had been working on assumptions which failed to carry across from the macro- to the microphysical domain, among them the separability principle and the putative existence of discrete measurable values for each particle.

There is an irony here which has not been lost on defenders of the standard (Bohr-derived) Copenhagen view. Einstein's purpose was to prove that orthodox QM theory *must* be 'incomplete' since it entailed consequences that went clean against any physically and logically consistent interpretation of the evidence. But the upshot of all his strenuous endeavours – so the story runs – was to demonstrate the strictly inescapable conflict that arose between quantum mechanics and a 'classical' worldview based on Einstein's conception of local realism. Thus, according to orthodox QM, there could be no possible procedure – given the uncertainty relations established by Heisenberg – for obtaining precise simultaneous values of particle location and momentum. However, as we have seen, Einstein countered with the EPR challenge: why not perform one kind of measurement on particle A of a separated pair and the other kind of measurement on particle B? This would get around the uncertainty problem precisely by appealing to orthodox QM with its theory of remote simultaneous anti-correlation. That is to say, the experimenter could determine *both* values for *both* particles by measuring each with respect to just one value (either location or momentum), and hence deducing what must be the case with regard to the other. However, Bohr responded to all these arguments by pointing out certain unnoticed complications in the proposed experimental set-up, factors which entailed an element of doubt as to whether the results could indeed be attained (as Einstein would have it) by precise measurement of objective properties and values, or whether – as Bohr claimed – they still left room for some observer-induced interference effect.[9]

Thus EPR was intended to have the force of a classic *reductio ad absurdum* argument directed against the very premises and logic of orthodox QM theory.[10] In Alastair Rae's succinct formulation:

> [t]his showed how quantum physics requires that a property, such as the polarization of a photon, could be measured at a distance by measuring the polarization of a second photon that had interacted with the first some time previously. If it is inconceivable that this measurement could have interfered with the distant object, it follows that the first photon must have possessed the measured property *before* the measurement was carried out. As the property measured can be varied by the experimenter adjusting the distant apparatus, EPR [Einstein, Podolsky and Rosen] concluded that all physical properties (in our example values of polarizations in all possible directions) must be 'real' *before* they are measured, in direct contradiction to the Copenhagen interpretation.[11]

However, this is where the irony finally struck home according to Bohr and his followers. For if indeed it is the case – as claimed by orthodox QM – that the particles will *always* exhibit anti-correlation no matter what sort of measurement is made (e.g. with respect to just which of their various spin-components), then the state of B at any given time must depend upon the momentary choice of parameter for measuring A or vice versa, rather than resulting from its intrinsic properties, causal history, 'real' spin-value, or whatever. For Einstein the whole purpose of the EPR thought experiment was to show that such properties *must* exist – and that such values *could* in principle be known or determined – quite apart from the inherent vagaries attaching to this notion of quantum measurement. After all, what sense could it make to talk of 'measuring' the location, momentum, or spin-component of a particle if any values thus arrived at were entirely an artefact of the measurement process itself? For Bohr, on the other hand, this simply went to show that Einstein and his colleagues had not yet grasped the extent to which quantum mechanics undermined their entire 'classical' worldview.

Einstein put the case for ontological realism in a well-known statement concerning the EPR proof. 'If, without in any way disturbing the system, we can predict with certainty (i.e., with probability equal to unity) the value of a physical quantity, then there exists an element of physical reality corresponding to this physical quantity.'[12] To which Bohr replied – as before – by rejecting Einstein's basic realist postulate that any such prediction could be made (or any such measurement performed) 'without in any way disturbing the system'. But he also came up with an additional argument which seemed to preclude any possible revision or adjustment of quantum theory so as to accommodate Einstein's thesis. For, ironically enough, it was the EPR paper – or a problematic aspect of it as remarked upon by Bohr – that was widely thought to have undermined the case for any consistent construal of quantum mechanics in keeping with a local-realist ontology or worldview. Thus, according to Bohr, it followed from the basic principles of quantum physics that any act of observation–measurement carried out on particle A of the separated pair would actually in itself *decide or determine* the result thus achieved, rather than establish an 'objective' (observer-independent) state of that particle which could then – at least in principle – be fully accounted for in causal-explanatory terms, i.e. as a result of its previous interactions, its consequent range of (measured or deduced) locations, momenta, spin-orientations, and so forth. Quite simply – though to Einstein unthinkably – the act of measurement was what *brought it about* that the particle 'possessed' this or that value, a value that in no way pertained to it prior to the choice of measurement parameter or setting of the spin-detector.

Thus, according to Bohr, the EPR thought experiment had in fact come up with the strongest evidence yet for abandoning any form of 'classical' (local) realism and acknowledging the existence of remote simultaneous (faster-than-light) particle interaction. For if the EPR thesis held good and was yet to be rendered compatible with basic QM theory, then surely it must follow that the act of observation–measurement on particle A determined not only that particle's value for any given parameter but *also the value for particle B* at precisely that moment or precisely that point in its space–time trajectory. And, moreover, since any pair of values thus obtained must be thought of as depending on the kind of measurement performed (e.g. the spin-detector setting) then it also followed – *contra* the local-realist precept of Einstein and his colleagues – that any adequate theory had to make room for the instant propagation of observer-induced effects over arbitrary space–time distances. 'Common cause' or 'in-the-source' explanations were ruled out by the fact that these effects were produced by momentary settings (or switchings) of the measurement apparatus and could therefore not be traced back to some antecedent causal history. In which case, of course, there was no escaping the conflict between quantum mechanics and the central claim of special relativity, i.e. that nothing could travel faster than light since this was the absolute invariant value with reference to which one had to assign all particular (localized) space–time coordinates and frameworks.

In short, the upshot of EPR was to pose this whole issue between Einstein and Bohr in the sharpest possible terms. *Either* there was something fundamentally wrong with the quantum theory, something that went beyond differences of interpretation and required that every previous advance in the field – such as Einstein's 1905 theory of photons, or light-quanta – should now be subject to wholesale revision. *Or* (as it seemed to Bohr) Einstein would have to abandon his ground, accept these unwelcome consequences of the EPR case, and acknowledge the 'completeness' of orthodox QM in so far as it precluded any viable alternative account. So the only line of argument open to those who rejected the orthodox (Copenhagen) approach was one that would somehow need to make room *within a realist ontology* for such 'realistically' unthinkable phenomena as superluminal remote interaction or nonlocal causality. In short, they would do much better to adopt the empiricist line of least resistance and give up the quest for a theory that could only be had at such (to them) unacceptable cost.

Thus Einstein was faced with a conceptual dilemma in the strictest sense of that term. On the one hand, special relativity required that causal influences could not be propagated faster than the speed of light, in which case he would need to explain – impossibly – how and why those correlations occurred (as predicted by orthodox QM) in the absence of any such 'spooky' superluminal force. Then again, he might adopt the alternative view that the measuring apparatus (i.e. the spin-detector) exerted some influence on the quantum 'system' so that the appearance of remote simultaneous interaction between particles was an artefact of the experimental set-up, and could thus be interpreted as posing no threat to local realism or the separability-principle. Yet this argument would plainly be at odds with Einstein's more basic realist conviction, i.e. his insistence that measured values must pertain to the objective, observer-independent properties of physical systems, rather than resulting – as the orthodox theory would have it – from the act of observation or the kind of measurement carried out. So if these were indeed the only alternatives, then either way it seemed that Einstein's position was in conflict with quantum mechanics. And since the latter was so strongly borne out by the best observational evidence to hand, then surely it must follow, according to Bohr, that Einstein was wrong in striving to uphold any version of the classical realist theory with respect to quantum phenomena.

So the EPR paper had failed in its purpose – so the orthodox community maintained – with respect to the supposed 'incompleteness' of quantum theory as currently understood. That is, it had shown that QM required the existence of faster-than-light interaction between widely separated particles and that this went against all the known 'laws' of classical physics, as well as contravening special relativity. Although clearly intended as a *reductio ad absurdum*, Einstein's argument could none the less be seen as proof that there existed no possible interpretation of quantum mechanics that would satisfy *both* the well-established quantum results *and* the requirements of a local-realist ontology as laid down by Einstein in his series of dialogues with Bohr. In which case, local realism would have to go – at least as concerned subatomic phenomena – since it came into conflict with a quantum theory whose predictive power, empirical warrant, and sheer formal elegance were such as to justify even the most far-reaching changes to our basic (commonsense-intuitive) notions of reality.

II

This whole debate was given fresh life – and the issues considerably sharpened – with the publication in 1964 of a paper by J.S. Bell, which specified precisely the conditions that would have to be met (and the consequences that would have to be taken on board) by any realist theory which sought to avoid such a conflict with orthodox QM.[13] I must here summarize some highly complicated arguments and hope that the interested reader will pursue them through the relevant source material. What Bell showed by application of an ingenious statistically based proof was the fact that no 'hidden-variables' theory – that is, no theory premised on the existence of some unknown property or deep further fact with respect to quantum phenomena – could *both* satisfy the QM predictions *and* avoid the postulate of nonlocal interaction or remote superluminal 'action-at-a-distance'. The hidden-variables theory was developed by David Bohm who agreed with Einstein that orthodox QM was 'incomplete' since it failed to deliver an adequate ontology in keeping with the basic principles of scientific realism.[14] More specifically, it failed to explain just how and why the wavefunction 'collapsed', i.e. underwent the crucial change from a wave-like distribution of probabilities in Hilbert space to a determinate wave or particle form as required (or perhaps brought about) by the localized act of observation–measurement. Hence all the well-known conceptual problems – most graphically figured in the 'superposed' alive-

and-dead predicament of Schrödinger's cat – that arose when physicists tried to explain at what point that transition occurred, and whether it involved the conscious intervention of a human (or maybe feline) observer.[15]

It was chiefly in order to avoid these problems that de Broglie had proposed his pilot-wave theory, according to which the wavefunction did not provide a complete description of the quantum system but rather acted as a guide for the particle, thus allowing the assignment of determinate values – e.g. of location or momentum – at every stage in the process.[16] This account would be compatible with orthodox QM in so far as it required that those values be established by use of Schrödinger's equation and the standard quantum formalism. However, it would break with that theory by maintaining that any results thus achieved had to do with objective properties or coordinates of the particle, rather than taking on this or that value as and when subject to measurement. Thus, the limits laid down by Heisenberg's uncertainty principle should be viewed as epistemological in nature, that is to say, as pertaining to our limited powers of precise observation, rather than construed ontologically as somehow pertaining to whatever it is in quantum-physical 'reality' that eludes such classical treatment.

On Bohm's account (following de Broglie) those limits apply *only if* one accepts the Bohr–Heisenberg theory according to which there is just no way of assigning such objective values, at least as concerns the subatomic domain where the wavefunction specifies whatever may be known or reliably predicted concerning quantum phenomena. Otherwise it will seem that the theory as it stands is most probably incomplete with regard to some explanatory factor – some as yet undiscovered 'hidden variable' – that would yield both a realist interpretation and a means of resolving the various quantum paradoxes. Peter Holland puts this case in the following passage from his book *The Quantum Theory of Motion*, by far the most detailed and vigorous defence of Bohm's theory in recent years.

> The fact that the centre of a packet moves along a well-defined orbit as if it were a particle of mass *m* does not demonstrate that there *is* such a particle pursuing that orbit. It is only in the causal interpretation that we can consistently claim that the classical-like motion of a packet when dispersion may be neglected is, in fact, the approach to the classical limit, since one starts by assuming the particle trajectory.[17]

That is to say, it would avoid the single most problematic feature of orthodox QM, the issue of where – and by what kind of agency – the transition occurs from a state of superposed (e.g. wave *and* particle) probabilities to a state where the wavefunction has 'collapsed' so as to produce a determinate (wave *or* particle) measurement. For it is this problem that has lately given rise to the most extraordinary flights of quantum-theoretical conjecture, among them the so-called 'many-minds' and 'many-worlds' interpretations, both of which entail some far-reaching (not to say mind-boggling) revisions to our basic concepts of physical reality.[18] At any rate – before those interpretations were proposed – it was already the firm belief of Einstein, Schrödinger, de Broglie, and Bohm that there must be a better, more adequate account that would show the dilemmas to have had their source in some defect of the orthodox theory.

It may be useful at this point to quote at greater length from the EPR paper so as to clarify the main issues and provide a more specific context for discussing Bell's theorem and its implications. 'In a complete theory', the authors maintain,

> there is an element corresponding to each element of reality. A sufficient condition for the reality of a physical quantity is the possibility of predicting it with certainty, without disturbing the system. In quantum mechanics, in the case of two physical quantities described by non-commuting operators, the knowledge of one precludes the knowledge of the other. Then either (1) the description of reality given by the wave function of quantum mechanics is not complete or (2) these two quantities cannot have simultaneous reality. Consideration of the problem of making predictions concerning a system on the basis of measurements made on another system that had previously interacted with it leads to the result that if (1) is false then (2) is also false. One is thus led to conclude that the description of reality as given by a wave function is not complete.[19]

However, it was in response to just such arguments – mostly inspired by the EPR paper – that Bell came up with his provocative theorem concerning the interlinked phenomena of particle spin and quantum nonlocality. What he showed, in brief, was that 'no hidden-variable theory which preserves locality and determinism is capable of reproducing the predictions of quantum physics for the two-photon experiment'.[20] Bell's reasoning to this conclusion may be summarized as follows. (1) EPR excluded the idea of remote simultaneous causal interaction, or 'spooky action-at-a-distance'. (2) This entailed – *contra* Bohr – that it could not be the act of measurement carried out on particle A that somehow influenced or determined the state of particle B as measured. (3) However, it was always possible to take different readings on particle A or vary the parameter so as to produce a whole range of different measurements for different spin-components. In which case, (4) it followed from the QM anti-correlation rule that particle B would always be found to possess precisely the opposite value for any given parameter at any given time. But it also follows – from (1) and (2) above – that this cannot be a matter of some causal influence or remote linkage between the two particles that produces the observed results. Thus, (5) according to EPR, it must be the case that each particle possesses an entire range of objectively existent properties (i.e. values for every parameter) *before* any measurement is carried out and *quite apart* from the particular experimental set-up that produces this or that measured result.

Yet it is at just this point, so Bell maintains, that the EPR argument runs into trouble. For if it is true – as required by quantum mechanics – that anti-correlation will always obtain no matter which parameter is 'chosen' (that is to say, no matter what result is produced by insertion of a spin-detector that 'decides' between various possible outcomes), then any momentary change of measurement setting for particle A will also momentarily decide the outcome for particle B were a measurement performed upon it with respect to the same spin-component or parameter. Moreover, this consequence is all the more difficult to take on board if one subscribes to an EPR-type hidden-variables theory which endorses local realism. Such a theory rejects any notion of the two particles as forming a quantum-mechanical system wherein both values are jointly affected by an act of measurement on either. But it is then confronted with the problem of explaining 'realistically' just how – on what alternative construal – those particles can somehow exhibit the properties predicted by QM and overwhelmingly confirmed by experiment. For, as we have seen, common-cause (or 'in-the-source') explanations cannot cope with the QM requirement that anti-correlation must be somehow brought about by momentary switchings of the measurement apparatus quite aside from any previous causal history pertaining to the two particles. Thus, any hidden-variables theory will need to make room for quantum nonlocality, at least in so far as it accepts those results as being operationally valid. And this problem is sharpened by the fact that, in keeping with its own realist criteria, there must be some objective (non-measurement-dependent) property of the particles which underlies and explains QM phenomena. For of course such a theory cannot have resort – like Bohr's purely instrumentalist account – to the argument that nothing more is required in the way of ontological commitments or depth-explanatory hypotheses.[21]

On this view the hidden-variables theory was a piece of otiose 'metaphysical' baggage which produced all sorts of unnecessary problems with an otherwise perfectly adequate method for performing the relevant calculations. Any question about the 'reality' underlying quantum results or measurements was a question that need not (and should not) be raised, given the impossibility – in Bohr's view – of finding an adequate descriptive language or conceptual framework. As Euan Squires puts it:

> the Copenhagen interpretation and the prevailing fashion in philosophy, which inclined to logical positivism, were mutually supportive. The only things that we are allowed to discuss are the results of experiments. We are not allowed to ask, for example, which way a particle goes in the interference experiment. The only way to make this a sensible question would be to consider *measuring* the route taken by the particle. This would give us a different experiment for which there would not be any interference. Similarly, Bohr's reply to the alleged demonstration of the incompleteness of quantum theory, based on the EPR experiment, was that it was meaningless to speak of the state of the two particles prior to their being measured.[22]

This claim was reinforced by Bell's demonstration of the problems which confronted any hidden-variables theory that also subscribed to a local-realist ontology. For such a theory would always necessarily entail a 'violation of Bell's inequality': a far greater (more precisely predictable) degree of anti-correlation between separated particles than could possibly occur were it not for the existence of some causal link – some system of remote simultaneous interaction – which ensured that the measurements would turn out in accordance with the standard quantum predictions. Yet of course it was just this point that Einstein and his EPR co-authors had seized upon as proving that orthodox quantum mechanics *must* be in some sense 'incomplete' if it required the introduction of far-fetched hypotheses at odds with the most basic principles of scientific realism, not to mention those of special relativity. Bell brought the issue to a head by devising an ingenious thought experiment – along with a rigorous mathematical proof procedure – which specified the conditions that would have to be met by any theory consistent with the evidence. His results have most often been taken as supporting Bohr's, rather than Einstein's position with regard to the EPR paper and its bearing on the quantum 'completeness' issue. Thus, according to Squires, 'any theory which is local must contradict some of the predictions of quantum mechanics', so that '[t]he world can either be in agreement with quantum theory or it can permit the existence of a local theory; both possibilities are not allowed'.[23] In which case – so it is often inferred – the orthodox (Copenhagen) interpretation must be right since there exists such a weight of statistical evidence in its favour. More precisely, the hidden-variables theory will lose much of its intuitive appeal if the promise of a more 'complete' (i.e. causal-realist) explanation has to be offset against the heavy cost of abandoning the EPR locality claim and thus readmitting 'spooky action-at-a-distance'.

As I have said, Bell's results were originally obtained by devising a suitable thought experiment – a variation on the EPR set-up – and then applying mathematical techniques in order to establish the strictly inescapable conflict between quantum mechanics and local realism. At this point it is worth going into more detail as to just how his reasoning differed from that of the EPR authors and just why it posed what many have thought to be a strictly unanswerable challenge to the realist case. One major difference in the Bell set-up is that the two polarizers (i.e. detectors or measuring devices) are arranged obliquely, not in parallel. That is to say, they are *neither* perfectly aligned *nor* set at precise right angles, in both of which EPR-type cases the existence of remote anti-correlation between particles could still be put down to some common-cause factor or explained in terms of their previous interaction. After all, as David Lindley remarks,

> [i]f you measure the first electron to be up, then you know the second must be down. But if you measure the second electron with a horizontal Stern–Gerlach magnet, that definite state translates into an indeterminate 'half-left, half-right' state, so that the second spin

measurement has an equal chance of coming out either way – just as it would for an isolated electron that you knew nothing about. This version of an EPR experiment doesn't seem to take you into interesting territory. It's just another example of quantum uncertainty: measure one thing, and you have complete ignorance of another.[24]

However, this situation changes sharply when it is asked what would happen if the two polarizing devices were set up at an intermediate (say 45°) angle as in Bell's proposed thought experiment. For we here have to do with a different kind of probability reckoning, one which effectively rules out the claim that both particles have objective values for every spin-dimension and hence that any uncertainty must be a matter of the limits placed upon our powers of observation–measurement. Rather, it results (or must be thought to result) from the *intrinsically* probabilistic character of quantum-mechanical systems and also from the way that probability values are somehow momentarily transmitted – in keeping with orthodox QM predictions but beyond any otherwise standard range of statistical expectation – between the separated particles.

Thus, to summarize, one can point to three main distinguishing features of Bell's thought experiment as compared with EPR. First, there is the use of a delayed-choice technique, i.e. the insertion of a polarizing device that 'decides' which spin-component to measure *after* the particles have commenced on their divergent paths; second, the deployment of oblique or intermediate measurement angles; and third, the adoption of statistical and probability-based methods in order to determine whether those QM predictions are indeed borne out as against the claims of local realism. Lindley again provides a clear statement of the case – unlike many writers whose descriptions tend to become rather fuzzy at this point – so I shall cite his commentary at length.

Let's say the first electron goes through a vertical magnet, and comes out up, so that the second must be in a down state. What happens now if this down electron passes through a Stern–Gerlach magnet set at forty-five degrees from the vertical? There can only be two possible outcomes: the electron must come out in one of the two directions defined by the magnetic field, which we can call northeast and southwest. But the probabilities of these two outcomes are not equal .... In fact, a down electron going through a magnet set on a northeast–southwest angle has about a 15 percent chance of coming out northeast and correspondingly an 85 percent chance of coming out southwest ... Bell's insight was to realize that this is a potentially telling intermediate case. The measurement of an up state for the first electron does not tell you with certainty what the outcome of a northeast-southwest measurement on the second electron will be, but neither does it leave you with a purely random, fifty-fifty result. What we have is a measurement on the second electron which is probabilistic (since both outcomes are possible) but that is also influenced by the measurement of the first electron (since the probabilities of those two outcomes are not equal).[25]

In brief, Bell's theorem has to do with the kinds of statistical finding that might be expected if one averaged over the results produced by many such delayed-choice experiments on the basis of standard, well-proven methods for calculating relative probabilities. It is no longer a matter of perfect, 100 per cent anti-correlation as strictly required by the EPR set-up where the polarizing magnets are arranged in parallel or at right angles, thus excluding the prospect of such measured deviations from a statistical norm. Rather, Bell's theorem shows that the extent of anti-correlation should not exceed certain specified limits just so long as there is nothing in the nature of quantum phenomena that contravenes the basic EPR premise, that is, the local-realist veto on any idea of superluminal interaction between widely separated particles. Yet if the quantum predictions are consistently applied, then they must be taken to impose a non-negotiable choice between (1) accepting the truth of quantum mechanics, or (2) accepting the truth of local realism and hence the 'incompleteness' of orthodox QM theory.

We are now better placed to understand precisely how EPR/Bell-type thought experiments differ from those originally conducted by Einstein and his colleagues in response to Bohr and the proponents of orthodox QM. The EPR case can be represented as an argument of the form: assuming that local realism holds, and given that the evidence appears to gainsay it on a certain (orthodox QM) construal, then necessarily that construal must be flawed and the evidence requires some alternative (non-orthodox) interpretation along local-realist lines. Where Bell's theorem sharpens the issue is by showing that the QM observational-predictive results are in conflict not only with local realism but also with some fairly basic and non-controversial methods for averaging-out over experimental data of the kind here in question. Indeed, that conflict can be shown to arise on *any* interpretation of quantum mechanics, that is, any account which accepts the empirical evidence along with the basic quantum formalisms. In other words, the violation of Bell's inequality leaves no choice but to acknowledge some form of nonlocal interaction between widely separated particles *whatever one's position with regard to the issue between realism and anti-realism.*

Thus, it can still be maintained – as by Bohm and indeed by Bell himself – that a realist interpretation is preferable in so far as it makes better sense of the measurement problem and moreover gives causal-explanatory content to an otherwise purely instrumentalist approach of the orthodox QM type. After all, as Holland pointedly remarks,

> nonlocality seems to be a small price to pay if the alternative is to forego any account of objective processes at all (including local ones). Also, it is inconsistent to deny the logical possibility of a pictorial representation of the phenomena, and then lay down conditions for what such a picture should consist of when one is produced.[26]

However, it is clear that Bell's argument creates large problems for anyone who espouses the kind of local-realist and broadly 'classical' worldview that Einstein set out to defend, and which still provides the framework for our dealings with macrophysical reality, whether at an everyday-commonsense or a practical-scientific level. What makes the violation of Bell's inequality such a very tough nut for the realist to crack is the fact that his theorem depends so little on the technicalities of this or that quantum-theoretical approach, and applies so widely on the basis of a few fairly simple algebraic calculations. 'The result seems inescapable', Lindley writes, 'and yet quantum mechanics contradicts it. Bell knew perfectly well that this contradiction existed; that was precisely the point of his theorem. His insight was in realizing that this contradiction could tell you something interesting about the workings of quantum mechanics.'[27]

III

Such thought experiments have played a large role in the development of QM theory, starting out with Planck's conjectures about black-body radiation and carried on through the famous series of debates between Einstein and Bohr.[28] Beyond that, of course, their history stretches right back to Galileo's classic thought-experimental proofs – mostly refutations of received scholastic wisdom – as applied to mechanics, gravitational effects, and other macrophysical phenomena.[29] What is so striking about these speculative arguments, in the quantum domain as elsewhere, is the fact that they reveal an implicit commitment to ontological realism even when (as with Bohr) they seem to come out clean against any realist construal of the evidence. That is to say, such arguments would lack all probative force were it not for the belief that any results obtained through consistent reasoning on hypothetical cases must also reveal what *would* be the upshot if the same experiment were actually conducted under controlled laboratory conditions. Thus Bohr implicitly takes it for granted, in his replies to Einstein, that by raising thought-experimental objections to the realist construal of quantum mechanics (e.g. as regards the impossibility of performing simultaneous measurements of position and momentum), he is also proving that Einstein's version of the experiment would encounter just such physical limits – as predicted by the orthodox theory – if somehow carried out 'in reality'. Indeed, it was only the restrictions imposed by currently available laboratory apparatus (restrictions on the speed of switching devices, spin-polarizers, observational instruments, etc.) which prevented such results from being achieved at the time.

I should not wish to claim that this implied ontological commitment on the part of orthodox theorists like Bohr amounts to a kind of transcendental argument against their position or in favour of an EPR-type deduction to the 'incompleteness' of orthodox QM and the need for a causal-realist account in keeping with Einstein's postulates. All the same, it is a point worth bearing in mind as we move to the next stage in this debate where the predicted 'violations' of Bell's inequality themselves became subject to physical testing with the advent of more advanced laboratory equipment. Various such experiments were performed from the early 1970s on, culminating in the best-known series that were carried out with remarkable precision by a team of French physicists (Aspect, Graingier and Roger) and have since been repeated on numerous occasions with a high degree of statistical-confirmatory warrant.[30] What they involved, very briefly, was a set-up of exactly the kind hypothesized by Bell with particles – in this case photons – whose polarization could be measured at any point in their trajectory so as to determine the number of coincident (anti-correlated) counts. The spin-detectors could be adjusted with great rapidity – some hundred million times per second – so as to measure the entire range of values for both particles and do so, moreover, in such a way (by switching momentarily between channels) that any results thus achieved must be a product of simultaneous remote correlation and could not be explained in terms of the particles' previous history, individual properties, or 'objective' (pre-measurement) polarization. Once again those results turned out to exhibit an impressive conformity with orthodox QM predictions and an equally striking violation of the kinds of coincidence rate that might be expected on a local-realist construal, one requiring that the particles should each possess a range of integral values irrespective of whatever measurement was performed at any given time. In Squires' words:

> [t]o demonstrate how effectively these results violate the Bell inequality, and hence forever rule out the possibility of a local realist description of the world, the authors measured explicitly at the angles where the violation was maximum .... A particular quantity S which according to the Bell inequality has to be negative, but which according to quantum theory has to be $0.118 + 0.005$ is measured to be $0.126 + 0.014$. It is very clear that quantum theory and not locality wins.[31]

In other words, it seemed that the proof (or the statistically preponderant case) for these remote quantum effects was such as could not possibly be explained unless on the premise – so repugnant to Einstein – of faster-than-light 'communication' between separated particles.

In subsequent experiments, Aspect and his colleagues sought to remove any remaining doubt of whether these results might not be subject to some alternative construal in accordance with local realism. One such possible line of counter-argument was that the spin-detectors might be 'communcating' with each other (i.e. somehow acting in concert so as to decide the joint measurement outcome) *before* the particles arrived. In that case the results might be seen as an artefact of the experimental set-up, thus avoiding any need to postulate 'messages' passing between them at superluminal velocity. I shall cite Squires again – at some length – since he offers a clear and detailed account of the experiment in question. 'In order to eliminate this possibility', he writes, it is necessary to arrange that the orientations are 'chosen' after the photons have been emitted. Clearly the time involved is too small to allow the rotation of mechanical measuring devices, so the experiment had two spin-detectors at each side, with pre-set orientations, and used switching devices to deflect the photons into one or the other detector. The switches were independently controlled at random. Thus, when the photons were emitted, the orientations that were to be used had not been decided .... The result ... was again in complete agreement with quantum theory, and in violation of the Bell inequality .... The experiments we have described confirm this feature of the quantum world [i.e., nonlocality]; no longer can we forget about it by pretending that it is simply a defect of our theoretical framework.[32]

There were further features of the Aspect experiments which appeared to block every avenue for an alternative (local-realist) construal. Among them was a test which varied the distance between the two detectors so as to determine whether – as maintained by one version of the hidden-variables theory – the wavefunction might spontaneously reduce (i.e. assume a determinate value for each measurement parameter) before reaching a detector. That is, it would do so simply as a function of the time required to traverse that distance, the latter exceeding the time limit for its 'collapse' into one or other of the discrete states (or spin-values) as subsequently measured on arrival. However, according to the Aspect results, '[e]ven when the separation was such that the time of travel of the photons was greater than the lifetime of the decaying states that produced them (which might conceivably be expected to be the timescale involved in such an effect), there was no evidence that this was happening'.[33] In other words, these results could not be accounted for in terms of some intrinsic probability (i.e. spontaneous decay rate) thought of as pertaining to each particle prior to the act of measurement. Thus, again it appeared that the predictions of orthodox QM were strongly borne out by experiment and, moreover, that any hidden-variables theory could match them only at the price of admitting simultaneous nonlocal interaction.

I have cited Squires on this topic since his book provides an uncommonly clear exposition of Bell's theorem and its consequences while also acknowledging the extent of their conflict with the basic principles of scientific realism. After all – as the EPR authors maintained – there need not be anything in the least 'spooky' about the fact of anti-correlation between remote particles just so long as this fact can be causally explained by application of the conservation law, i.e. that any two particles with their source in a singlet-state and with zero joint angular momentum will always exhibit a sum-total zero value when measured thereafter at any point on their divergent paths of travel. Thus, for instance, take the situation of a blindfolded subject who is presented with a box which she knows to contain two 'anti-correlated' billiard balls, red and white. She then removes one of them, throws it away, and remains unsure which one it was *until she takes off the blindfold* and discovers that the white ball is still there in the box, thus proving beyond doubt that the red ball *must* have been the one she took out. This analogy may seem simplistic but it captures the basic set of assumptions – ontological realism plus space–time locality relative to the speed of light – which motivated the original EPR paper. Where the problems arise is with Bell's demonstration (empirically confirmed by Aspect's experiments) that in quantum mechanics there is just no fact of the matter until it is decided – through random switching of the spin-polarizer – what values obtain for the two particles and also what *shall have been* their values up to that point from the time of emission.

Orthodox QM gets around this difficulty by taking an instrumentalist line, that is, by adopting the philosophy propounded by thinkers from Berkeley to Mach. On this view the proper business of physical science is to 'save appearances' by accepting the results of empirical observation, devising the simplest possible theory to accommodate those data, and eschewing the quest for causal explanations of a realist ('metaphysical') kind.[34] Thus, according to Bohr, there is simply no answer to the question how and where the wavefunction 'collapses' so as to produce determinate results at the point of measurement.[35] Such questions are ill-framed in so far as they adopt a descriptive language that works well enough for observable objects or events but which cannot be applied to the quantum domain since it imposes a wholly inappropriate conceptual apparatus or explanatory scheme. In classical (Newtonian) mechanics the assumption was that one could – at least in principle – specify the state of any given system by assigning values of position and momentum to all its component parts. The motion of a particle could then be determined by applying Newton's Second Law (acceleration = force ÷ mass), thus producing a unique set of values that predicted its position and momentum at all future times. In orthodox QM, on the other hand, it is the wavefunction – as specified by Schrödinger's equation – that defines the state of the system *so far as it can possibly be known*, and which permits the assignment of probability values rather than determinate (classical) values of space–time location and momentum. As regards location, '[t]he relation between the wavefunction and the probability is very simple: the probability is proportional to the square of the magnitude of the wavefunction [and] does not depend in any way on the angle of the wavefunction'. As regards momentum, conversely, 'this is related to the angle [and is] proportional to the rate at which the angle of the wavefunction varies with the point of space'.[36] The deployment of Schrödinger's equation is analogous to the deployment of Newton's Second Law because, as Squires points out, 'it allows the wavefunction to be uniquely determined at all times if it is known at some initial time. Thus quantum mechanics is a deterministic theory of wavefunctions, just as classical mechanics is of position.'[37]

However, this analogy proves to have sharp limits as soon as one asks the kind of question that Bohr ruled out: the question of how and when – at what precise stage in the measurement process – the wavefunction somehow collapses and thus produces determinate values of space–time location or momentum. Indeed, it is another great irony that Schrödinger should have produced the very formalism which enabled Bohr and the proponents of orthodox QM to reject any interpretation (such as the hidden-variables theory) that sought to reconcile quantum mechanics with a 'classical' realist ontology. For it was Schrödinger also who joined with Einstein in arguing that the orthodox model *must* be 'incomplete' if it failed to resolve the EPR paradox and provide some adequate means of explaining why macrophysical objects (like the famous cat-in-a-box) were not likewise subject to quantum probability, superposition, observer-induced wavefunction collapse, and so forth.[38] Squires himself shares this sense of dissatisfaction with a theory – orthodox QM – which decrees that we *cannot or should not* raise such questions on pain of either contradicting the well-proven quantum observational-predictive results or engendering further 'metaphysical' problems and paradoxes. All the more so since, as a practising physicist, he is aware of the enormous success of quantum mechanics in 'explaining' a range of otherwise inexplicable phenomena, among them findings that have given rise to some of the most remarkable advances in present-day physics. Thus – to take just a few striking examples – quantum theory alone makes it possible to 'account for' the classically anomalous features of black-body radiation and the photoelectric effect; to 'explain' chemical bonding in terms of subatomic structure; to 'understand' the working of transistors, silicon chips, and other such microelectronic devices from which there emerged the revolution in modern communications technology; and to 'comprehend' such recently discovered phenomena as superconductivity and superfluidity through the effect of low temperatures in producing a low-energy quantum state where electrons condense (i.e. lose the normal repulsive force that exists between particles with equivalent charge) and thus make possible an energy flow without resistance or loss. Given these successes – and a great many more besides – it seems well nigh unthinkable that quantum mechanics could turn out to rest on some huge mistake concerning its own conceptual foundations or the nature of quantum 'reality'.

Yet there is a reason for placing those queasy quotation marks around words such as 'explain', 'understand', 'comprehend', and 'reality' when used in this context. For it is precisely the problem with orthodox QM – a problem (that is) for all but its hard-line advocates – that it deprives such terms of any real explanatory content. On this view, we have everything required of an adequate theory or interpretation when we apply the standard quantum formalisms, obtain a probability value as yielded by the Schrödinger equation, and then go on to compare the results with those achieved through empirical observation or measurement. But in that case, so its critics maintain, the word 'interpretation' is itself being redefined in quantum-instrumentalist terms, i.e. as involving no claim to understand what is *really going on* beyond the requirements of statistical warrant, empirical adequacy, or predictive confirmation. This is surely hard to square with the above-cited evidence of its great – indeed unequalled – success as a physical theory that has managed not only to 'explain' such a range of classically unexplained phenomena but also to inspire the development of technologies undreamt of before the advent of quantum mechanics. At the very least there is a problem in upholding the standard Copenhagen line on this issue while proclaiming – as orthodox theorists frequently do – the extent to which QM has been instrumental in bringing those advances about. For such claims are 'instrumentalist' in a sense wholly opposed to the usual, somewhat specialized philosophy-of-science usage of the term. That is to say, they involve a strong supposition that any theory (or interpretation thereof) that yields scientific or technological progress will do so by providing a better, more adequate grasp of the real-world operative features – microstructural attributes, causal dispositions, law-governed regularities, etc. – which make such progress possible.[39] In which case clearly there is something awry about a theory (orthodox QM) that erects the non-availability of any such realist or causal-explanatory account into a high point of a priori doctrine.

As I have said, its chief rival in terms of present-day QM debate is the de Broglie–Bohm 'hidden-variables' theory, which (at least until recently) was ignored or marginalized by exponents of the orthodox view. This theory embodies a thoroughgoing realist outlook with respect both to particles (taken as possessing objective, observer-independent values throughout their trajectory) and to fields (taken as guiding those particles through the action of a pilot-wave that determines their position and momentum at every stage). Moreover, it has proved capable of meeting the challenge of spin-$\frac{1}{2}$ multipath or delayed-choice experiments by postulating the existence of a spinor wave, a 'new type of physical field [in Peter Holland's summary] propagating in spacetime that exerts an influence on a particle moving within it'.[40] On this account it is the spinor wave that carries information concerning such values as internal angular momentum, that is to say, those further properties of the particle (besides position and momentum) that are commonly thought most resistant to any such construal. However, the case can best be understood in connection with the classic two-slit experiment, which first gave rise to the theory of wave–particle dualism and hence – via EPR and subsequent debates – to the widely held idea of quantum mechanics as requiring a radical break with all forms of objectivist or causal-realist thinking. For if indeed it is possible to interpret that experiment in accordance with Bohm's hidden-variables theory, then there is strong presumptive warrant for rejecting the orthodox (Copenhagen) view with respect to those other, more refined or sophisticated variants.

Holland once again states the issue with admirable clarity and force, so I shall cite him at length as a reference point for further discussion.

> The statistical interpretation of the wavefunction is in accord with experimental facts. An interference pattern on a screen is built up by a

series of apparently random events, and the wavefunction correctly predicts where the particle is most likely to land over an ensemble of trials. Yet the interpretation of the wavefunction which ascribes to it a purely statistical significance is not forced upon us by the experimental results .... On the contrary, one may take the view that the characteristic distribution of spots on a screen which build up an interference pattern is evidence that the wavefunction indeed has a more potent physical role than a mere repository of information on probabilities, for how are the particles guided so that statistically they fall into such a pattern? Such a question is naturally ruled out by the purely probabilistic interpretation. But the latter is appropriate only if we wish to reduce physics to a kind of algorithm which is efficient at correlating the statistical results of experiments. If we wish to do more, and attempt to understand the experimental results as the outcome of a causally connected series of individual processes, then we are free to enquire as to the further possible significance of the wavefunction (beyond its probabilistic aspect), and to introduce other concepts in addition to the wavefunction.[41]

Bohm's theory is thus premised on the realist assumption that any adequate account of QM phenomena will indeed 'do more' than establish a high degree of predictive correlation or empirical warrant. That is to say, it will work on the joint principles that (1) the reality underlying those phenomena might always turn out to exceed or transcend our current methods of empirical verification, and (2) this entails a method of inference to the best causal-explanatory theory consistent with the evidence to hand. Where orthodox QM falls short of that aim is in resting content with a highly developed and sophisticated formal approach – one that has undeniably passed all the tests for predictive–observational accuracy – while offering no guidance as to how those results might be given some genuine (i.e. substantive and not merely formal) content. For this would mean breaking with the orthodox veto on any interpretation – such as Bohm's – which oversteps the limits of empirical warrant. Where Bohm's theory is at its strongest, conversely, is in putting up a realist interpretation of the evidence which *on principle* rejects this self-denying ordinance and instead takes scientific theories to be warranted by their jointly observational, predictive, *and* causal-explanatory power. If this entails going 'beyond' the evidence – strictly or empirically construed – then it cannot be accounted a fault in Bohm's theory except from the opposing (orthodox QM) standpoint. Thus, as Holland remarks, '[s]cience would not exist if ideas were only admitted when evidence for them exists. One cannot after all empirically prove the completeness postulate. The argument in favour of the trajectory lies elsewhere, in its capacity to make intelligible a swathe of empirical facts'.[42]

IV

Some philosophers, Bas van Fraassen among them, would reject this whole line of argument in favour of a 'constructive empiricist' approach with no ontological commitments beyond what is given as a matter of direct observational warrant.[43] On this view – closely akin to old-style logical positivism – it is simply unnecessary to posit the existence of recondite subatomic particles that can be 'observed' only with the aid of advanced instrumentation, yet which happen to play an explanatory role in our best scientific theories. Rather, we should adopt an agnostic stance, continue our practice of 'referring' to those objects whenever there is occasion to do so, but construe that practice always in terms of empirical warrant or conformity with the evidence currently to hand. Thus, according to van Fraassen, it is the aim of an adequate scientific theory to save empirical appearances without any need for ontological underpinnings in the realist or causal-explanatory mode. 'To be an empiricist', he asserts,

> is to withhold belief in anything that goes beyond the actual, observable phenomena, and to recognise no objective modality in nature .... [I]t must involve throughout a resolute rejection of the demand for an explanation of the regularities in the observable course of nature, by means of truths concerning a reality beyond what is actual and observable, as a demand which plays no role in the scientific enterprise.[44]

I have written elsewhere about the problems that arise for any such approach if one takes a longer term view of the history of science, a view that allows for convergence on truth as a matter of inference to the best explanation.[45] To support this claim, one could instance the way in which various once unobservable entities – e.g. molecules and atoms – have often started out as speculative 'posits' of the kind that van Fraassen describes, but have then acquired strong realist credentials through the development of more refined observational techniques coupled with more advanced explanatory theories concerning their structure, interactive capacities, causal powers, and so forth.[46]

Van Fraassen meets such arguments part-way by stretching the term 'observable' to cover what *could* be described under optimal conditions by the best-placed human observers. Still, there is the obvious objection – raised by Ian Hacking and others – that science has various techniques for extending the limits of human observation (from radio telescopes to electron microscopes), and also various means of checking their accuracy to a degree of precision far beyond that attainable by the naked eye.[47] Also, it is hard to see any reason – anthropocentric prejudice apart – for restricting the scientific object-domain to just those entities and events that happen to fall within the range of unaided *human* perceptual grasp. For there are many things that elude even the most sensitive or sharp-eyed human observer simply through the limits imposed by our physical constitution, perceptual apparatus, modes of cognitive processing, etc. C.J. Misak makes the point – following Paul Churchland – when she lists some of the ways in which an object or event may lie beyond the furthest limits of unaided human perception. Thus:

> it may not be spatially or temporally placed so that we can observe it; it may be too small, too brief, or too protracted; it may lack the appropriate energy, being too feeble or too powerful to permit us to discriminate it; it may fail to have the appropriate wavelength or mass; or it may fail to feel the relevant forces which our sensory apparatus exploits.[48]

Churchland has a nice supporting argument when he asks how van Fraassen's doctrine might apply to beings who were rooted to the spot like trees – say Douglas Firs – and whose epistemic modalities obliged them to draw a very different line between the 'merely unobserved' and the 'downright unobservable'. 'It may help', he suggests, 'to imagine here a suitably rooted arboreal philosopher named ... Douglas van Firrsen, who, in his sedentary wisdom, urges an antirealist scepticism concerning the spatially very *distant* entities postulated by his fellow trees'.[49] In other words, there is something decidedly parochial – not to say myopic – about fixing the limits of genuine knowledge at just that point where human observers must cease to rely on their highly restricted powers of direct observation.

Now it might well appear, on the evidence so far, that quantum mechanics is one branch of science where van Fraassen's programme of constructive empiricism has a fair claim to be the best, most sensible approach when confronted with the kinds of interpretative problem thrown up by the EPR paper and Bell's theorem. That is, it would seem fully justified to adopt an agnostic stance with regard to the 'reality' of quantum phenomena which exhibit such a deep (perhaps intrinsic) resistance to treatment in the realist or causal-explanatory mode. Such is van Fraassen's argument in his book *Quantum Mechanics: an empiricist view*, in which he follows a basically positivist line in rejecting 'the seductive temptation of metaphysical realism', or the idea that there are certain fundamental questions about science – such as those concerning the existence of subatomic entities or the status of causal explanations – 'which the philosopher can answer speculatively by positing abstract, unobservable, or modal realities'.[50] Thus, QM provides him with an ideal test case for the claim that philosophy of science goes too far – oversteps the limit of reputable scientific method – when it raises ontological issues or enquires into the putative reality 'behind' appearances. Rather, it should seek to save those appearances, in good empiricist fashion, by refusing to enter such otiose 'metaphysical' debates and resting content with the best observational data to hand.

Van Fraassen's is a highly sophisticated line of argument which surveys the whole range of interpretative options and by no means ignores the counter-proposals advanced by advocates of a realist approach. Indeed he proposes a modal interpretation which claims to represent a significant advance on the standard Copenhagen doctrine and also to provide a more 'complete' physical theory in something like the sense required by Einstein and Bohm. All the same, his thought is still much indebted to that old-style verificationist doctrine according to which the only truth claims admissible in science are those arrived at through logical analysis as applied to observational data or empirical findings. Besides, it is far from clear why a constructive empiricist like van Fraassen should feel any need to reconcile his approach with a Bohm-type hidden-variables theory premised on the 'incompleteness' of orthodox quantum mechanics. Hence the kinship between van Fraassen's 'Copenhagen Variant of the Modal Interpretation' and Bohr's many statements to the general effect that quantum theory must adopt a strictly empiricist approach and eschew all attempts to describe or explain the so-called 'quantum world'. For Bohr, quite simply, '[t]here is no quantum world. It is wrong to think that the task of physics is to find out how nature is. Physics concerns what we can say about nature'.[51] On this view – endorsed with certain reservations by van Fraassen – there is no point in seeking a more 'complete' (i.e. realist or causal-explanatory) theory of the kind proposed by physicists such as Einstein, Schrödinger, and Bohm.

Of course Bohr's statement leaves some room for different understandings of 'what we can say about nature', since a realist could well come back with the argument that we can say a lot more – and 'about nature' in a far stronger (depth-explanatory) sense – than Bohr wished to maintain. Indeed, Bohr's thoughts are often so fuzzily expressed that it is hard to make out just where he takes the line to fall between ontological and epistemological issues, or the underlying reality of quantum phenomena (whatever that could mean on his account) and the limits of human understanding as applied to those same phenomena.[52] Van Fraassen is himself highly critical of wholesale anti-realist doctrines that extrapolate too easily from the quantum realm to that of macrophysical objects and events. After all, it is precisely his point to uphold this distinction between humanly observable objects (over which we can quantify with empirical confidence) and other, more elusive entities – such as quarks or maybe electrons – of which we had better say with due caution that they figure in our current best scientific theories but should none the less be treated as convenient posits whose ontological status remains undecided. Still it is the case that van Fraassen's doctrine of constructive empiricism tends very often to blur that line by generalizing from problem cases (e.g. the current more speculative posits of subatomic particle theory) to an argument against realist or causal-explanatory theories as applied to the macrophysical domain.[53] That is to say, his thesis in the latter regard – that 'empirical adequacy' is the best we can reasonably hope for – gains a good deal of its persuasive force from the idea of quantum mechanics as having problematized all our most basic conceptions of knowledge, truth, and reality. Only in a climate of widespread scepticism *vis-à-vis* those 'classical' conceptions could such a thesis present itself as really nothing more than a sensible refusal to overstep the limits of good scientific practice.

Indeed, van Fraassen's stance may appear quite moderate by comparison with other current forms of anti-realist or ontological-relativist thinking. Nevertheless it is a doctrine that rejects some major tenets of the scientific outlook that prevailed (with occasional dissenting voices) from Galileo to Einstein, and which embodies the working faith of most physical scientists, if not philosophers and historians of science. On this view – rejected by van Fraassen – it is the business of scientific theories not only to save empirical appearances and match predictions with results but also (in Squires' words) 'to explain observed phenomena and to understand the nature of what exists'.[54] Of course there are some notable precedents for the broadly instrumentalist approach to issues concerning the scope and limits of scientific knowledge. On the one hand are those thinkers – a diverse and variously motivated company from Berkeley to Mach, Duhem and Bohr – who have adopted a phenomenalist standpoint and consistently refused to speculate on whatever 'reality' might lie beyond or behind the empirical evidence.[55] (Karl Popper traces the relevant prehistory from a strongly opposed realist standpoint in his book *Quantum Theory and the Schism in Physics*.[56]) On the other may be counted scientists like Newton and the early Einstein – himself much influenced by Mach – who expressly renounced the quest for causal or depth-explanatory hypotheses but whose actual methods and thought procedures tell a very different story. In Einstein's case the conversion from Machian instrumentalism to causal realism was noted with regret – understandably so – by Bohr and others in the orthodox QM camp who considered it a strange lapse into old 'metaphysical' ways of thinking. To Einstein, conversely, his early position now appeared to have been just a brief unfortunate lapse from the standards and aims of proper scientific enquiry.[57]

No doubt this attitude was strongly reinforced by his debates with Bohr and his deep dissatisfaction with the failure to apply such standards – as Einstein saw it – among proponents of orthodox QM theory. However, it is only on the crudest of reductive psychobiographical accounts that his reaction appears just the product of brooding *ressentiment* in an erstwhile pioneer of quantum physics overtaken by new developments. Rather, it expresses the basic realist conviction that there exist components of reality on whatever scale – from microphysical structures to the rotation of galaxies – which are not directly (humanly) observable but which possess a range of objective or determinate features quite apart from the various contingent limits of our own sensory, perceptual, or cognitive equipment. Sometimes it is a matter of relying on other, indirect or technologically assisted means of observation – such as electron microscopes or radio telescopes – along with the kinds of theoretical understanding that enable us to use that technology and interpret the results. Ian Hacking, in his book *Representing and Intervening*, argues strongly in support of this claim and against what he sees as the absurdly narrow (anthropocentric) idea that the limits of unaided sensory perception are also the limits of what properly counts as genuine scientific knowledge.[58] There are truths that we might be incapable of *ever* coming to know on account of our innate constitution as creatures whose faculties and reasoning powers are well adapted to life on our own physical scale or within our particular biological evolutionary niche.

On this view there is nothing in the least surprising about the fact that we experience problems of conceptual as well as perceptual grasp when thinking about certain, e.g. quantum-physical or astrophysical events which lie far beyond the middle-range dimensions of our normal spatio-temporal cognitive framework. All the same those problems should not be taken – in orthodox QM or 'constructive empiricist' fashion – as imposing some ultimate limit on the scope for genuine scientific knowledge. That is, we can come up with well-supported theories that exceed the best current evidence (narrowly construed), but which still lay claim to a high measure of realist and causal-explanatory warrant. For it is in just this way that sciences like particle physics have advanced from an early phase of pure speculation to a stage of theoretically informed conjecture as regards the existence of atoms, nuclei, protons, electrons, etc., and thence to a point where they are able to explain an impressive range of phenomena – such as atomic valence or chemical bonding – which would otherwise lack any adequate scientific account.

V

This makes it odd – to say the least – that so many orthodox QM theorists have elected to follow Bohr and adopt an instrumentalist stance which on principle rejects the very possibility of a realist (e.g. Bohm-type hidden-variables) interpretation. After all, as Rae very pointedly remarks,

> [t]he success of the matter–wave model did not stop at the atom. Similar ideas were applied to the structure of the nucleus itself which is known to contain an assemblage of positively charged particles, called protons, along with an approximately equal number of uncharged neutrons .... Nowadays, even 'fundamental' particles such as the proton and neutron (but not the electron) are known to have a structure and to be composed of even more fundamental particles known as 'quarks'. This structure has also been successfully analysed by quantum physics in a similar manner to those of the nucleus and the atom, showing that the quarks also possess wave properties. But modern particle physics has extended quantum ideas even beyond this point. At high enough energies a photon can be converted into a negatively charged electron along with an otherwise identical, but positively charged, particle known as a positron, and electron–positron pairs can combine into photons. Moreover, exotic particles can be created in high-energy processes, many of which spontaneously decay after a small fraction of a second into more familiar stable entities like electrons or quarks.[59]

I have quoted this passage at length because it brings out very strikingly the tension that exists between any account of quantum mechanics that adheres to the orthodox veto on realist or causal-explanatory talk, and any account – such as Rae's – which acknowledges the extent of its contribution to our better understanding of microphysical reality. It is worth noting also that the passage finds room for differing degrees of ontological commitment with regard to the various particles mentioned and their role *vis-à-vis* the best current theories of subatomic structure. (His book was first published in 1986, but the point would hold good for any updated version of the argument based on more recent research.) Thus, for instance, whereas the nucleus is *known* to be made up of neutrons and protons, the former uncharged and the latter possessing a positive charge which is balanced by an equal number of surrounding (negatively- charged) electrons, when it comes to those particles *known as* 'quarks', there is at least some measure of doubt as to their precise ontological status and hence their claim to occupy a place in our best explanatory theories. But in other respects – as compared, say, with those transient 'exotic particles' produced in high-energy accelerators – quarks can be considered as belonging in the company of 'more familiar stable entities such as electrons'.

Then again, we are warranted in referring to anti-particles 'known as' positrons (positively charged electron counterparts) in so far as they fulfil certain basic symmetry requirements and appear to explain just how it is that photons can undergo the kinds of transformation described in the above-cited passage from Rae. Still there is a difference between cases like this – which involve some degree of hypothetical conjecture on the basis of other, more 'familiar' results – and cases (such as that of the proton–neutron structure of the nucleus) where those results can be directly applied. All the same, we are justified in granting more credence (i.e. a higher probability weighting) to the positron hypothesis than we should be as concerns the existence of other, presently more elusive or recondite particles which play a role in the most advanced speculative theories of present-day physics. With respect to these entities we had much better say that their existence is still a moot question, though it becomes more probable – or less a matter of pragmatic-instrumental convenience – with each new result that can best be explained by building them into our favoured ontological scheme. What this amounts to is a version of the basic realist principle: that the truth of scientific theories is decided by the way things stand in reality, rather than those theories deciding what shall count as true according to our presently accepted notions of reality or acculturated habits of belief. Aristotle was the first to enounce this principle, and it remains the touchstone of realist philosophies in quantum physics as elsewhere. Such is the reasoning behind Bohm's theory and its justification for espousing a viewpoint which posits the existence of objective (observer-independent) values of particle position and momentum. In Holland's words:

> [i]t is the assumption of a corpuscle which transforms quantum mechanics

into a theory of matter having substance and form. The pure wave dynamics described by Schrödinger's equation does not yield any account of which result is actually realised in an individual measurement operation. The wavefunction collapse hypothesis only gains physical content if actual coordinates for the collapsed system are posited. Since the point at which these are introduced in the chain of connected physical systems is arbitrary, the only consistent assumption is that they are well defined all along.[60]

Instrumentalists take the opposite view, i.e. that there is no legitimate appeal to anything beyond the current best evidence as given by empirical methods of enquiry or criteria of predictive warrant. Such is the standard Copenhagen 'interpretation' of quantum mechanics, one that effectively debars all attempts to interpret the quantum formalisms aside from their purely instrumental yield as a matter of observation and measurement. What is thereby excluded is any prospect of advancing beyond that stage to the point where it becomes possible to achieve a more adequate (realist or causal-explanatory) account of quantum phenomena. Hence the 'unspoken contradiction' – as Holland describes it – 'at the heart of quantum physics: physicists do want to find out "how nature is" and feel they are doing this with quantum mechanics, yet the official view which most workers claim to follow rules out the attempt as meaningless!'.[61]

This is one argument for scientific realism: the analogy with previous developments in the history of science – e.g. the atomist hypothesis from the ancient Greek materialists, through Dalton, to present-day particle physics – where erstwhile hunches or pieces of inspired guesswork have matured into powerful explanatory hypotheses and thence into theories of capable of testing through more advanced (technologically assisted) means of observation or experimental warrant. The other chief argument is more basic to the realist case though also, by its nature, more a matter of ultimate ontological commitment and hence always open to various kinds of long-familiar sceptical response. Realism in this sense has to do with asserting the existence of verification-transcendent truths, i.e. truths for which as yet we may possess no means of proof or ascertainment, but which none the less hold quite aside from our present limited state of knowledge and therefore determine the truth-value of any statements we might make concerning them.[62] Thus, for instance, one could state a vast number of hypotheses (or candidate truths) about history, geography, remote astrophysical events, the subatomic structure of matter, and so forth, that lie beyond the bounds of verification for various contingent or non-contingent reasons. It might be merely that evidence is lacking, or that the historical records haven't survived, or that we don't have sufficiently powerful radio telescopes, or electron microscopes with high enough powers of resolution. Then again, it might be that we lack the scientific knowledge or depth of theoretical grasp to interpret certain puzzling phenomena (such as quantum nonlocality or the wave–

particle dualism) for which we have strong experimental warrant but as yet no adequate explanation. At the limit – epistemologically speaking – it could even be the case that there were aspects of reality that lay beyond reach of human understanding on account of some intrinsic deficit in our powers of conceptual grasp. After all, we can imagine that there might exist intellects better adapted than our own to comprehend matters that we find deeply mysterious, just as – to the best of our knowledge – non-human animals are incapable of grasping the truths of elementary number theory or Newtonian celestial mechanics.

Each of these arguments has considerable force as applied to issues in the interpretation of quantum theory. Thus there is good reason to think that our present state of knowledge regarding quantum phenomena is at roughly the stage that had been reached by the mid-nineteenth century regarding the atomist-molecular theory of matter. That is to say, it involves the construction of hypotheses that are well borne out by a range of predictive, indirect-observational and theoretical results but which as yet lack any adequate explanation in causal-realist terms. At this stage the best (most rational) attitude for physicists and philosophers to adopt is one of qualified instrumentalism, or a willingness to work with the theory as it stands while acknowledging its limits and keeping an open mind with respect to alternative accounts – such as Bohm's – that hold out the prospect of a fuller, more complete understanding.[63] Thus, according to Holland,

> Bohm showed conclusively by developing a consistent counterexample that the assumption of completeness ..., a notion that pervaded practically all contemporary quantal discourse, was not logically necessary. One *could* analyse the causes of individual atomic events in terms of an intuitively clear and precisely definable conceptual model which ascribed reality to processes independently of acts of observation, *and* reproduce all the empirical predictions of quantum mechanics .... It is thus very much a 'physicist's theory' and indeed puts on a consistent footing the way in which many scientists think instinctively about the world anyway.[64]

No doubt there are problems with Bohm's hidden-variables theory, among them its complex mathematical structure and its need to assign a realist interpretation to components of the standard model (e.g. linear operators in Hilbert space) which offer less resistance when treated in a purely instrumentalist fashion.[65] However, this argument should not be taken as ruling out the prospect of a future advance that would either vindicate Bohm's theory – perhaps in modified form – or manage to resolve those problems within some alternative realist and causal-explanatory framework. At any rate, there seems little merit in a doctrine, such as orthodox QM, which leaps so quickly from the limits of present-day knowledge to the presumed limitations of knowledge in general or to various highly problematical consequences concerning quantum phenomena.

Endnotes

1 See for instance P.C.W. Davies, *Other Worlds* (London: Dent, 1980); John Gribbin, *In Search of Schrödinger's Cat: quantum physics and reality* (New York: Bantam Books, 1984); David Lindley, *Where Does the Weirdness Go? why quantum physics is strange, but not so strange as you think* (London: Vintage, 1997); John Polkinghorne, *The Quantum World* (Harmondsworth: Penguin, 1986).

2 See especially J. Alberto Coffa, *The Semantic Tradition from Kant to Carnap: to the Vienna Station* (Cambridge: Cambridge University Press, 1991).

3 Peter Holland, *The Quantum Theory of Motion: an account of the de Broglie–Bohm causal interpretation of quantum mechanics* (Cambridge: Cambridge University Press, 1993), p. 26.

4 See especially Tim Maudlin, *Quantum Non-Locality and Relativity: metaphysical intimations of modern science* (Oxford: Blackwell, 1993) and Michael Redhead, *Incompleteness, Nonlocality and Realism: a prolegomenon to the philosophy of quantum mechanics* (Oxford: Clarendon Press, 1987).

5 For a good introductory account of these phenomena, see Alasdair I.M. Rae, *Quantum Physics: illusion or reality?* (Cambridge: Cambridge University Press, 1986); also Euan Squires, *The Mystery of the Quantum World*, 2nd edn. (Bristol & Philadelphia: Institute of Physics Publishing, 1994).

6 See Note 4, above; also Albert Einstein, *Relativity: the special and the general theories* (London: Methuen, 1954); J.R. Lucas and P.E. Hodgson, *Spacetime and Electro-Magnetism* (Oxford: Clarendon Press, 1990).

7 See Maudlin, *Quantum Nonlocality and Relativity* (op. cit.).

8 See especially A. Einstein, B. Podolsky and N. Rosen, 'Can Quantum-Mechanical Description of Reality be Considered Complete?', *Physical Review*, series 2, Vol. 47 (1935), pp. 777–80; Niels Bohr, article in response under the same title, *Physical Review*, Vol. 48 (1935), pp. 696–702; Bohr, 'Conversation with Einstein on Epistemological Problems in Atomic Physics', in P.A. Schilpp (ed.), *Albert Einstein: philosopher–scientist* (La Salle: Open Court, 1969), pp. 199–241; also Arthur Fine, *The Shaky Game: Einstein, realism, and quantum theory* (Chicago: University of Chicago Press, 1936); Don Howard, 'Einstein on Locality and Separability', *Studies in the History and Philosophy of Science*, Vol. 16 (1985), pp. 171–201; J.A. Wheeler and W.H. Zurek (eds.), *Quantum Theory and Measurement* (Princeton, NJ: Princeton University Press, 1983); and entries under Note 4, above.

9 See Fine, *The Shaky Game* (op. cit.); also Niels Bohr, *Atomic Theory and the Description of Nature* (Cambridge: Cambridge University Press, 1934 and *Atomic Physics and Human Knowledge* (New York: Wiley, 1958); John Honner, *The Description of Nature: Niels Bohr and the philosophy of quantum physics* (Oxford: Clarendon Press, 1987); Henry J. Folse, *The Philosophy of Niels Bohr: the framework of complementarity* (Amsterdam: North-Holland, 1985); Dugald Murdoch, *Niels Bohr's Philosophy of Physics* (Cambridge University Press, 1987).

10 See Note 8, above.

11 Rae, *Quantum Physics: illusion or reality?* (op. cit.), p. 50.

12 Einstein, Podolsky and Rosen (op. cit.), p. 778.

13 See J.S. Bell, *Speakable and Unspeakable in Quantum Mechanics: collected papers on quantum philosophy* (Cambridge: Cambridge University Press, 1987); also James T. Cushing and Ernan McMullin (eds.), *Philosophical Consequences of Quantum Theory: reflections on Bell's Theorem* (Notre Dame, IN: University of Notre Dame Press, 1989) and entries under Note 4, above.

14 See especially David Bohm, *Causality and Chance in Modern Physics* (London: Routledge & Kegan Paul, 1957); David Bohm and B.J. Hiley, *The Undivided Universe: an ontological interpretation of quantum theory* (London: Routledge, 1993); also David Z. Albert, 'Bohm's Alternative to Quantum Mechanics', *Scientific*

American, No. 270 (May 1994), pp. 58–63; F.J. Belinfante, *A Survey of Hidden Variable Theories* (Oxford: Pergamon Press, 1973); S.H. Bhave, 'Separable Hidden Variables Theory to Explain the Einstein–Podolsky–Rosen Paradox', *British Journal for the Philosophy of Science*, Vol. 37 (1986), pp. 467–75; James T. Cushing, *Quantum Mechanics: historical contingency and the Copenhagen hegemony* (Chicago: University of Chicago Press, 1994); Peter Holland, *The Quantum Theory of Motion* (op. cit.).

15 Erwin Schrödinger, *Letters on Wave Mechanics* (New York: Philosophical Library, 1967); also – in popularizing vein – John Gribbin, *In Search of Schrödinger's Cat* (op. cit.).

16 Louis de Broglie, *Physics and Microphysics* (New York: Harper & Row, 1960).

17 Holland, *The Quantum Theory of Motion* (op. cit.), p. 271.

18 See for instance P.C.W. Davies and J.R. Brown (eds.), *The Ghost in the Atom* (Cambridge: Cambridge University Press, 1986); B. de Witt and N. Graham (eds.), *The Many-Worlds Interpretation of Quantum Mechanics* (Princeton, NJ: Princeton University Press, 1973); David Deutsch, *The Fabric of Reality* (Harmondsworth: Penguin, 1997); Wheeler and Zurek (eds.), *Quantum Theory and Measurement* (op. cit.); E.P. Wigner, *The Scientist Speculates*, ed. I.J. Good (London: Heinemann, 1962).

19 Einstein, Podolsky and Rosen (op. cit.), p. 778.

20 Rae, *Quantum Physics: illusion or reality?* (op. cit.), p. 36.

21 For further discussion see entries under Notes 4, 13, 14 and 18, above; also Evadro Agazzi (ed.), *Realism and Quantum Physics* (Amsterdam & Atlanta, GA: Rodopi, 1997); David Z. Albert, *Quantum Mechanics and Experience* (Cambridge, MA: Harvard University Press, 1993); Bernard D'Espagnat, *Veiled Reality: an analysis of present-day quantum-mechanical concepts* (Reading, MA: Addison-Wesley, 1995); Max Jammer, *The Philosophy of Quantum Mechanics* (New York: Wiley, 1974); Henry Krips, *The Metaphysics of Quantum Theory* (Clarendon Press, 1987); A. Sudbury, *Quantum Mechanics and the Particles of Nature* (Cambridge: Cambridge University Press, 1986); Bas C. van Fraassen, *Quantum Mechanics: an empiricist view* (Clarendon Press, 1992).

22 Squires, *The Mystery of the Quantum World* (op. cit.), p. 118.

23 Ibid, p. 98.

24 Lindley, *Where Does the Weirdness Go?* (op. cit.), p. 131.

25 Ibid, pp. 131–2.

26 Holland, *The Quantum Theory of Motion* (op. cit.), p. 67.

27 Lindley, *Where Does the Weirdness Go?* (op. cit.), p. 137.

28 See Notes 8 and 9, above.

29 See James Robert Brown, *The Laboratory of the Mind: thought experiments in the natural sciences* (London: Routledge, 1991) and *Smoke and Mirrors: how science reflects reality* (Routledge, 1994); also Paul Davies, 'The Thought that Counts: thought-experiments in physics', *New Scientist*, May 6th 1995, pp. 26–31 and Roy Sorensen, *Thought Experiments* (New York: Oxford University Press, 1992).

30 A. Aspect, P. Graingier and C. Roger, 'Experimental Realization of the E–P–R Paradox', *Physical Review*, Vol. 48 (1982), pp. 91–4; also entries under Notes 2 and 11, above.

31 Squires, *The Mystery of the Quantum World* (op. cit.), p. 99.

32 Ibid, p. 101.

33 Ibid, pp. 101–2.

34 See for instance Pierre Duhem, *To Save the Phenomena: an essay on the idea of physical theory from Plato to Galileo*, trans. E. Dolan and C. Maschler (Chicago: University of Chicago Press, 1969); Michael Gardner, 'Realism and Instrumentalism in Nineteenth-Century Atomism', *Philosophy of Science*, Vol. 46 (1979), pp. 1–34; Ernst Mach, *The Science of Mechanics: a critical and historical account of its development*, trans. T.J. McCormack (La Salle, IL: Open Court, 1960); C.J. Misak, *Verificationism: its*

history and prospects (London: Routledge, 1995); Hans Reichenbach, *Experience and Prediction* (University of Chicago Press, 1938); Wesley C. Salmon, *Four Decades of Scientific Explanation* (Minneapolis: University of Minnesota Press, 1989); Bas C. van Fraassen, *The Scientific Image* (Oxford: Clarendon Press, 1980) and *Quantum Mechanics: an empiricist view* (op. cit.).

35 See Notes 8 and 9, above.

36 Squires, *The Mystery of the Quantum World* (op. cit.), p. 24.

37 Ibid, p. 24.

38 See Note 15, above; also – for a brief introductory account – John Gribbin, 'A Tale of Two Kitties', *New Statesman and Society*, 7th April 1995, pp. 45–6.

39 See for instance D.M. Armstrong, *What is a Law of Nature?* (Cambridge: Cambridge University Press, 1983); J. Aronson, R. Harré and E. Way, *Realism Rescued: how scientific progress is possible* (London: Duckworth, 1994); Roy Bhaskar, *A Realist Theory of Science* (Leeds: Leeds Books, 1975); Rom Harré and E.H. Madden, *Causal Powers* (Oxford: Blackwell, 1975); Wesley C. Salmon, *Scientific Explanation and the Causal Structure of the World* (Princeton, NJ: Princeton University Press, 1984); Peter J. Smith, *Realism and the Progress of Science* (Cambridge University Press, 1981); Michael Tooley, *Causation: a realist approach* (Blackwell, 1988).

40 Holland, *The Quantum Theory of Motion* (op. cit.), p. 379.

41 Ibid, p. 66.

42 Ibid, p. 25.

43 See van Fraassen, *The Scientific Image* and *Quantum Mechanics: an empiricist view* (Notes 34 and 21, above).

44 Van Fraassen, *The Scientific Image* (op. cit.), p. 202.

45 Christopher Norris, 'Anti-Realism and Constructive Empiricism: is there a (real) difference?' and 'Ontology According to van Fraassen: some problems with constructive empiricism', in *Against Relativism: philosophy of science, deconstruction and critical theory* (Oxford: Blackwell, 1997), pp. 167–95 and 196–217.

46 See Gardner, 'Realism and Instrumentalism in Nineteenth-Century Atomism' (op. cit.); also J. Perrin, *Atoms*, trans. D.L. Hammick (New York: Van Nostrand, 1923) and Mary Jo Nye, *Molecular Reality* (London: MacDonald, 1972).

47 Ian Hacking, *Representing and Intervening: introductory topics in the philosophy of natural science* (Cambridge: Cambridge University Press, 1983).

48 Misak, *Verificationism: its history and prospects* (op. cit.), p. 169.

49 Paul M. Churchland, 'The Ontological Status of Observables: in praise of the superempirical virtues', in P.M. Churchland and C.M. Hooker (eds.), *Images of Science: essays on realism and empiricism, with a reply from Bas C. van Fraassen* (Chicago: University of Chicago Press, 1985).

50 Van Fraassen, *Quantum Mechanics* (op. cit.), p. 481.

51 Cited in Bell, *Speakable and Unspeakable in Quantum Mechanics* (op. cit.), p. 142.

52 See entries under Note 9, above.

53 See Norris, 'Ontology According to Van Fraassen' (op. cit.).

54 Squires, *The Mystery of the Quantum World* (op. cit.), p. 123.

55 See Note 34, above.

56 Karl R. Popper, *Quantum Theory and the Schism in Physics* (London: Hutchinson, 1982).

57 See Fine, *The Shaky Game* (op. cit.); also various contributors to Paul A. Schilpp (ed.), *Albert Einstein: philosopher-scientist* (La Salle, IL: Open Court, 1969).

58 Hacking, *Representing and Intervening* (op. cit.).

59 Rae, *Quantum Physics: illusion or reality?* (op. cit.), pp. 14–15.

60 Holland, *The Quantum Theory of Motion* (op. cit.), p. 350.

61 Ibid, p. 9.

62 See especially William P. Alston, *A Realist Conception of Truth* (Ithaca, NY: Cornell University Press, 1996) for a vigorous statement of this alethic-realist case as opposed to various epistemic approaches.

63 For a general defence of weighted probability estimates with regard to scientific realism, see Michael Devitt, *Realism and Truth*, 2nd edn., revised (Oxford: Blackwell, 1987).

64 Holland, *The Quantum Theory of Motion* (op. cit.), p. 17.

65 See entries under Note 14 above, especially Cushing, *Quantum Mechanics: historical contingency and the Copenhagen hegemony*.

2 Quantum theory and the logic of anti-realism

I

Although realism comes in a good many present-day forms and varieties, I shall take it that the single most salient distinction is that between epistemic and alethic realism, as argued by William P. Alston in his recent survey of the field. Thus, according to Alston, the alethic conception 'implies that (almost always) what confers a truth value on a statement is something independent of the cognitive-linguistic goings on that issued in that statement, including any epistemic status of those goings on'.[1] That is to say, it avoids the sorts of confusion that typically arise when philosophers (whether realists or anti-realists) equate truth with warranted assertability, or restrict it to whatever we can justifiably claim to know concerning some particular object-domain or field of enquiry. The chief problem with epistemic approaches – at least from a realist viewpoint – is that they open the way to sceptical arguments which deny the knowability and hence the very existence of objective or verification-transcendent truths. Thus for any statement S of the disputed class (i.e. one that is presently undecidable as regards its truth-value) there is simply no question, so the anti-realist argues, of asserting that S must be *either* true *or* false with respect to some knowledge-independent state of affairs which obtains quite apart from the limits imposed by our perceptual apparatus, conceptual equipment, restricted information sources, etc. Rather, we should count it as failing to meet the basic conditions for warranted assertability, namely (1) that it possess adequate verification criteria, and (2) that those criteria are fully satisfied by the best evidence to hand.

Epistemic realism invites this sceptical response by making truth dependent on our state of knowledge at any given time and knowledge dependent on our various (always fallible) sources of evidence. Alethic realism rejects that approach and puts forward the case that truth is indeed verification-transcendent in so far as it pertains to objective matters of fact – or valid conjectures in logic or mathematics – which in no way depend on our possession of decisive evidence or our ability to produce the relevant kind of proof. On this view the truth of such statements is determined by how things stand in reality (or as a matter of objective logico-mathematical warrant) quite aside from what we happen to know or the extent of our available proof procedures. After all, is not the concept of objective truth *presupposed* in any account we can give of the difference between genuine knowledge and evidentially warranted items of belief? Thus (Alston again):

> if we say that the most we can do with respect to, for example, a theoretical statement in science is to adduce evidence that makes it more or less probable but cannot exhibit conclusively verifying evidence, by virtue of what do we make the judgement that the evidence available to us is insufficient to strictly verify? We must have some grasp of what is being asserted by the statement that goes beyond any possible evidence we can compile.[2]

The trouble with epistemic realism is that it runs straight into the verificationist trap by reducing truth to the limited compass of presently attainable human knowledge. For the alethic realist, conversely, truth is a property of just those statements that assert some veridical fact about the world or some theorem (say) in mathematics or logic whose validity depends not at all on our happening to know the relevant proof or means of verification. From their point of view, the epistemic fallacy is one that so severely weakens the realist case as to leave it prone to all manner of infection by sceptical and relativist arguments.[3] In the context of quantum-physical debate, alethic realism is basically the position that Einstein defends against Bohr, and which finds perhaps its clearest expression in the following passage by d'Espagnat:

> [T]he definition of the notion of objective state is absolutely general (for a realist) in the sense that it is completely theory independent. Anybody who believes in realism feels that he understands such a notion, even if he doesn't know a word of physics (either classical or quantum). In that sense the concept of objective state is logically prior (again for a realist) to any quantum theoretical concept.[4]

D'Espagnat goes on to compare this quantum-related conception of 'objective states' with Boltzmann's hypothesis concerning the existence of molecules, one that likewise went beyond the current observational evidence but did so on strong theoretical grounds and also on the alethic-realist assumption that the status of such hypotheses was a matter of objective (verification-transcendent) truth. Thus, '[t]he fact that Boltzmann's idea finally "worked" after all, although it was at first subjected to the very objection here discussed, shows in a quite convincing way that the objection in question is in fact a mere prejudice'.[5]

This case is challenged by anti-realists of various persuasions, among them philosophers such as Michael Dummett, who reject it on the logico-semantic grounds that we could never be in a position to know – or to state as a matter of truth – that there exist certain truths for which we possess no proof or means of ascertainment.[6] Thus, according to Dummett, it is strictly nonsensical (whether in mathematics, the natural sciences, history, or any other discipline) to suppose that there must be some truth of the matter to us unknown or perhaps unknowable for the above-cited sorts of reason. In mathematics this would apply to cases such as that of Goldbach's conjecture – that every even number greater than two is the sum of two odd primes – which are borne out to the limits of current computational power but are still not ultimately proven despite their strong (indeed overwhelming) force of intuitive self-evidence.[7] In astrophysics or subatomic particle theory, it would endorse an instrumentalist line by rejecting any realist construal of those various hypothetical objects, processes, or events that play a role in our current best scientific theories. (Thus, for instance, it would deny the reality of the neutrino at that time when its existence was strongly predicted through the measurement of energy loss and no other explanation could be found that was consistent with the basic conservation laws.) More than that, it would deny on principle that there must be some fact of the matter – albeit to us presently unknown – as regards their existence and the truth-value of any statements we might make concerning them.[8] Other more homely instances include the ascription of certain qualities (e.g. courage) to human individuals who have not been called upon to display those qualities in action, or such historical statements as 'it must be either true or false that event X occurred at location Y at some time during the period T', where we haven't any means of settling the issue one way or the other.[9] In each case, according to Dummett, if we lack adequate grounds or criteria for judgement then we cannot properly be said to *understand* what is required in order for the statement in question to count as determinately true or false.

This argument derives partly from his acceptance of the Wittgensteinan doctrine that 'meaning is use', i.e. the thesis that our understanding of statements is manifest only in our ability to use them in the right sorts of context and with the right sorts of evidence or assertoric warrant.[10] Thus whenever such evidence is lacking – as in the above-cited instances – then plainly (according to Dummett) we cannot have an adequate grasp of what those statements mean and are hence in no position even to assert that they must be either true or false in keeping with the principle of bivalent (distributed) truth- and falsehood-values. From which Dummett concludes that there are not only gaps in our knowledge but also 'gaps in reality', that is to say, pseudo-statements which appear to assert something meaningful concerning historical events, or subatomic structures, or other such putative realia but which in fact 'concern a region of reality which is simply indeterminate'.[11] This applies not only in cases where the event in question is mythic, fictitious, or counter-factual, or where the structure is one that belongs to a realm of far-fetched hypothetical conjecture. Rather, according to Dummett, 'whenever a statement is true, it must be possible ... in principle, for *us* to know that it is true, that is, for beings with our particular restricted observational and intellectual facilities and spatio-temporal viewpoint'.[12] Only if it meets these criteria can a statement be considered apt for evaluation in terms of 'warranted assertability', or be treated as a proper candidate for ascription of bivalent truth–falsehood status. In which case reality extends just as far as we possess adequate knowledge of it, or at any rate the means of acquiring such knowledge within the limits imposed by our sensory equipment, cognitive apparatus, powers of intellectual grasp, and so forth. Beyond that there can only be 'gaps in reality' corresponding to the gaps – or the areas of indeterminacy – in those various statements that we are prone to make while not in possession of the relevant criteria or methods of verification.

This will surely strike the realist as a cautionary instance of what goes wrong when philosophy takes the Wittgensteinian turn towards a wholesale doctrine of meaning-as-use, one which denies that any statement can be meaningful – or any truth-claim intelligibly count as such – except in so far as it conforms to existing (communally sanctioned) modes of linguistic expression. Thus, for Dummett, '[t]he meaning ... of a statement cannot be, or contain as an ingredient, anything which is not manifest in the use made of it, lying solely in the mind of the individual who apprehends that meaning'.[13] Of course this derives from Wittgenstein's condign reflections on the impossibility that there should exist any such thing as a 'private language', i.e. a language that would somehow make sense only to its solitary user, and would therefore involve no shared conventions or communal modes of understanding.[14] Up to a point one can accept Dummett's analogous argument with regard to the criteria of intelligibility for truth claims in the natural sciences, mathematics, history, and elsewhere. After all, these claims have to be couched in a language that is subject to the normal requirements for shared communicative grasp, plus various other, more specialized requirements pertaining to the discipline or subject area concerned. However, Dummett goes far beyond this – in company with other anti-realists – when he asserts that we simply *cannot make sense* of the idea of verification-transcendent truths, or that nothing can count as a candidate item (such as the possible existence of certain elusive subatomic particles or the possible truth of Goldbach's Conjecture) unless we are already in possession of the relevant evidence or an adequate proof procedure.[15] Yet it seems paradoxical in the extreme, not to say absurd, that mathematicians who work long and hard to obtain a proof of that conjecture must be thought not to have understood it – and thus not to have known what they were looking for – until (if ever) the proof is finally achieved. The same applies to those other areas of enquiry, from particle physics to historical research, where Dummett takes the anti-realist argument to apply. For in each case the very nature of any such enquiry is to postulate a certain gap between reality and our beliefs concerning it, one that shows up in anomalous data, unlooked-for results, theoretical quandaries, lacunae in the extant historical record, explanatory shortcomings, conflicts between theory and observation, and so forth. For Dummett these can only be 'gaps in reality' since 'reality' *just is* the sum total of those statements we can make with adequate reason on the basis of our present-best knowledge.

To the realist this will seem a strictly preposterous idea, one that inverts the dependence relation or order of priority between objective truth and knowledge as a matter of justified true belief. Where Dummett goes wrong (after Wittgenstein) is in arguing that the relevant criteria – for truth, evidence, assertoric warrant – must *either* be located in a shared structure of linguistic-interpretative norms *or* 'in the mind of the individual who apprehends that meaning'. Since the latter is clearly not an option, it must therefore be the case that validity conditions are synonymous with those for meaningful utterances in general, that is to say, with the capacity to manifest a grasp of the way that certain statements are properly used in certain contexts of utterance. Thus, for instance (Dummett's example), we are justified in thinking that someone has an adequate grasp of the concept 'square' if they show themselves capable of discriminating between square and non-square objects, and also – crucially – applying the word 'square' to square objects and not to circles, triangles, trapezoids, etc. Or again (my example), we are justified in thinking that a physicist knows what she is talking about if she uses a term such as 'wave/particle superposition' in a context – that of quantum physics – where it manifests a knowledge of the relevant issues. However, this example should also give pause to anyone who follows Dummett (or Wittgenstein) in adopting a language-based verificationist approach whereby truth can simply drop out in favour of 'warranted assertability', the latter defined in accordance with the principle of meaning-as-use. For in that case any problems of interpretation – of attempting to explain just what it is that produces such puzzling quantum-mechanical phenomena – must be treated as problems that need not (and should not) arise for anyone who possesses the relevant concepts, i.e. anyone who discusses these matters in a well-informed plausible way. It is not just 'truth' that drops out in a somewhat technical (logico-semantic) sense of that term, but also the basic realist premise that our beliefs are rendered true or false by the way things stand in reality rather than by the way things happen to appear under this or that currently favoured description.

Thus the upshot of Dummett's anti-realist argument, as applied to quantum mechanics, is precisely to endorse the instrumentalist line taken by Bohr and other adherents of the orthodox Copenhagen view.[16] On this account issues of quantum 'reality' are strictly beside the point so long as the statistical predictions hold – or the equations work out – and we can find some language in which to describe (if not explain) the observed phenomena. In Bohr's words, '[t]here is no quantum world. There is only an abstract quantum mechanical description. It is wrong to think that the task of physics is to find out how Nature *is*. Physics concerns what we can say about Nature.'[17] For Dummett, as I have said, the anti-realist argument must be taken to apply right across the board – to mathematics, the natural sciences, history, and so forth – and not just to problem cases like that of quantum mechanics. Nevertheless, one might suggest that his argument derives a good deal of its seeming plausibility from the existence of such long-standing unresolved problems at the heart of advanced scientific enquiry. This conjecture is borne out by the way that other philosophers – some of whose arguments I shall examine later – have put forward a range of anti-realist or ontological-relativist proposals by analogy with quantum mechanics as a test case for claims concerning the scope and limits of scientific knowledge in general.

One could cite any number of passages from Dummett where the connection is not made explicitly but where it figures in the background as a source of enhanced credibility. Such is his above-mentioned claim about 'gaps in reality', taken to follow in strict (anti-realist) logic from the existence of equivalent gaps or lacunae in our present state of knowledge. Thus, 'there are meaningful statements which we can understand and whose truth or falsity we can therefore conceive of establishing, but for which, nevertheless, the question whether they are true or false has no answer; they concern a region of reality which is simply indeterminate.'[18] (This is actually a soft version of the argument, since elsewhere – as in the case of Goldbach's conjecture – Dummett seems to think that we cannot even understand [or meaningfully claim to understand] such statements unless we already have some definite evidence or adequate proof-procedure.) However, this idea can only seem remotely plausible if there is thought to be some ultimate ontological sense in which reality *itself* might be 'indeterminate', rather than our own best knowledge or theories concerning it. At which point the realist will surely respond that there is no warrant for any such claim since (1) it involves a straightforward confusion between ontological and epistemological issues, (2) it ignores the existence of alternative (e.g. Bohm-type hidden-variable) theories which purport to resolve at least some of these quantum problems,[19] and (3) – *pace* Wittgenstein – the 'limits of my language' may indeed be (in some sense) the 'limits of my world' but are *not* for that reason coextensive with 'the world' or with reality *sans phrase*.

Dummett is obliged to draw this conclusion in keeping with his anti-realist precept that it cannot make sense to say of any statement for which (as yet) we possess no means of verification that its truth value is none the less objectively decided by the way things stand quite apart from our present state of knowledge. Quantum theorists in the Copenhagen camp arrive at it on similar grounds, i.e. that what counts as 'truth' or 'reality' cannot be conceived as existing beyond our present-best means of observation plus the various predictive successes obtained by application of the standard formalisms. Thus here, as with Dummett, any 'gaps' in our understanding – like the various well-known puzzles and paradoxes entailed by orthodox QM – must be thought of as concerning 'a region of reality which is simply indeterminate'. Most instructive is the way that 'indeterminate' slides across from the sense 'neither true nor false so far as we can presently determine' to the sense 'intrinsically or of its very nature possessing no determinate truth-value'. For that slide is the basic enabling move for a good many forms of current anti-realist thinking, whatever their particular field of application.

II

For the realist such arguments amount to nothing more than a re-run of long familiar sceptical themes that go back to Berkeley, receive something less than an adequate rejoinder from Kant, and are then taken up by a diverse company of empiricists, positivists, verificationists, and proponents of Dummett-style anti-realism in the jointly Wittgensteinian and logico-semantic mode. However, it is the merest of parochial illusions to suppose that 'man is the measure', or that reality should somehow be expected to conform to our perceptions and conceptions of it. Kant may have argued that these were the a priori forms of all possible experience and knowledge, and hence that scepticism made no sense just so long as we remained within the bounds thus set for human understanding.[20] However, this argument is open to three main objections: first, that Kant was decidedly premature in hitching his a priori claims to the supposedly absolute, self-evident truths of Euclidean geometry and Newtonian physics;[21] second, that he thereby undercut the grounds for *any* such 'transcendental' deduction to the truth of our current best theories; and third, that Kant's critical philosophy – against his avowed intention – spawned a whole range of later anti-realist or sceptical-relativist arguments by placing 'reality' in a noumenal realm beyond reach of phenomenal cognition.

For the sceptic such arguments serve to make the point that human knowledge is inherently fallible and that 'truth' is nothing more than the way things appear to creatures with our particular range of sensory inputs, cognitive capacities, intellectual powers, etc.[22] For others – those of a realist persuasion – what it shows is that Kant went the wrong way around in answering the sceptical case of empiricists like Hume. That is to say, his response took the form of an argument from the impossibility of justifying knowledge on empirical grounds to the necessity of introducing a priori concepts and categories, and thence to the idea that reality and truth *just are* (for all humanly relevant purposes) what we make of them according to those same concepts and categories. Thus Kant may have thought to hold the line against scepticism by declaring himself an 'empirical realist' as regards the existence of a mind-independent (but to us unknowable) object-domain and a 'transcendental idealist' with respect to the conditions of possibility for human experience and knowledge. But in that case – so anti-realists urge – why not give up the otiose fiction of a noumenal 'thing-in-itself' and just admit that the way things appear to us is the way things are so far as we can possibly know, and hence – the crucial step in this argument – the way things are *tout court*? At which point the realist will again respond that this is a false extrapolation from Kant's critical doctrine of the faculties but one to which Kant laid himself open by failing to draw a sharp enough distinction between epistemological and ontological issues. That is, he made things easy for the sceptic by resting his case too much on the (supposed) a priori powers and limits of human understanding, and too little on the way that reality can kick back – so to speak – and force us to rethink what had hitherto counted as established or self-evident truth.

Such is the basic realist principle whether in epistemology and philosophy of science or as concerns our everyday process of belief adjustment when things fail to turn out in accordance with previous expectations. However, there is a yet more crucial point on which realists and anti-realists divide, namely (as I have said) the issue of verification-transcendent truths and whether it can make sense to suppose that such truths obtain quite apart from any limits on our proof procedures, sources of evidence, means of ascertainment, etc. Here the realist will typically assert that there exist all manner of contingent truths about the world – and likewise all manner of necessary truths about immaterial 'objects' such as numbers, sets, classes, theorems, or possible (as yet unaccomplished) mathematical proofs – which depend not at all on our current best methods for finding them out. In support of this claim, she will adduce two arguments: first, our knowledge of the growth of knowledge (i.e. that we now know a great many things that were once unknown), and second – following from this – the massive unlikelihood that we have now reached a consummate stage of advance in various scientific, historical, mathematical, and other fields of enquiry where truth *just is* what we take it be on the evidence presently to hand. For is it not absurd, the realist will surely protest, to think that such truths are somehow dependent on what we happen to believe concerning them? It is as if to say that our lack of determinate knowledge on a whole range of issues – from remote astrophysical events to the prehistory of sentient life forms, or what Napoleon had for breakfast on the eve of Waterloo – necessarily entails that there just cannot be any objective truth of the matter.

To adopt this line is somewhat like maintaining – on the currently fashionable 'strong anthropic' principle – that the universe and all its constituent properties (subatomic and molecular structures, gravitational constants, laws of dynamics, conservation, matter–energy conversion, and so forth) *must* be intrinsically knowable to creatures such as ourselves since we represent the high-point of sentient awareness and the end to which all those properties evolved. Or again, no less anthropocentrically, they are the products of a cosmic design – call it 'God' – whose purpose it was to create such a world whose intelligibility is proof of his existence and also of our own uniquely privileged place in the order of divine creation. For the realist, conversely, we had much better settle for some version of the 'weak' anthropic principle, that is, the argument that creatures like ourselves are pretty well equipped to pursue all kinds of reliable (truth-conducive) methods of enquiry since these are just the methods that have enabled us to survive and flourish in our particular physical environment or evolutionary niche.[23] Still, it is important to be clear that this argument cannot work if it reduces 'truth' to purely and simply a matter of 'warranted assertability', or to what counts for us – for creatures with our particular range of sensory inputs, cognitive interests, information sources, conceptual capacities, etc. – as pragmatically 'good in the way of belief'. For if there is one great lesson of evolutionary theory (whether in the life sciences or as applied by extension to epistemological issues), it is the lesson that survival-value is determined by objectively existent real-world conditions, rather than those conditions being somehow 'selected' in accordance with human aims, priorities, purposes, values, or desires.[24]

This is why the realist maintains the existence of verification-transcendent truths, as opposed to the idea – paraphrasing Wittgenstein again – that the limits of our knowledge are also the limits of our world. More precisely, they may be the limits of 'our' world in the epistemological (Kantian) sense that we cannot consistently or logically deny the truth of what counts for us as a matter of definite truth. However, there is no legitimate route from this self-evident point about the logical grammar of belief to the anti-realist argument that it cannot make sense to postulate truths (or conceivable states of reality) beyond our present-best powers of understanding or evidential warrant. For it is abundantly clear from the history of science – as even relativists like Kuhn are hard put to deny – that some beliefs have later shown up as erroneous *precisely on account* of inaccurate observation, limited knowledge, theoretical shortcomings, the influence of received ideas, metaphysical worldviews, fixed doctrinal attachments, and so forth.[25] One response is a sceptical meta-induction to the effect that, since most of our scientific 'knowledge' to date has turned out to be false, therefore we can have no reason – natural prejudice apart – to think that our current theories and truth-claims will fare any better. The other is to take this lesson on board but turn it around in support of the realist case. Thus the fact that we can argue from the falsehood (or the partial validity) of past theories is evidence for the existence of truth-standards that by definition transcend the criteria by which those theories were originally judged and which themselves later proved inadequate. Moreover, applying this same line of argument to our present situation, we can see that the sceptical meta-induction is capable of a quite different outcome. For if a good many current beliefs go the same way, then this can only occur through the future development of scientific theories that explain – among other things – just why those beliefs came up against certain recalcitrant data, theoretical anomalies, or limits to their own explanatory power.

In response the sceptic may argue that we are then confronted with an endless procession of beliefs, truth-claims, and theories, none of which could ever claim anything more than warranted assertability according to prevalent ideas of what counts as viable scientific knowledge. However, that argument will lose its force if one takes the realist's cardinal point: that at any stage in the history of science there *are and must be* 'verification-transcendent' truths, that is to say, objective features of reality and truths concerning them for which (as yet) there exist no adequate means of proof or ascertainment. Only thus can we explain why knowledge should ever make progress or meet any challenge to its present-best powers of conceptual-explanatory grasp.

J.S. Bell puts the case with typical forthright vigour when he remarks on the 'complacency' of pragmatist or instrumentalist beliefs – like that of Bohr – which equate truth with what works for present observational-predictive purposes, and which thereby avoid any deeper questioning as to the reality behind quantum appearances. This approach, he writes,

> has undoubtedly played an indispensable role in the evolution of contemporary physical theory. However, the notion of the 'real' truth, as distinct from a truth that is presently good enough for us, has also played a positive role in the history of science. Thus Copernicus found a more intelligible pattern by placing the sun rather than the earth at the centre of the solar system. I can well imagine a future phase in which this happens again, in which the world becomes more intelligible to human beings, even to theoretical physicists, when they do not imagine themselves to be at the centre of it.[26]

Such is the case for scientific realism in its widest, most generalized ontological form. What is further required if this case is to gain some epistemological purchase is an argument from inference to the best explanation, that is to say, a detailed descriptive account of particular phases in the history of science – e.g. particle physics – where realism eventually replaced instrumentalism as by far the most convincing candidate hypothesis by which to explain observational results.[27] On this account the growth of scientific knowledge with respect first to atoms, then electrons and photons, then quarks, muons, or neutrinos is precisely a growth in the justifiability of statements concerning their objective existence and possession of certain known attributes, electrical charges, causal powers, interactive capacities, and so forth.

Thus for instance, in the case of atoms, it was reasonable to hold an instrumentalist position just so long as they remained unobservable and had not yet acquired a crucial role in explaining a vast range of otherwise unaccountable phenomena in physics, chemistry, and the life sciences.[28] However, this attitude became less reasonable – or more a matter of standing prejudice – as evidence mounted in favour of the realist hypothesis, from Dalton's atomic theory of the elements to Mendeleev's periodic table and Perrin's famous series of experiments which offered something close to direct observational proof.[29] Of course there were still some eminent physicists – Ernst Mach and the early Einstein among them – who continued to maintain an instrumentalist line and who regarded the realist interpretation as a kind of metaphysical extravagance.[30] With regard to electrons, likewise, it remained possible to treat their existence as an open question even after the discovery of the cathode-ray phenomenon and Thomson's explanation of it as resulting from the negatively charged particles emitted by a heated metal wire in a vacuum. But this view was upheld very much against the odds as physics moved on and produced ever more detailed depth-explanatory accounts of the subatomic structure of matter. Hence Rutherford's model of the atom as a positively charged nucleus which contained most of the atomic mass and was surrounded by electrons of equivalent inverse charge. These he envisaged as orbiting the nucleus under the force of electrical attraction just as the planets orbited the Sun according to the laws of gravity. And clearly, if this model lays claim to something more than a vaguely metaphoric or impressionistic content, then its truth must be a matter of there *actually existing* such entities as 'atoms', 'nuclei', and 'electrons', which finally decide the truth-value of any statements or hypotheses concerning it.

At this point the instrumentalist will no doubt remark that the realist should take no comfort from Rutherford's theory since it ran into various well-known problems which at length gave rise to quantum mechanics and the retreat from any kind of naive 'metaphysical' realism. Thus it followed from Maxwell's field equations that the orbiting electrons would radiate energy through the propagation of electromagnetic waves. In which case – according to the classical laws of momentum and mass–energy conservation – they would be sure to slow down, enter a series of lower orbits, and eventually collapse into the nucleus. One solution, proposed by Bohr, explained how this 'solar-system' model might remain stable at least as applied to the simplest case, that of the hydrogen atom which included just a single electron. But it proved incapable of extension to more complicated subatomic structures and also lacked any means of determining precisely the conjoint values of mass and momentum that produced stability in the hydrogen atom. So it was – in response to these difficulties with the early Bohr model – that he and others (including de Broglie) proposed the radically alternative theory according to which both matter and energy existed only in quantized form, i.e. as wave–particle 'packets' whose state at any given time was a function of their 'jumping' from one discrete level to another in keeping with certain invariant values determined by the Schrödinger equation. This offered a solution to the problem of atomic collapse, since it explained why the orbiting electron could not fall below a certain energy level and hence below a certain value of angular momentum in its travel around the nucleus.[31] Also, it promised to resolve the problem of black-body radiation, which had first given rise to Planck's 1900 quantum conjecture concerning the discrete (non-continuous) character of energy emission. That is to say, it explained why the energy levels produced by a light-emitting source within the confines of a closed and totally absorbing (black-walled) container would *not* – as predicted by classical physics – very rapidly increase to the point of producing some quite cataclysmic event. For in this case also the highest energy level could not be greater than that allowed for by the highest state into which the wave–particle system could 'jump' according to the laws of quantum mechanics as derived from Schrödinger's equation.

III

So what are we to make of all this in the context of the realist versus instrumentalist debate? As we left it two paragraphs above, the realist had taken such entities as 'atoms' and 'electrons' to possess a strong claim to objective existence by virtue of their role as indispensable components in various theories of subatomic structure which proved highly fruitful of further, more detailed or depth-explanatory research. To which the instrumentalist predictably responded by pointing out the problems that arose with models – like those of Rutherford and the early Bohr – where 'atom' and 'electron' were realistically construed as referring expressions which occurred in statements with determinate truth-values, rather than as terms of descriptive convenience which entailed no such further ontological commitments. Moreover, so it is argued, these problems led on to the development of a theory – quantum mechanics – whose immense *instrumental* or observational–predictive success went along with its resistance to any kind of realist construal. Thus the upshot of this story is to drive home the lesson that realism will always undermine its own case by offering examples which then turn out to constitute decisive counter-examples from an instrumentalist viewpoint.

In its generalized (post-Kantian) epistemological version the argument runs as follows: that if truth is indeed that which pertains to a realm of objective and mind-independent reality, then by very definition it lies beyond the limits of any possible human understanding. At which point supposedly the sceptic wins hands down, or – where this argument is not pushed through to its ultimate sceptical conclusion – the Machian instrumentalist, Dummett-type anti-realist, or 'constructive empiricist' after Bas van Fraassen's methodological prescription.[32] Van Fraassen's argument to this effect is perhaps best captured in the following colourful passage:

> If I believe a theory to be true and not just empirically adequate, my risk of being shown wrong is exactly the risk that the weaker, entailed belief will conflict with actual experience. Meanwhile, by avowing the stronger belief, I place myself in the position of being able to answer more questions, of having a richer, fuller picture of the world .... But, since the extra opinion is not additionally vulnerable, the risk is – in human terms – illusory, and *therefore so is the wealth*. It is but empty strutting and posturing, this display of courage not under fire and avowal of additional resources that cannot feel the pinch of fortune any earlier.[33]

In its orthodox QM version the argument runs along much the same lines but with the further (ontological) twist: that 'reality' itself has somehow turned out to elude, resist, or subvert any claims that the realist can make concerning it. Such is at any rate the version to be found in a great many statements by Bohr and others who espouse an instrumentalist approach.

Still, there is a case for realism in quantum mechanics which counters both these lines of argument by observing simply – on the strength of past evidence – that the existence of objective truths about the world has never depended on science's claim to have got them right at some particular stage in the development of scientific thought. It is not just (as van Fraassen thinks) that realism offers us a 'richer, fuller picture of the world' but also that it takes due account of the crucial distinction between ontology and epistemology, or the truth of our theories conceived as a matter of objective correspondence with the ways things stand in reality and what we are currently disposed to believe on the best evidence to hand. As Nicholas Rescher succinctly puts it, the anti-realist's standard argument from error (e.g. with respect to Rutherford's model of the atom) is not so much an argument against scientific realism as 'against the ontological finality of science as we have it'.[34] Thus it may very well be the case that by far the greater part of our presently accredited 'knowledge' will eventually prove either false, inadequate, or restricted in its range of application. But this argument has absolutely no bearing – *pace* van Fraassen – on the issue of whether there exist objective truths about objects, events, or real-world states of affairs that determine the truth-value of our statements concerning them, quite apart from our present (maybe limited) sources of evidence or means of verification. To suppose otherwise, with anti-realists and proponents of orthodox QM, is to adopt just the kind of dogmatic sceptical *parti pris* that led the church authorities in Galileo's time to insist that he interpret the heliocentric hypothesis as an instrumental fiction devoid of substantive cosmological import.[35]

In that instance, of course, there were urgent doctrinal and ideological reasons for maintaining an instrumentalist line. But a similar case has been argued with respect to the establishment of orthodox QM at a time and in a socio-historical context – that of Germany after the first World War – that favoured the emergence of new scientific ideas with a certain irrationalist appeal, or at any rate theories which promised a break with old, presumptively discredited norms of *Wertfrei* objective method.[36] When applied elsewhere – as by 'strong' sociologists of knowledge – such approaches are mostly aimed to discredit any truth-based or realist conception of science. That is to say, these theorists routinely reject the distinction between 'context of discovery' and 'context of justification', taking it that sociological explanations go all the way down and that scientific truth *just is* whatever counts as such according to prevalent (ideologically motivated) interests, values, priorities, etc.[37] What is distinctive about the qualified form of this argument as applied to QM theory is that it strengthens the case for a realist interpretation by suggesting that the orthodox (Copenhagen) line gained credence more as a result of extraneous – socially and historically determined – factors than by virtue of any proven superiority in theoretical or causal-explanatory terms.[38]

However, it is clear that no amount of socio-cultural background research could settle this issue between the realist and anti-realist (or instrumentalist) interpretations. To suppose that it could is effectively to endorse the anti-realist case, since it gives the last word to a sociology of knowledge that treats all scientific truth-claims and theories as products of – or as 'relative to' – the norms and values that happen to prevail within some given cultural belief system. Rather, as I have said, any adequate defence of a realist approach will have go by way of an inference to the best explanation based partly on evidence from the history and philosophy of science which bears out the case for objective (verification-transcendent) truths, and partly on the progress in our detailed understanding of subatomic structures achieved by application of QM principles. As concerns the latter, the realist can assemble an impressive list of discoveries which at any rate lengthens the odds against any hard-line instrumentalist view. Rae sets them out in summary form and describes them, moreover, from the standpoint of practising physicist, so I can do no better than cite his account.

> The successes of quantum physics are not confined to atomic or subatomic phenomena. Soon after the establishment of the matter–wave hypothesis, it became apparent that it could be used to explain chemical bonding. For example, in the case of a molecule consisting of two hydrogen atoms, the electron waves surround both nuclei and draw them together with a force that is balanced by the mutual electrical repulsion of the positive nuclei to form the hydrogen molecule. These ideas can be developed into calculations of molecular properties, such as the equilibrium nuclear separation, which agree precisely with experiment. The application of similar principles to the structure of condensed matter, particularly solids, has been just as successful. Quantum physics can be shown to account for the fact that some solids are insulators while others are metals that conduct electricity and others again – notably silicon and germanium – are semiconductors. The special properties of silicon that allow the construction of the silicon chip with all its ramifications turn out to be direct results of the existence of electron waves in solids. Even the exotic properties of materials at very low temperatures, where liquid helium has zero viscosity and some metals become superconductors devoid of electrical resistance, can be shown to be manifestations of quantum behaviour.[39]

Of course there are those of an anti-realist persuasion to whom such arguments will seem completely wide of the mark. From their point of view, any attempt to deduce the truth of scientific realism from a mere listing of (supposed) realia and (presumed) explanatory achievements in this or that field of research is a hopeless endeavour that will always fall prey to the standard sceptical response. It is rather like G.E. Moore's confident belief that all one had to do in order to refute the sceptic and vindicate commonsense realism was to raise one hand, point to it with the other, and thereby prove – beyond reasonable doubt – that there existed at least two mind-independent, real-world, physical objects.[40] Moreover, as we have seen, the sceptic can always come back against sophisticated versions of the realist case by remarking that these are likewise subject to a form of self-refuting paradox. For if reality is indeed 'objective' (i.e. verification-transcendent) in the sense thus required, then there is – so it is argued – no possible way for us to gain knowledge of it or avoid being driven to the sceptical conclusion bequeathed by philosophers like Kant.

It should be apparent to anyone who has perused even a small part of the literature devoted to this topic that the debate is set up on just such terms as to ordain that conclusion in advance and ensure that the sceptic will always trump the realist if he or she accepts those terms.[41] Some philosophers have thought to avoid the whole problem by declaring – like the later Wittgenstein – that certain beliefs about truth and reality go so deep into our 'form of life' that they cannot be questioned (or called to epistemological account) without lapsing into incoherence.[42] Thus Wittgenstein talks about 'hinge' propositions on which everything else turns, since to doubt them is to undermine everything else that could possibly count as a meaningful statement or a well-founded item of belief.[43] Others, Hilary Putnam among them, have recommended a form of 'internal realism' that accepts the framework-relative character of all theories and truth-claims but sees this as no problem so long as we adopt the right sorts of framework for various purposes.[44] However, both 'solutions' – Putnam's more explicitly – amount to just a kind of naturalized Kantianism relieved of all that surplus 'transcendental' baggage but still providing no answer to the sceptic's challenge.[45] For it is scarcely an answer to be told (as by Wittgenstein) that there is no point raising questions of validity with regard to our various practices – in mathematics, the natural sciences, historical understanding, social customs, religious belief, or whatever – since these are so firmly embedded in our 'language-games' or cultural 'life-forms' as to constitute rules for intelligible discourse in the area concerned.[46] After all, we might conceivably adopt some other set of 'hinge' propositions or beliefs that entailed reinterpreting just about anything, from the 'rules' of elementary number theory to the basics of physical science. Nor does it help very much to be told, as by Putnam, that we can carry on talking about 'truth' and 'reality', though always with the express proviso that such talk must be construed in framework-relative or pragmatist ('good-for-some-particular-purpose') terms. Here again, the sceptic can always respond that this is just scepticism under a different and somewhat less alarming name, that is to say, a strategic fall-back position that counts objective reality a world well lost for the sake of avoiding all those old post-Kantian epistemological dilemmas.

One could multiply examples of this pattern of retreat among recent philosophers, especially those (such as Putnam himself) who started out as convinced causal realists in philosophy of science and language, and then backed off to the point of espousing an 'internal realist' position that is anti-realist and cultural-relativist in all but name.[47] Then again, there are others, like P.F. Strawson, who once took a scaled-down ('descriptivist' rather than 'prescriptivist') Kantian approach to epistemological issues, but who have lately come around to a 'naturalist' perspective which sees no hope of any answer to the sceptic's challenge except by following Hume's advice to keep philosophy separate from our everyday forms of unreflective commonsense practice.[48] On this view such problems arise only as a product of hyperinduced philosophic doubt, and can hence be relied upon simply to vanish – or lose their disturbing force – as soon as we return to the natural attitude. However, there is a fairly obvious sense in which these various purported 'solutions' to scepticism miss the whole point of the sceptic's argument, or respond to it only by shifting ground in a way that leaves the problem firmly in place. That is to say, they either take the route 'beyond' Kant and the perceived dead-end of transcendental idealism by adopting a Moore-like commonsense outlook that flatly refuses to engage such issues, or follow Wittgenstein's lead in declaring that everything is perfectly in order with our language games, practices, life forms, etc., and hence that the sceptic can only be deluded in labouring his or her pointless philosophical scruples.[49]

My point in all this is that many influential philosophers – even those (like Putnam) with a well-developed interest in the history of science – have shown an increasingly marked unwillingness to argue from our knowledge of the growth of scientific knowledge to the case for realism *vis-à-vis* the objects of that knowledge. As I have said, this case must ultimately rest on some form of inference to the best explanation, which in turn involves a number of premises – the regularity of nature, the validity of inductive reasoning, the existence of a mind-independent 'external world' – which sceptics since Hume have routinely denied or at any rate refused to grant as a basis for realist counter-arguments. However, there is no reason to suppose that such arguments necessarily run out at this point or that, since the sceptic cannot be answered on his or her chosen ground, therefore the realist has no choice but to adopt a line of least resistance as suggested in various ways by Wittgenstein, Putnam, or Strawson. On the contrary, the strongest case for scientific realism is that which starts out from particular examples of the growth in knowledge typically achieved through a deeper (causal-explanatory) account of objects, events, processes, properties, microstructural features, etc. For such advances would themselves lack any remotely plausible explanation were it not for the fact that the object terms and predicates in a valid scientific theory can be taken as referring to (or quantifying over) a real-world physical object domain and its various integral attributes.[50]

Now of course this argument can be turned around by someone who asserts – following Quine – that 'to be is to be the value of a variable', but who interprets that statement (again after Quine) in ontological-relativist terms, i.e. as showing that our various putative realia must always be construed as framework-relative or internal to a given ontological scheme.[51] Indeed, Quine takes quantum mechanics as a striking illustration of the radical change that might always be forced upon our basic conception of 'reality' – not to mention our accepted criteria for what counts as a valid scientific theory – by some new development in the physical sciences.[52] However, this still begs the realist's question of how we can explain advances in knowledge *including those achieved by application of quantum theory* except on the premise that our theories are rendered objectively true or false according to whether they capture some aspect of a belief-independent (or non-scheme-relative) reality. For otherwise, as Putnam himself once argued, it would be nothing short of a miracle that there existed such a vast range of technologies that worked *just as if* the current best scientific theories were true, but which – it might very well turn out – functioned on entirely different (to us unknown or maybe unknowable) principles.[53]

Putnam presents an interesting test case here since he has swung right across from an early commitment to causal realism in philosophy of science, epistemology, and philosophical semantics to his later view (as summarized above) that 'reality' and 'truth' are always relative to some given conceptual framework. It is my impression from reading his work on issues in quantum theory – especially the issue concerning alternative or deviant quantum logics – that these have been a large factor in Putnam's conversion to a fig leaf variety of so-called 'internal' realism.[54] On the one hand, he agrees with the Quine of 'Two Dogmas' that discoveries in physical science (such as wave-particle dualism or kindred quantum phenomena) may force us to envisage a change in the 'ground-rules' of logical thought, i.e. the adoption of a non-bivalent or many-valued logic. On the other hand, Putnam is well aware of the obvious objection that to change the logical rules in response to some perceived empirical anomaly is to open the way to all manner of evasive or shuffling compromise solutions. This objection was raised very forcefully by Popper[55] and also, improbably enough, by Feyerabend in his early writings on quantum philosophy. Thus Feyerabend: 'this sly procedure is only one (the most "modern" one) of the many devices which have been invented for the purpose of saving an incorrect theory in the face of refuting evidence and ... consistently applied, it must lead to the arrest of scientific progress and to stagnation'.[56] Yet according to orthodox QM – which Putnam accepts as having set the terms for debate – one is confronted with a choice between revising certain ground rules of logic and revising (or abandoning) some of our most basic notions of physical reality. In which case it is not hard to see why Putnam should have adopted the welcome escape route of an 'internal' (framework-relative) approach that refused on principle to prejudge the issue between these different lines of response to the problems of interpreting quantum mechanics. And from here – as often happens – the way might well have seemed open to a generalized application of the theory which extended to issues of 'reality' and 'truth' in the wider (macrophysical) domain.

IV

There is a pertinent passage in Putnam's 1995 book *Pragmatism: an open question*, where he takes exactly this line in response to a well-known argument by Ian Hacking concerning the reality of subatomic particles.[57] Hacking's point is that we should be realists with respect to anything that we can 'manipulate', such as the electrons beamed through an electron microscope or the various entities that show up when observed under such conditions, or again, the subatomic particles deployed (or produced) in superconducting collider experiments.[58] His approach can thus be described as 'instrumentalist' in a certain sense of that term – i.e. that it treats particles as real in so far as they exert or respond to some instrumental power – but not in the more customary sense of treating their existence as a useful hypothesis adopted purely for the sake of descriptive or explanatory convenience. The passage to which Putnam takes exception is one where Hacking recounts an experiment involving the gradual change of electrical charge on a supercooled niobium ball. 'Now how does one alter the charge on the niobium ball? "Well, at that stage", said my friend, "we spray it with positrons to increase or decrease the charge". From that day forth I've been a scientific realist. *So far as I'm concerned, if you can spray them they are real.*'[59]

I shall need to quote Putnam's response at some length since it brings out very clearly the way in which orthodox QM is taken to constitute a powerful (even a knock-down) argument against scientific realism. 'What does it mean', Putnam asks, 'to believe that "they" [i.e. the positrons] are "real"?'

> If it means that one believes that there are *distinct things* called 'positrons', then we are in trouble – a *lot* of trouble – with the theory. For the theory – quantum field theory – tells us that positrons do not in general have a definite *number*! In the particular experimental set-up Hacking is describing, they do have a definite number, perhaps, but it would be quite possible to set up an experiment in which one 'sprayed' the niobium ball, not with three positrons, and not with four positrons, but with a *superposition of three and four positrons*. And elementary quantum mechanics already tells us that we cannot think of positrons as having *trajectories*, or as being, in general, *reidentifiable*.[60]

Now Hacking is perfectly aware of all this – as his book and other writings show – and can well do without the italicized lessons in elementary quantum theory. The point of his argument is *not* to deny (if I can turn this strategy round for a moment) that QM places large problems in the way of any realist ontology based on the idea of positrons as *distinct things* having a *definite number* and thus capable of being *reidentified* from one observation to the next. To ignore those problems – or adopt that ontology despite them – would amount to the claim that quantum mechanics rested on a huge mistake and that all the various interpretations were so much wasted effort. At any rate, Hacking's instrumentalist realism is committed to no such ontology of discrete particles and no such avoidance of the issue concerning quantum superposition.

Nor can he fairly (or seriously) be charged with holding the belief that positrons are 'real' in the classical sense that each possesses a unique and well-defined 'trajectory' enabling us to track it across different space–time locations. This is indeed a problem for scientific realism and one that Putnam too easily brushes aside with his offhand appeal to quantum field theory as having settled the issue for anyone sufficiently up-to-date on these matters. After all, the reidentification criterion has been taken by many philosophers – from Kant to Strawson – as among the most basic conditions of possibility for knowledge and experience in general. Thus, according to Strawson, the perdurance of 'individuals' (objects and persons) from one spatio-temporal location to the next is one of those necessary presuppositions without which we could make no sense of ourselves or the physical world.[61] We could also make no sense of mathematics since, for instance, the truth of '1 + 1 = 3' could be shown by placing two rabbits together for a certain period of time and then discovering that there were three, or the truth of '1 + 1 = 1' likewise proven by finding just one drop of water on a surface where two separate drops had previously existed side by side. Such examples are often brought up as evidence against empirically based theories of mathematics like that proposed by J.S. Mill. What rules them out as controverting the truths of elementary arithmetic is firstly our grasp that those truths are analytic, i.e. not subject to any form of empirical refutation, and secondly our knowledge that the extra rabbit and the missing drop can be explained in terms of certain events that befell certain otherwise perduring and numerically self-identical physical objects. For in cases like these – as with geometrical axioms or the conservation laws in physics – any conflict with 'the evidence' can only be construed as proof that there must have been something amiss with our methods for obtaining that evidence.

Now in quantum mechanics, as Hacking would scarcely deny, there are problems that elude any clear-cut settlement along these or similar lines. Chief among them is the reidentification problem: the impossibility of determining whether two distinct measurements of particle location or momentum carried out at separate points on its assumed trajectory should be interpreted as yielding values for the 'same' particle or for two different particles that happened to show up in the right spatio-temporal vicinity. For this is just the point of Putnam's put-down rejoinder to Hacking: that in quantum field theory such questions cannot properly arise since any values thus obtained are dependent on the kind of measurement performed and the way that the wavefunction 'collapses' so as to yield those values. More precisely, it is the wavefunction that assigns a certain probability to certain outcomes (e.g. values of particle location or momentum) but only in so far as these represent a sampling from the range of alternative experimental set-ups and hence of alternative measurement results. Thus, if we are to speak of quantum 'reality', then that reality is the wavefunction itself, or the complex equations that describe it, as distinct from any localized manifestation in particle- or wave-like form. From which it follows (according to Putnam) that any idea of particles as 'really' existing – let alone as reidentifiable – cannot be sustained in conjunction with the findings of quantum field theory.

It is tempting, at this point, to draw comparisons with the waterdrop case (mentioned above) and to suggest that the issue of numerical identity with respect to subatomic particles might be resolved along similar lines. Alan Musgrave appears to have some such solution in mind when he offers the example of two particles (say an electron and a positron) that carry opposite charges and which thus cancel out – or cease to 'exist' in whatever one takes to be the operative sense of that term – on the instant that they come into contact. Thus, concerning the waterdrops,

> '2 + 2 = 4' is not a generalisation about what will happen if you count two things, and two more, then physically amalgamate them, then perhaps wait a while, and then count what you have. Nor do those physicists who talk about types of subatomic particles which mutually annihilate one another if brought together think that when this happens one and one had failed to equal two – rather, they say that some particles have disappeared.[62]

This is in the context of Musgrave's argument that certain truths – e.g. those of elementary arithmetic or the physical conservation laws – are simply not such as could ever be subject to revision under pressure of empirical counter-evidence. Of course there will be differences of view as to where exactly the line falls between truths that are purely analytic or definitional (as with mathematics on one interpretation), those that are synthetic a priori (mathematics and the conservation laws according to Kant), and cases – like the waterdrop instance – where empirical observation is just what allows us to explain away anomalous results that would otherwise (impossibly) constitute such counter-evidence. However, Musgrave is here making the larger point: that the identity criterion for physical objects (i.e. what allows us to pick them out from one observation or measurement to the next) is a *sine qua non* both for scientific realism and also for the application of mathematics to physics and the other natural sciences. That is to say, if we lacked this criterion then we could have no means of distinguishing contingent facts-of-observation from necessary truths-of-reason, or grasping why some non-arithmetical fact – having to do with the behaviour of liquids under certain conditions – must be taken to explain how the two drops of water added up to one.

Whether these principles can possibly extend from the macro- to the microphysical realm is a chief point at stake between realists and instrumentalists with respect to QM phenomena. Musgrave and Putnam are squarely opposed on this issue, the former supposing that it still makes sense to consider particles as possessing both a history (or trajectory) and unique identity criteria, the latter that no such supposition is warranted given what we know – or may justifiably conjecture – on the basis of quantum field theory. Thus Musgrave takes it as self-evident that in the case of mutual annihilation 'some particles have disappeared', whereas Putnam treats it as evidence of

Hacking's quaint realist beliefs that he should carry on talking about positrons as if they were reidentifiable objects with definite locations, momenta, trajectories, points of disappearance, etc. For the chief lesson of quantum field theory is that we cannot, as in classical mechanics, distinguish any one particle from any other in terms of their respective positions or coordinate values from one measurement to the next. According to this theory, in Squires' words, 'they are described by a wavefunction which tells us the probability of finding an electron at one place and an electron at another place; in no way are the two electrons distinguished'.[63] In which case – so it seems – there can be no question of establishing numerical identity criteria such as Hacking requires in order to support a realist ontology, and which Musgrave invokes (by analogy with the waterdrop case) to avoid confusion between matters of necessary truth and matters of empirical warrant. If there is any ultimate reality 'behind' quantum appearances, it is that of the wavefunction itself, construed as distributing probability values for the outcome of this or that particle measurement. Thus, 'quantum mechanics is a deterministic theory of wavefunctions, just as classical mechanics is of positions.'[64] However, the realist can extract small comfort from this given that the wavefunction exists 'at all points of space' and must be taken as yielding determinate values only as and when such a measurement is carried out.

The prospects for realism appear still bleaker if one takes account of relativistic quantum field theory, that is, the theory which claims to reconcile orthodox QM with Einstein's special relativity. This involves a procedure of 'second quantization' described by Squires in terms that bring out its extreme resistance to any realist construal reliant on maintaining identity criteria for particles. 'In the transition from classical to quantum mechanics', he writes,

> variables like position changed from being definite to being uncertain, with a probability distribution given by a wavefunction, i.e. a (complex) number depending upon position. In relativistic quantum field theory we have a similar process taken one stage further; the wavefunctions are no longer definite but are uncertain, with a probability given by a 'wavefunctional'. This is again a (complex) number, but it depends upon the wavefunction, or, in the case where we wish to talk about several different types of particle, upon several wavefunctions, one for each type of particle .... The total number of particles of a given type is not a fixed number. Thus the theory permits creation and annihilation of particles to occur, in agreement with observation.[65]

There is an irony here comparable to that which overtook Einstein's attempt, in his debates with Bohr, to establish the 'incompleteness' of orthodox QM theory and hence the case for an alternative realist or hidden-variables account.[66] Just as that paper gave rise to problems that appeared to undermine its own argument, so likewise the attempt to square QM with special relativity generates a further development – relativistic quantum field theory – according to which any talk of 'particles' as possessing determinate space–time coordinate values is rendered more problematic.[67]

Thus when Squires poses the rhetorical question, 'do we learn anything in all this which might help us with the nature of reality?' he is obliged to answer 'apparently not', since the best current theory for removing the conflict between orthodox QM and realism (relativistically construed) is one that lands us with problems undreamt of on Einstein's account. 'If, in our previous, non-relativistic, discussion, we regarded the wavefunction as a part of reality, we now have to replace this with the wavefunctional, which is even further removed from the things we actually observe. The wavefunctions have become a part of the observer-created world, i.e. things that become real only when measured.'[68] That is to say, every attempt so far to achieve a workable reconciliation between realism and orthodox QM has had the upshot of pushing such a wished-for resolution yet further out of sight. I should emphasize again that Squires is himself very far from endorsing the anti-realist or instrumentalist position and devotes a large part of his book to canvassing various possible alternative (e.g. Bohm-type hidden variables) theories.[69] In fact he clearly regards it as something of a scandal that instrumentalism has remained the default theory over such a long period in so fundamental a discipline of scientific thought. Nevertheless, as the above passages make clear, he is also acutely aware of the problems that stand in the way of any realist interpretation. From which the reader might well be tempted to conclude that Putnam is right – as against Hacking and Musgrave – in his belief that quantum mechanics undermines every possible argument for realism *vis-à-vis* the ultimate nature of 'reality'.

V

I put the case in this sharply paradoxical (not to say nonsensical) form because it brings out the strain that so often results when philosophers routinely argue from the presumptive *truth* of orthodox QM theory to the presumptive supersession of all those concepts – such as the existence of a quantum reality beyond phenomenal appearances – which they *cannot do without* in the process of defending their own anti-realist position. This is all the more evident when Putnam extrapolates from quantum mechanics to its supposed implications for the realist case with respect to objects and events in the non-quantum (macrophysical) domain. Thus he takes it – along with adherents to the orthodox QM line – that such phenomena as quantum superposition and particle nonlocality must constitute a large and most likely an insuperable problem for any conception of physical reality as existing apart from some given descriptive framework, ontological scheme, system of measurement, etc. Hence (I have suggested) Putnam's adoption of a theory – so-called 'internal realism' – which appears little more than a notional alternative to Kuhnian anti-realist or paradigm-relativist talk. And yet there are passages elsewhere in the same book where Putnam comes out quite explicitly against such ideas and indeed reverts to something very like his earlier causal-realist approach to issues of knowledge and truth. Thus for instance:

> Suppose a terrestrial rock were transported to the moon and released. Aristotle's physics clearly implies that it would fall to the earth, while Newton's physics gives the correct prediction (that it would stay on the moon, or fall to the surface of the moon if lifted and released). There is a certain magnificent indifference to *detail* in saying grandly that Aristotle's physics and Newton's are 'incommensurable'.[70]

Of course Putnam is choosing his words with some care so as not to let the realist cat too far out of the instrumentalist bag, as by saying that Newtonian physics 'gives the correct prediction', rather than that it provides the true (objectively valid) causal-explanatory account. Nevertheless, the main force of his argument resides in its offering support for the case that the truth of scientific theories is a function of their actually describing and explaining certain real-world physical objects and events, rather than their merely conforming (as the 'internal realist' Putnam would have it) to some given paradigm or conceptual scheme. For it does indeed require a majestic 'indifference to detail' – as well as an indifference to basic standards of valid reasoning on the evidence – for the Kuhnian to cite such well-worn examples as bearing out the sceptical-relativist thesis of paradigm 'incommensurability'.[71]

So there is, we may conclude, dubious warrant for extending the claims of quantum anti-realism beyond the microphysical domain. The case of Schrödinger's unfortunate cat – 'superposed' between life and death until some observer opens the box and thereby collapses the wavefunction – should at least give pause to anyone tempted in this direction.[72] But then, what justification is there for taking the opposite line – in company with (among others) Einstein, Schrödinger, and Bohm – and supposing that the QM problems and paradoxes might yet be resolved in some manner consistent with the tenets of scientific realism? At an anecdotal level Squires recounts having attended a party thrown by the physicists at CERN (Conseil Européen pour la Recherche Nucléaire) to celebrate their recent discovery of the W boson. 'There would have been no excuse for the party', he remarks, 'if the scientists had not been convinced they had shown that these particles, predicted by theorists, really existed. Science is concerned with *discoveries*, not *inventions*!'[73] However, such an argument will scarcely convince orthodox QM theorists, instrumentalists, anti-realists, van-Fraassen-type 'constructive empiricists', or anyone who – like Putnam – would think it good enough reason to celebrate that the physicists had managed to devise (or invent) a descriptive framework consistent with the theorists' current best predictions. What is needed is clearly something much stronger by way of principled justification for supposing that there could (indeed *must*) exist a more adequate realist interpretation of QM phenomena than anything presently available. I have suggested several such lines of argument in the course of this book so far and will now re-state them in summary form.

Most crucial to any version of the realist case is that which asserts the existence of objective or verification-transcendent truths. In other words, there exist many features of reality that lie beyond our knowledge or present-best powers of understanding, but which none the less obtain quite apart from what we happen to think or believe. Moreover – so this argument holds – they determine the truth value of any speculative statements or hypotheses that we might advance concerning them *even if* we possess no adequate means of ascertaining their truth or falsehood. To think otherwise, the realist maintains, is to adopt an absurdly parochial and anthropocentric view of what might be the case in reality, as distinct from what we fallible enquirers have managed to discover up to now. Thus Squires:

> long before the human brain gave names to their constituents, there were protons and neutrons binding together to form nuclei. The way we describe reality is dependent upon the human brain and is therefore subject to the brain's limitation; it is an unnecessary arrogance on our part to assume that reality is itself subject to similar limitations.[74]

Anti-realist construals of quantum mechanics – Bohr's most prominent among them – trade heavily on the kinds of post-Kantian epistemological dilemma that result from adopting a representationalist approach to issues of knowledge and truth.[75] From here it is but a short step to Putnam's 'internal realism', arrived at by abandoning Kant's 'metaphysical' framework, i.e. his transcendental arguments from the conditions of possibility for experience in general, and admitting as many different versions of reality as there are different frameworks (or conceptual schemes) under which to interpret them.

Clearly this goes along very well with the orthodox QM doctrine, according to which – on Bohr's account – there is no point raising issues of quantum 'reality' since the best we can do when confronted with paradoxes like wave–particle superposition is deploy alternate 'complementary' (framework-relative) descriptions which enable us to avoid such unwanted ontological commitments.[76] However, it is an argument which all too readily equates the limits of our present-best knowledge with the way things somehow mysteriously are in the quantum-physical domain. After all, as Rachel Wallace Garden remarks,

> [i]t is surely sensible to see the quantum peculiarities as products of weak description rather than as incomprehensible features of the world. It is surely more reasonable to suppose that our theory is inadequate rather than to argue that reality itself is bizarre .... We should accept quantum mechanics as the most successful theory we have at present, while setting out to develop a new and better theory of reality.[77]

At any rate, there is something decidedly perverse about a theory – orthodox QM – which refuses on principle to accept the possibility of any such 'new and better' alternative. All sorts of (often contradictory) arguments tend to get mixed up here, from the notion that quantum 'reality' is intrinsically so strange that our theories and descriptions cannot hope to encompass it, to the claim that QM descriptions are the best to be had – even if themselves paradoxical – and hence must be construed as 'realistically' describing a quantum world that stubbornly defies rational comprehension.

Now we can return to Musgrave's point about the confusions engendered by a failure to distinguish empirical facts-of-observation from necessary truths such as those that apply in mathematics, logic, and certain core regions of physical science, e.g. the conservation laws. Anti-realists very often seem to forget the extent to which this distinction carries over into quantum-theoretical debate and provides a large measure of common ground even for opposed interpretations. Thus, for instance, it is essential to Bell's theorem – as likewise to the original Einstein–Podolsky–Rosen thought experiment – that the law of conservation of spin angular momentum should apply without exception in quantum mechanics just as in the macrophysical domain.[78] That is, such reasoning takes for granted that the two divergent particles will always have inversely correlated spin-values since the singlet pair from which they derived had combined spin-value zero, in which case their conjoint value at any point thereafter will necessarily be measured as zero. Therefore – Einstein and his colleagues maintained – one could disprove orthodox QM by obtaining two simultaneous measurements for different parameters on particles A and B which would also yield determinate (inversely-correlated) values for B and A. It would then be possible – in theory at least – to specify *both* values for *each* particle and thereby vindicate the realist case (against Heisenberg's uncertainty principle) that those values were obtained objectively for the system as a whole and were no way dependent on this or that choice of measurement parameter.

Bell's theorem threw a paradox into this line of argument by proving that if 'reality' was consistent with the well-established quantum formalisms then the weight of statistical evidence for the degree of anti-correlation went far beyond anything accountable in terms of 'classical' (EPR) realism. What it showed, in short, was that '[t]he world can either be in agreement with quantum theory or it can permit the existence of a local theory; both possibilities are not allowed'.[79] This case was supported by subsequent laboratory experiments which confirmed Bell's theorem by establishing a clear discrepancy between the anti-correlation effects brought about by momentary switches of the measuring apparatus and the kinds of result that might have been predicted according to classical conservation laws. Hence the irony, as Squires describes it, that

> Einstein believed in local realism (as we do); quantum theory seemed to
> deny such a belief and was therefore considered by Einstein to be

incomplete. The EPR thought experiment was put forward as an argument, in which the idea of locality was implicitly used, to support this view. We now realize, however, that the experiment actually demonstrates the impossibility of there being a theory which is both complete and local.[80]

However, this should not obscure the fact that Bell's calculations and those applied in interpreting the Aspect results are themselves dependent – no less than EPR – on a range of distinctly 'classical' assumptions, among them the existence of a physical object-domain which, however puzzling its details, permits such experiments to be carried out and conclusions to be drawn from them. Nor are these assumptions so trivial or self-evident as to count for nothing either way as regards the debate between realist and instrumentalist QM theories. For they concern the basic realist contention – basic, that is, to the entire development of modern science from Galileo to Einstein and beyond – that the world exhibits various physical properties some of which are matters of empirical observation, whereas others (like the conservation laws) are treated as synthetic a priori, and others again turn out to have the character of necessary (i.e. mathematical and logical) constants.

I have argued that this applies just as much to quantum mechanics – on whatever interpretation – as to scientific theories in the macrophysical or 'classical' domain. That is, QM avails itself of various strictly indispensable premises which entail both a realist ontology (what would *have to be the case* in physical reality if this or that theory were correct), and a constant appeal to mathematical procedures – such as those involved in Heisenberg's or Schrödinger's equations – as a source of *precise and informative* knowledge concerning that physical reality. In this respect there is a striking resemblance between Galileo's famous dictum that 'the book of nature is written in the language of mathematics' and the way that quantum mechanics has evolved through progressively more refined applications of the established QM formalisms.

Take for instance the procedure of 'second quantization', as described above, and its deployment of Schrödinger's equation twice over so as to provide a mathematical basis for relativistic quantum field theory. From one point of view – that of commonsense realism – this creates yet further problems for any realist construal of quantum mechanics since it seems to entail that the 'reality' in question is entirely a product of complex numbers or abstruse mathematical formalisms. In so far as these latter are conceived in instrumentalist terms (i.e. as convenient heuristic devices for achieving a match between predictive and observational data), they will of course not satisfy the realist's ontological or causal-explanatory requirements. Thus, in P.H. Eberhard's words,

quantum theory is generally considered only as a tool to make predictions, not as a description of the mechanisms behind quantum phenomena.

Mathematical entities, such as wave functions and density matrices which are used to define quantum states, depend not only on characteristics of the quantum system itself, but on our knowledge of it as well. Models have been sought that would have features that are only characteristic of the system and are disentangled from the mathematical description of our information about it. These features could be called objective reality.[81]

But from another point of view this appears as just one more stage in the progressive mathematization of the physical sciences which began with Galileo and continued right down to the equations of special and general relativity. At each point in this process the resulting theory was one that conflicted with various tenets of a commonsense-realist, straightforwardly intuitive, or anthropocentric view of the world.[82] Yet at each point also the theory turned out to describe and explain some aspect of reality which could not have been thus described and explained except for the extraordinary power of mathematics to deliver accurate knowledge of that world on every scale from the laws of celestial mechanics to the structure of subatomic particles. Thus commonsense 'realism' and scientific 'realism' are two very different things, subject to different criteria, and tending to diverge most sharply during periods of radical scientific theory change.

What is perhaps most unusual about the case of quantum mechanics is the extent to which this divergence has opened up *within* the physics community and given rise to a conflict of interpretations that has lasted for almost a century. However, we need not be driven to conclude either that the conflict is unresolvable – perhaps (as Bohr sometimes suggests) on account of the inherent strangeness or 'irrationality' of quantum phenomena – or that any eventual resolution of it will take the form of a theory so mathematically complex as to exclude a realist interpretation. For, as I have said, this ignores a decisive counter-argument; namely, that certain 'abstract' quantum formalisms – such as Schrödinger's equation – have managed not only to describe and predict but also to *explain* a great range of otherwise puzzling physical phenomena. Among them are the properties of chemical bonding, of semiconductors and (more recently) of superconductivity and superfluidity, properties that either would not have been discovered or would remain wholly mysterious were it not for the singular explanatory power of the matter–wave hypothesis arrived at by use of Schrödinger's equation. Of course such quantum-based explanations must fail to satisfy the causal realist who – like Einstein and Schrödinger himself – requires something more of a complete physical theory. But there is no compelling reason (anti-realist or orthodox QM prejudice aside) to suppose that such a theory cannot be achieved or even, should it lie beyond human grasp, that the limits of our knowledge must therefore represent the very nature of quantum reality.

Endnotes

1 William P. Alston, *A Realist Conception of Truth* (Ithaca, NY: Cornell University Press, 1996), p. 84.

2 Ibid, p. 126.

3 For further discussion in various contexts, see Roy Bhaskar, *Scientific Realism and Human Emancipation* (London: Verso, 1986); Michael Devitt, *Realism and Truth*, 2nd edn (Oxford: Blackwell, 1986); Frank B. Farrell, *Subjectivity, Realism, and Postmodernism: the recovery of the world in recent philosophy* (Cambridge: Cambridge University Press, 1994); Richard L. Kirkham, *Theories of Truth* (Cambridge, MA: M.I.T. Press, 1992); Penelope Maddy, *Realism in Mathematics* (Oxford: Oxford University Press, 1990); Mark Platts (ed.), *Reference, Truth and Reality: essays on the philosophy of language* (London: Routledge & Kegan Paul, 1980); Gerald Vision, *Modern Anti-Realism and Manufactured Truth* (London: Routledge, 1988).

4 B. d'Espagnat, 'Nonseparability and the Tentative Descriptions of Reality', in W. Schommers (ed.), *Quantum Theory and Pictures of Reality* (Berlin: Springer Verlag, 1989), pp. 89–168; p. 103.

5 Ibid, pp. 103–4.

6 See Michael Dummett, *Truth and Other Enigmas* (London: Duckworth, 1978); Richard L. Kirkham, 'What Dummett Says about Truth and Linguistic Competence', *Mind*, Vol. 98 (1989), pp. 207–24; Michael Luntley, *Language, Logic and Experience: the case for anti-realism* (London: Duckworth, 1988); N. Tennant, *Anti-Realism and Logic* (Oxford: Clarendon Press, 1987); Alan Weir, 'Dummett on Meaning and Classical Logic', *Mind*, Vol. 95 (1986), pp. 465–77; Timothy Williamson, 'Knowability and Constructivism: the logic of anti-realism', *Philosophical Quarterly*, Vol. 38 (1988), pp. 422–32; Kenneth P. Winkler, 'Scepticism and Anti-Realism', *Mind*, Vol. 94 (1985), pp. 46–52; Crispin Wright, *Realism, Meaning and Truth* (Oxford: Blackwell, 1987) and *Truth and Objectivity* (Cambridge, MA: Harvard University Press, 1992).

7 See Dummett, *Truth and Other Enigmas* (op. cit.); also *Frege: philosophy of language* (London: Duckworth, 1973); *Elements of Intuitionism* (Oxford: Oxford University Press, 1976); *The Interpretation of Frege's Philosophy* (Duckworth, 1981).

8 For further discussion see Michael Gardner, 'Realism and Instrumentalism in Nineteenth-Century Atomism', *Philosophy of Science*, Vol. 46 (1979), pp. 1–34.

9 Dummett, *Truth and Other Enigmas* (op. cit.).

10 See Ludwig Wittgenstein, *Philosophical Investigations*, trans. G.E.M. Anscombe (Oxford: Blackwell, 1953); *On Certainty*, ed. G.E.M. Anscombe and G.H. von Wright (Blackwell, 1969); *Lectures on the Foundations of Mathematics*, ed. C. Diamond (Chicago: University of Chicago Press, 1976).

11 Dummett, 'The Metaphysics of Verificationism', in L.E. Hahn (ed.), *The Philosophy of A.J. Ayer* (La Salle, IL: Open Court, 1992), p. 146.

12 Dummett, 'What Is a Theory of Meaning?', in G. Evans and J. McDowell (eds.), *Truth and Meaning: essays in semantics* (Oxford: Clarendon Press, 1976), p. 100.

13 Dummett, *Truth and Other Enigmas* (op. cit.), p. 216.

14 Wittgenstein, *Philosophical Investigations* (op. cit.), Sections 242–315.

15 See Note 6, above.

16 See for instance Niels Bohr, *Atomic Theory and the Description of Nature* (Cambridge: Cambridge University Press, 1938) and *Atomic Physics and Human Knowledge* (New York: Wiley, 1958); also – for a range of viewpoints on the realism/anti-realism issue – James T. Cushing, *Quantum Mechanics: historical contingency and the Copenhagen hegemony* (Chicago: University of Chicago Press, 1994); Bernard D'Espagnat, *Veiled Reality: an analysis of present-day quantum concepts* (Reading, MA: Addison-Wesley, 1995); John Honner, *The Description of Nature: Niels Bohr and the philosophy of quantum physics* (Oxford: Clarendon Press, 1987); Max Jammer, *Philosophy of Quantum*

Mechanics (New York: Wiley, 1974); Alastair I.M. Rae, *Quantum Physics: illusion or reality?* (Cambridge University Press, 1986); Euan Squires, *The Mystery of the Quantum World*, 2nd edn (Bristol and Philadelphia: Institute of Physics Publishing, 1994).

17 Bohr, cited in J.S Bell, *Speakable and Unspeakable in Quantum Mechanics: collected essays on quantum philosophy* (Cambridge: Cambridge University Press, 1987), p. 142.

18 Dummett, 'The Metaphysics of Verificationism' (op. cit.), p. 146.

19 See especially David Bohm, *Causality and Chance in Modern Physics* (London: Routledge & Kegan Paul, 1957); David Bohm and B.J. Hiley, *The Undivided Universe: an ontological interpretation of quantum theory* (London: Routledge, 1993); also David Z. Albert, 'Bohm's Alternative to Quantum Mechanics', *Scientific American*, No. 270 (May 1994), pp. 58–63; F.J. Belinfante, *A Survey of Hidden Variable Theories* (Oxford: Pergamon Press, 1973); Cushing, *Quantum Mechanics: historical contingency and the Copenhagen hegemony* (op. cit.).

20 Immanuel Kant, *Critique of Pure Reason*, trans. N. Kemp Smith (London: Macmillan, 1964).

21 See Kant, *Metaphysical Foundations of Natural Science*, trans. J. Ellington (Indianapolis: Bobbs-Merrill, 1979); also Gordon G. Brittan, *Kant's Theory of Science* (Princeton, NJ: Princeton University Press, 1978) and Michael Friedman, *Kant and the Exact Sciences* (Hemel-Hempstead: Harvester, 1992).

22 For recent discussions, see A.C. Grayling, *The Refutation of Scepticism* (London: Duckworth, 1985); Christopher Hookway, *Scepticism* (London: Routledge, 1992); Alan Musgrave, *Common Sense, Science and Scepticism: a historical introduction to the theory of knowledge* (Cambridge: Cambridge University Press, 1993); John Watkins, *Science and Scepticism* (Princeton, NJ: Princeton University Press, 1984).

23 See Roger Penrose, *The Emperor's New Mind: concerning computers, minds and the laws of physics* (London: Vintage, 1990); also J. Barrow and F.J. Tipler, *The Anthropic Cosmological Principle* (Oxford: Oxford University Press, 1991).

24 See for instance Peter Muntz, *Our Knowledge of the Growth of Knowledge* (London: Routledge & Kegan Paul, 1985).

25 Thomas S. Kuhn, *The Structure of Scientific Revolutions*, 2nd edn (Chicago: University of Chicago Press, 1970); also Ian Hacking (ed.), *Scientific Revolutions* (Oxford: Oxford University Press, 1981).

26 Bell, *Speakable and Unspeakable in Quantum Mechanics* (op. cit.), p. 125.

27 See for instance J.L. Aronson, 'Testing for Convergent Realism', *British Journal for the Philosophy of Science*, Vol. 40 (1989), pp. 255–60; Gilbert Harman, 'Inference to the Best Explanation', *Philosophical Review*, Vol. 74 (1965), pp. 88–95; Peter Lipton, *Inference to the Best Explanation* (London: Routledge, 1993); Wesley C. Salmon, *The Foundations of Scientific Inference* (Pittsburgh, PA: University of Pittsburgh Press, 1967); Peter J. Smith, *Realism and the Progress of Science* (Cambridge: Cambridge University Press, 1981).

28 See Gardner, 'Realism and Instrumentalism in Nineteenth-Century Atomism' (op. cit.).

29 J. Perrin, *Atoms*, trans. D.L. Hammick (New York: van Nostrand, 1923); also Mary Jo Nye, *Molecular Reality* (London: MacDonald, 1972); Josef M. Jauch, *Are Quanta Real? a Galilean dialogue* (Bloomington, IN: Indiana University Press, 1973); Wesley C. Salmon, *Scientific Explanation and the Causal Structure of the World* (Princeton, NJ: Princeton University Press, 1984).

30 See Ernst Mach, *The Science of Mechanics* (La Salle, IL: Open Court, 1984); also various contributions to P.A. Schilpp (ed.), *Albert Einstein: philosopher–physicist* (Open Court, 1994), especially Einstein's 'Autobiographical Notes', pp. 1–95; Bas van Fraassen, *The Scientific Image* (Oxford: Oxford University Press, 1980) and *Quantum Mechanics: an empiricist view* (Oxford University Press, 1991).

31 For further discussion see Peter Forrest, *Quantum Metaphysics* (Oxford: Blackwell, 1988); R.I.G. Hughes, *The Structure and Interpretation of Quantum Mechanics* (Cambridge, MA: Harvard University Press, 1989); Dugald Murdoch, *Niels Bohr's Philosophy of Physics* (Cambridge: Cambridge University Press, 1987); A. Sudbury, *Quantum Mechanics and the Particles of Nature* (Cambridge University Press, 1986).

32 See van Fraassen, *The Scientific Image* and *Quantum Mechanics* (Note 30, above).

33 Van Fraassen, 'Empiricism in the Philosophy of Science', in P.M. Churchland and C.A. Hooker (eds.), *Images of Science: essays on realism and empiricism, with a reply from Bas C. van Fraassen* (Chicago: University of Chicago Press, 1985), p. 255; cited in C.J. Misak, *Verificationism: its history and prospects* (London: Routledge, 1995), p. 165.

34 Nicholas Rescher, *Scientific Realism: a reappraisal* (Dordrecht: D. Reidel, 1987), p. 61.

35 For further discussion from an instrumentalist viewpoint, see Pierre Duhem, *The Aims and Structure of Physical Theory*, trans. Philip Wiener (Princeton, NJ: Princeton University Press, 1954) and *To Save the Phenomena: an essay on the idea of physical theory from Plato to Galileo*, trans. E. Dolan and C. Maschler (Chicago: University of Chicago Press, 1969). Karl Popper takes a different, highly critical view in his book *Quantum Theory and the Schism in Physics* (London: Hutchinson, 1982). See also Christopher Norris, 'Truth, Science and the Growth of Knowledge', in *Reclaiming Truth: contribution to a critique of cultural relativism* (London: Lawrence & Wishart, 1996), pp. 154–79.

36 See especially Paul Forman, 'Weimar Culture, Causality, and Quantum Theory, 1918–1927: adaptation by German Physicists and Mathematicians to a hostile intellectual environment', *Historical Studies in the Physical Sciences*, Vol. 3 (1971), pp. 1–115 and 'The Reception of an Acausal Quantum Mechanics in Germany and Britain', in S.H. Mauskopf (ed.), *The Reception of Unconventional Science, AAAS Selected Symposium*, No. 25 (1979), pp. 11–50; also Cushing, *Quantum Mechanics: historical contingency and the Copenhagen hegemony* (op. cit.).

37 See for instance Barry Barnes, *About Science* (Oxford: Blackwell, 1985); David Bloor, *Knowledge and Social Imagery* (London: Routledge & Kegan Paul, 1976); Steve Fuller, *Philosophy of Science and its Discontents* (Boulder, Colorado: Westview Press, 1989); Karin D. Knorr-Cetina, *The Manufacture of Knowledge: an essay on the constructivist and contextual nature of knowledge* (Oxford: Pergamon Press, 1981); Bruno Latour and Steve Woolgar, *Laboratory Life: the social construction of scientific facts* (London: Sage, 1979); Steve Woolgar, *Science: the very idea* (London: Tavistock, 1988).

38 See especially Cushing, *Quantum Mechanics* (op. cit.).

39 Rae, *Quantum Physics: illusion or reality?* (op. cit.), p. 14.

40 G.E. Moore, 'Proof of an External World', in T. Baldwin (ed.) *Selected Writings*, (London: Routledge, 1993), pp. 147–70; see also 'The Refutation of Idealism' and 'A Defence of Common Sense', pp. 23–44 and 106–33.

41 See especially the fine recent study by Michael Williams, *Unnatural Doubts: epistemological realism and the basis of scepticism* (Princeton, NJ: Princeton University Press, 1996).

42 See Note 10, above.

43 Wittgenstein, *On Certainty* (op. cit.), Sections 97–9, 341–4.

44 Hilary Putnam, *The Many Faces of Realism* (La Salle, IL: Open Court, 1987); *Realism With a Human Face* (Cambridge, MA: Harvard University Press, 1990); *Renewing Philosophy* (Harvard University Press, 1992).

45 See Alston, *A Realist Conception of Truth* (op. cit.) for a detailed critique of Putnam's 'internal realist' approach; also various contributions to Peter Clark and Bob Hale (eds.), *Reading Putnam* (Oxford: Blackwell, 1993).

46 See for instance David Bloor, *Wittgenstein: a social theory of knowledge* (New York: Columbia University Press, 1983); Derek L. Phillips, *Wittgenstein and Scientific Knowledge: a sociological perspective* (London: Macmillan, 1977); S. Shanker, *Wittgenstein and the Turning-Point in the Philosophy of Mathematics* (Albany, NY: State University of New York Press, 1987).

47 For the 'early' (causal-realist) Putnam, see for instance his *Philosophical Papers*, Vols 1 and 2 (Cambridge: Cambridge University Press, 1975). The shift towards 'internal realism' is already well under way by the time of *Realism and Reason: philosophical papers*, Vol. 3 (Cambridge University Press, 1982).

48 P.F. Strawson, *Individuals: an essay in descriptive metaphysics* (London: Methuen, 1959); *The Bounds of Sense: an essay on Kant's Critique of Pure Reason* (Methuen, 1966); *Scepticism and Naturalism: some varieties* (Methuen, 1985).

49 See Notes 22 and 41, above.

50 See entries under Note 3, above; also D.M. Armstrong, *Universals and Scientific Realism*, 2 vols. (Cambridge: Cambridge University Press, 1978); J.Aronson, R. Harré and E. Way, *Realism Rescued: how scientific progress is possible* (London: Duckworth, 1994); Roy Bhaskar, *Reclaiming Reality: an introduction to contemporary philosophy* (London: Verso, 1989); Michael Devitt, *Realism and Truth*, 2nd edn (op. cit.); Jarrett Leplin (ed.), *Scientific Realism* (Berkeley & Los Angeles: University of California Press, 1984); Karl Popper, *Realism and the Aim of Science* (London: Hutchinson, 1983); Peter J. Smith, *Realism and the Progress of Science* (op. cit.); M. Tooley, *Causation: a realist approach* (Blackwell, 1988).

51 W.V.O. Quine, 'Two Dogmas of Empiricism', in *From a Logical Point of View*, 2nd edn (Cambridge, MA: Harvard University Press, 1961), pp. 20–46; also *Ontological Relativity and Other Essays* (New York: Columbia University Press, 1969).

52 Quine, 'Two Dogmas of Empiricism' (op. cit.), p. 43.

53 See Note 47, above; also Putnam, 'Explanation and Reference', in G. Pearce and P. Maynard (eds.), *Conceptual Change* (Dordrecht: D. Reidel, 1973).

54 See especially Putnam, 'How to Think Quantum-Logically', *Synthèse*, Vol. 29 (1974), pp. 55–61 and other papers collected in *Mathematics, Matter and Method* (Cambridge: Cambridge University Press, 1979).

55 See Popper, *Quantum Theory and the Schism in Physics* (op. cit.).

56 Paul K. Feyerabend, 'Reichenbach's Interpretation of Quantum Mechanics', *Philosophical Studies*, Vol. XX (1958), p. 50.

57 Putnam, *Pragmatism: an open question* (Oxford: Blackwell, 1995).

58 Ian Hacking, *Representing and Intervening: introductory topics in the philosophy of natural science* (Cambridge: Cambridge University Press, 1983).

59 Ibid, p. 23.

60 Putnam, *Pragmatism* (op. cit.), p. 59.

61 See Strawson, *Individuals* (op. cit.).

62 Alan Musgrave, *Common Sense, Science and Scepticism* (op. cit.), p. 188.

63 Squires, *The Mystery of the Quantum World* (op. cit.), p. 43.

64 Ibid, p. 24.

65 Ibid, p. 110.

66 Albert Einstein, B. Podolsky and N. Rosen, 'Can Quantum-Mechanical Description of Reality be Considered Complete?', *Physical Review*, series 2, Vol. 47 (1935), pp. 777–80; Niels Bohr, article in response under the same title, *Physical Review*, Vol. 48 (1935), pp. 696–702; also Bohr, 'Conversation with Einstein on Epistemological Problems in Atomic Physics', in P.A. Schilpp (ed.), *Albert Einstein: philosopher-scientist* (La Salle, IL: Open Court, 1969), pp. 199–241; Arthur Fine, *The Shaky Game: Einstein, realism, and quantum theory* (Chicago: University of Chicago Press, 1936); Don Howard, 'Einstein on Locality and Separability', *Studies in the History and Philosophy of Science*, Vol. 16 (1985), pp. 171–201.

67 See especially Tim Maudlin, *Quantum Nonlocality and Relativity: metaphysical intimations of modern science* (Oxford: Blackwell, 1993); Michael Redhead, *Incompleteness, Nonlocality and Realism: a prolegomenon to the philosophy of quantum mechanics* (Oxford: Clarendon, 1987).

68 Squires, *The Mystery of the Quantum World* (op. cit.), p. 111.

69 See entries under Note 19, above.

70 Putnam, *Pragmatism* (op. cit.), p. 53.

71 Kuhn, *The Structure of Scientific Revolutions* (op. cit.).

72 Erwin Schrödinger, *Letters on Wave Mechanics* (New York: Philosophical Library, 1967); also John Gribbin, *In Search of Schrödinger's Cat: quantum physics and reality* (New York: Bantam Books, 1984).

73 Squires, *The Mystery of the Quantum World* (op. cit.), p. 124.

74 Ibid, p. 132.

75 On this aspect of Bohr's thinking, see especially Honner, *The Description of Nature* (op. cit.).

76 See Note 16, above; also Henry J. Folse, *The Philosophy of Niels Bohr: the framework of complementarity* (Amsterdam: North-Holland, 1985).

77 Rachel Wallace Garden, *Modern Logic and Quantum Mechanics* (Bristol: Adam Hilger, 1983); cited by Squires (op. cit.), p. 122.

78 See Notes 66 and 67 above: also J.S. Bell, *Speakable and Unspeakable in Quantum Mechanics: collected papers on quantum philosophy* (Cambridge: Cambridge University Press, 1987); James T. Cushing and Ernan McMullin (eds.), *Philosophical Consequences of Quantum Theory: reflections on Bell's Theorem* (Notre Dame, IN: University of Notre Dame Press, 1989).

79 Squires (op. cit.), p. 98.

80 Ibid, p. 102.

81 P.H. Eberhard, 'A Realistic Model for Quantum Theory with a Locality Property', in Schommers (ed.), *Quantum Theory and Pictures of Reality* (op. cit.), pp. 169–216; pp. 169–70.

82 For a detailed historical and philosophical account of these developments, see J. Alberto Coffa, *The Semantic Tradition from Kant to Carnap: to the Vienna Station* (Cambridge: Cambridge University Press, 1991).

3　Bell, Bohm and the EPR debate

A case for nonlocal realism

I

Philosophers have standardly drawn a distinction between truths of reason that are taken to hold necessarily or come what may, and matters of fact or contingent truths which hold with respect to the way things are in our particular world.[1] They have further distinguished analytic propositions whose truth is self-evident, since purely a function of their logico-semantic form, and synthetic a priori judgements – such as those of geometry and also mathematics, at least according to Kant – which are likewise self-evident, since they are presupposed by any knowledge we can have concerning that world.[2] These latter are purportedly in no way dependent on experience but must rather be thought of as defining the very conditions of possibility for knowledge in general. They are closely bound up with our primordial intuitions of space and time – the very forms of 'outward' and 'inner' experience – and extend to our capacity for grasping causal relations, perceiving the truth of geometrical theorems, and applying a priori concepts and categories to the manifold of sensuous perceptions.

Such was at any rate Kant's response to the sceptical argument of those (like Hume) who denied that there was anything more to our causal-explanatory theories and conjectures than a fixed habit of association between contiguous objects or events.[3] However, it was open to various kinds of renewed sceptical attack, not least on account of his failure to secure any necessary link between the forms and modalities of human understanding and the real-world (ontological) object domain to which they presumptively applied. For Kant, so it seems, there was no conflict between declaring himself on the one hand a 'transcendental idealist' in epistemological matters, and on the other an 'empirical realist' as concerned the existence of a mind-independent reality which *for that very reason* belonged a noumenal realm beyond our utmost powers of cognitive grasp.[4] However, this response to the sceptical challenge has turned out to spawn all manner of anti-realist or framework-relativist arguments which can plausibly claim to press Kant's case to its logical conclusion. The pattern was set early on by idealists like Fichte, who viewed reality as a 'posit' of the world-constitutive Ego which left no room for Kant's lingering and self-contradictory attachment to (so-called) empirical realism.[5]

Beyond that, the history is familiar enough, from Hegel's all-encompassing dialectics of Spirit to Schopenhauer's solipsist ruminations and Nietzsche's all-out sceptical assault on the ideas of truth, knowledge, and objectivity. But there is also a parallel (if somewhat less dramatic) story to be told about the way that Kant's dilemmas have been re-enacted by recent Anglo-American philosophers – Hilary Putnam among them – who for various reasons have felt themselves driven to adopt some version of 'internal' or framework realism.[6]

In any case, my main concern here is with how these issues have been taken up in epistemology and philosophy of science since the early years of this century. Thus it is often argued that Kantian foundationalism no longer retains credibility in the face of certain scientific developments – such as non-Euclidean geometry, relativity theory, and quantum mechanics – that have brought about a radical change in our thinking about what (if anything) should count as an instance of a priori knowledge.[7] Certainly few philosophers nowadays, except the hardiest metaphysicians, would venture along the Kantian path to a justification of the physical sciences based on philosophy's prior claim to deduce the very scope and limits of attainable knowledge.[8] Mostly they are content to accept the science-led process that has called such claims into doubt, or at any rate – like Strawson – to adopt a kind of scaled-down ('descriptivist') Kantianism that limits them to such elementary matters as establishing the criteria by which we can identify enduring particulars from one to another spatio-temporal context.[9] Even so, there is a question whether quantum theory has rendered that assurance problematic, a question to which one possible response is Strawson's subsequent retreat to a form of qualified Humean naturalism.[10] Indeed, one could argue that quantum theory is the latest stage in that process by which mathematics, geometry and the physical sciences have increasingly detached themselves from any idea of a priori knowledge and truth. On the orthodox QM view, it is the kind of physics that results when 'reality' itself is treated as a construct out of various complex mathematical formalisms – such as Planck's constant or Schrödinger's equation – which exist in a realm utterly remote from anything graspable in commonsense-intuitive or 'naive' realist terms.[11]

All of which appears to place large problems in the way of any argument that assumes particles to retain at least some vestige of discrete continuous identity from one measurement to the next, and that moreover assumes the 'book of nature' to be written 'in the language of mathematics', as it was for scientists from Galileo to Einstein. What underlies this twofold assurance is the belief that reality must possess an intelligible structure, that is to say, the kind of structure that can best be understood through a joint process of empirical observation and logico-mathematical reasoning. On one version of the QM story – the orthodox version according to Niels Bohr[12] – this classical worldview was finally put to the test when Einstein and his colleagues (Podolsky and Rosen) framed the EPR thought experiment with the aim of defending both a realist ontology and a rationalist conception of scientific method. From which it would follow that quantum mechanics was necessarily 'incomplete' in respect of some deep further fact (or 'hidden variable') whose discovery would bring it back into line with the requirements for an adequate physical theory. However, that argument failed, so the orthodox version goes, since when conjoined with certain other (strictly ineliminable) features of quantum theory it turned out to harbour implications – such as nonlocal causality or remote interaction between widely separated particles – which undermined its own chief premises.[13] Peter Gibbins offers a clear brief account of the EPR paradox as viewed by its original framers and then by those in the henceforth dominant 'orthodox' QM camp. Thus:

> Einstein showed that if it is admitted, as it is by the Copenhagen interpretation in one of its forms, that the act of making a measurement on a quantum system disturbs it, then this disturbance can be transmitted over large distances. Einstein rejected action-at-a-distance on principle and so considered that he had demonstrated the incompleteness of quantum mechanics .... [However], a deeper analysis of EPR due to J.S. Bell in the middle 1960s shows, so most philosophers of physics would say, that quantum mechanics is inconsistent with any hidden-variables theory that rejects action-at-a-distance, and further that quantum mechanics is itself a nonlocal theory. Experiments, though difficult ones to perform, can decide between quantum mechanics and any local hidden-variables theory. The consensus is that experiment has vindicated quantum mechanics and also refuted locality.[14]

Thus realism (at any rate local realism) is simply no longer an option for anyone who has grasped the point of Bohr's arguments *contra* Einstein and who accepts QM as strongly borne out on both statistical-predictive and observational grounds. *Tertium non datur*, it seems: either QM will prove to have been fundamentally mistaken while somehow coming up with all the right results, or we shall have to change our most basic ideas of what counts as an 'adequate physical theory'.[15]

II

However, there is a third alternative and one that is just as well supported by the evidence despite having been consistently sidelined – or dismissed out of hand – by proponents of the orthodox view. This case is well made by James T. Cushing in his recent book *Quantum Mechanics: historical contingency and the Copenhagen hegemony*.[16] What Cushing brings out to striking effect is the extent to which that orthodox view was imposed by a steadfast refusal, on the part of influential figures like Bohr and Heisenberg, to entertain even the possibility that any such alternative existed. And this despite the fact that at least one candidate – de Broglie's pilot-wave theory, later taken up by David Bohm's – was observationally equivalent to the standard account (that is to say, fully compatible with the quantum formalisms, predictions, experimental results, etc.) while offering the basis for a realist ontology and a causal explanation of quantum phenomena.[17]

Bohm's first statement of the realist case was in a 1952 paper which set out its basic principles as follows:

> The usual interpretation of the quantum theory is self-consistent, but it involves an assumption which cannot be tested experimentally, namely that the most complete possible specification of an individual system is in terms of a wave function that determines only probable results of actual measurement processes. The only way of investigating the truth of this assumption is by trying to find some other interpretation of the quantum theory in terms of at present 'hidden' variables, which in principle determine the precise behaviour of an individual system, but which are in practice averaged over in measurements of the type that can now be carried out .... [A]s long as the mathematical theory retains its present general form, this suggested interpretation leads to precisely the same results for all physical processes as does the usual interpretation. Nevertheless, the suggested interpretation provides a broader conceptual framework ... because it makes possible a precise and continuous description of all processes, even at the quantum level.[18]

Cushing argues that the Copenhagen view prevailed not so much through its intrinsic merits – or for want of any other viable interpretation – but chiefly on account of its advocates' high prestige and their rapid success in spreading the idea that quantum mechanics absolutely required a break with all previous ('classical') conceptions of scientific truth and method. Hence one of the main problems with that theory, namely its failure to specify the point of transition between micro- and macrophysical domains, or the point at which the necessary measure of allowance for effects of quantum indeterminacy gave way to the kinds of causal regularity observed in all other (scientific and everyday) contexts of enquiry. Such was indeed Schrödinger's objection when he framed his cat-in-a-box hypothesis by way of a classic *reductio ad absurdum* designed to test the Copenhagen theory on this issue of how and where the line should properly be drawn.[19]

As Cushing sees it there were just two possible answers to the question thus raised, i.e. the issue of quantum superposition – the cat as somehow momentarily both dead and alive – and whether this pertained to the very nature (the putative 'reality') of quantum-mechanical systems or rather to the limits of our current best understanding. This uncertainty might be put down to

> (a) *our state* of knowledge (quantum mechanics is incomplete) and (b) the *actual state* of the system (there is a sudden change upon observation). If we choose (a) (which is what Schrödinger feels we *must* do intuitively),

then quantum mechanics is *incomplete* (i.e. there are physically meaningful questions about the system that it cannot answer – *surely* the cat was *either* alive *or* dead *before* we looked). On the other hand, choice (b) saddles us with the measurement problem (and with a vengeance). The collapse of the wave packet becomes an actual *physical* process which must be explained.[20]

It seems to me that Cushing is right to think it something of a scandal that so many commentators – not only popularizing adepts but physicists and theorists of high distinction – have elected to grasp one or both horns of the Schrödinger dilemma and failed to take the point of his *reductio* argument.[21] That is, they have worked on the orthodox (Copenhagen) assumption that any problems or paradoxes here must be taken as somehow intrinsic to the nature of quantum-mechanical systems, and hence as debarring any possible appeal to standards of 'intuitive' (realist and causal-explanatory) thought whether at the quantum or the macrophysical level. This in turn leads on to a further problem with the claims of quantum mechanics *vis-à-vis* other well-attested observations and theories in the physical sciences. For it is a basic principle with regard to any new candidate-theory that it should *both* constitute a genuine advance in respect of some hitherto unexplained range of phenomena, *and* manage to incorporate previous findings within the limit of their own (henceforth restricted) ontological and causal-explanatory scope. The classic case here is of course that of Einstein's special and general relativity, interpreted *not* as a full-scale replacement for Newtonian theories of space, time, and gravitation, but rather as a more comprehensive unified account which conserved Newton's theory as a special-case instance valid for all space–time frameworks whose gravitational fields were weak and where maximal velocities were nowhere near the speed of light.[22] Thus Einstein's hypothesis respected the standard constraint, i.e. that 'a theory that supersedes a previous one whose domain of validity has been established must reduce to the old one in a suitable limit'.[23] However, this was not the case with quantum mechanics on the orthodox Copenhagen interpretation. For here – as witness Schrödinger's cat and a range of kindred paradoxes – there was simply no way of establishing the limit within which the quantum formalisms applied, and beyond which physicists could properly appeal to more familiar (realist and causal-explanatory) conceptions of scientific method.

So there seems little merit in maintaining a theory which generates intractable paradoxes of the kind pointed out by Schrödinger while *de jure* excluding any possible advance (such as that envisaged by Bohm) towards a better, more encompassing, or depth-explanatory account. All the more so since Bohm's interpretation is at no point in conflict with the standard theory as regards its undoubted record of success at the level of predictive and purely observational yield. What remains inexplicable on the orthodox view is just *why* the theory should have been so successful by its own instrumentalist criteria. According to Bohr, such demands are misconceived as soon as one passes from the macrophysical domain (where causality holds for all practical purposes) to the realm of subatomic phenomena (where it has to be redefined in irreducibly statistical or probabilistic terms). But here again there is the issue why one should prefer a theory which fails to explain where and how that transition occurs, and which moreover leaves it an ultimate mystery – a puzzle *in the very nature* of quantum 'reality' – as to what might occur at the moment when the wavepacket 'collapses' and brings about one or another measurement. On this account, as P.H. Eberhard describes it:

> quantum theory is just a set of mathematical rules to predict future observations. There is no mention of a 'reality' of the quantum system with features to be defined, described, or ruled by an equation of evolution between times at which we make observations. No attempt is made to describe that reality between measurements. Computations make use of a wave function ..., evolving according to the Schrödinger equation between observations, and collapsing instantaneously everywhere at the time every one of these observations is made. It expresses everything we can know of the system.[24]

In which case it can offer no adequate explanation of *why* Schrödinger's cat should not be thought of as existing in a superposed (dead-and-alive or neither-dead-nor-alive) state until the box is opened up for inspection and the wavepacket thereby collapsed into one or the other henceforth determinate state. In so far as Bohm both explains these phenomena and does so in accord with the best observational evidence, his account would appear to have much stronger claims by all the usual (hitherto reliable) standards of scientific theory construction. On the Copenhagen view, this desire for some 'deeper' understanding can only be 'a leftover (historically conditioned) prop from our classical (physics) worldview, one that must simply be exorcised'. On Bohm's account, conversely, 'probabilistic "explanations" produce no really satisfying understanding when a visualizable causal story is in principle blocked'.[25] That is to say, in so far as it erects this *de jure* barrier to causal hypotheses of whatever kind, quantum mechanics can scarcely claim to provide an 'explanation' (let alone a genuine 'understanding') of its own operative methods, formalisms, heuristic devices, instrumental fictions, or whatever.

Cushing goes various ways around in attempting to explain the near-hegemonic status of the orthodox Copenhagen version. One is a broadly sociological account which suggests the influence of cultural factors, in particular the strain of anti-rationalist feeling and the deep mistrust of overweening scientific certitudes in post-war Weimar Germany. This is *not* – and the point requires emphasis – the kind of vulgar-reductionist or 'strong' sociological account which would treat all scientific theories as equally explainable by reference to their cultural conditions of emergence or the various pressures and incentives at work in their original context of discovery. On this view the only acceptable approach is one that adopts a strictly non-partisan principle of symmetry between those theories that are nowadays accepted (by mainstream philosophers and historians of science) as having made some genuine and lasting contribution to the advancement of knowledge, and those that are seen as having failed in that regard, and hence as fit material for sociological or psychobiographical treatment.[26] I have argued elsewhere – along with other critics of the strong sociological line – that this approach is thoroughly misconceived and cannot begin to explain either the manifest achievements of the physical sciences to date or our knowledge of the growth of scientific knowledge.[27] However, Cushing is far from espousing such a grossly conflationist view of the contexts of discovery and justification, or of the two separate issues: (1) Why were so many physicists so quickly won over to the orthodox Copenhagen theory?, and (2) What were the merits (or failings) of that theory as a matter of scientific warrant quite apart from the beliefs, motives, or predisposed bias of those who accepted it? Indeed his argument pursues a course precisely opposite to that of the strong sociologists when they apply their all-purpose levelling approach. For it is Cushing's claim that the Copenhagen theory, if not ruled out in favour of Bohm's alternative account, must at any rate be seen as a weaker theory on various well-established scientific grounds, among them those of rational inference to the best (most adequate) causal explanation.

So where the strong sociologists seek to efface the distinction between true and false theories – thinking thereby to redress the injustice of 'mainstream' (Whiggish) historiography – Cushing has a quite different aim in view. Merely to place them on a par, sociologically speaking, would be to miss the whole point of his argument that the orthodox version has managed to prevail through factors very largely unrelated to its merits in the context of justification. On this account it is the very success of the Copenhagen theory that requires sociological explanation, whereas Bohm's causal-realist alternative should be seen as possessing a stronger claim on properly scientific grounds. In other words, Cushing maintains the two-contexts principle – thus holding the line against cultural relativism – but turns it back against the 'mainstream' (orthodox) interpretation of quantum mechanics. And he does so precisely in order to argue that social and cultural factors played a large (though not entirely determinant) role in securing that theory despite the various challenges mounted against it by theorists such as Einstein, de Broglie, and Bohm. Of course he is not suggesting – like the strong sociologists – that externalist accounts of this sort are sufficient to provide a *complete explanation* of why certain scientific communities were swung into accepting certain ideas in a given historical and socio-cultural context. For only on the most reductionist view could it be thought that the central themes and concerns of quantum-mechanical debate – uncertainty, complementarity, the wave-particle dualism, statistical probability, nonlocal interaction, and so forth – were indeed *nothing more* than specialized symptoms of a wider malaise in society or the body politic.

Other commentators – among them Paul Forman in a well-known series of articles – have pressed much further in this direction.[28] Thus Forman links the emergence of quantum mechanics, and the Copenhagen theory in particular, to the widespread movement of revolt against rationalist and causal-realist philosophies of science which were viewed at the time as having somehow contributed to the crisis of European (more specifically German) national identity. 'If the physicist were to improve his public image', Forman writes,

> he had first and foremost to dispense with causality, with rigorous determinism, that most universally abhorred feature of the physical world picture. And this, of course, turned out to be precisely what was required for the solution of those problems in atomic physics which were then at the focus of the physicist's interest.[29]

From which Forman concludes – in distinctly 'strong' sociological vein – that 'substantive problems in atomic physics played only a secondary role in the genesis of the acausal persuasion'; moreover, that 'the most important factor was the social-intellectual pressure exerted upon the physicists as members of the German academic community'.[30] Other theorist-historians of quantum mechanics, including Max Jammer, have offered a broadly similar diagnosis, citing the rise of various anti-rationalist movements ('contingentism, existentialism, pragmatism, and logical empiricism') as evidence that there existed a post-war cultural climate highly receptive to the kinds of argument advanced by Bohr, Heisenberg, and their colleagues.[31]

Cushing finds these claims persuasive up to a point, but marks his own distance from them by insisting that issues of scientific truth and method cannot be reduced, so to speak, *without remainder* to the currency of socio-cultural attitudes and beliefs. More precisely, '"internal" factors were most important for the emergence of the formalism of quantum mechanics, "external" ones for the nature of the interpretation that was accepted.'[32] That is to say, those formalisms were arrived at on the basis of well-defined operational constraints upon the range of possible measurements, findings, assignments of particle location or wave-like probability distribution, etc., discovered through experiment and theory. To treat them as externally (that is, socio-culturally or ideologically) determined would be to give up any claim – such as Cushing requires – to distinguish between those interpretations of quantum phenomena that were justified by the scientific evidence and those others which betrayed, at least arguably, a predisposed irrationalist or anti-realist bias. For it is just his point that the Copenhagen version, although perfectly consistent with the evidence, nevertheless went far beyond anything required by that evidence and did so – moreover – in a way that conflicted with a great many otherwise well-supported principles of scientific method and theory construction. Among the latter were the three major tenets of Einstein's case against Bohr; namely, 'the existence of an objective, observer-independent reality, the necessity for causal (essentially deterministic) explanations for physical processes, and the locality/separability of the physical world'.[33] It could scarcely be claimed – except by the strongest of strong sociologists – that any problem with these tenets raised by quantum mechanics must be attributed chiefly or solely to the influence of cultural or sociopolitical factors.

Cushing himself is quite clear that one or other of them must be abandoned in consequence of Bell's theorem and the various experiments since carried out in order to test Bell's hypothetical results.[34] What these appear to establish is the impossibility of sustaining a realist or Bohm-type hidden-variables theory that would not entail the existence of nonlocal causal effects, or the anti-correlation of particle-spin measurements conducted over arbitrary distances from source and involving some form of superluminal (faster-than-light) 'communication' between the separated particles. In short, one cannot conserve all three of Einstein's realist postulates while also conserving quantum theory, or allowing for its great (indeed unrivalled) success as a matter of probabilistic and predictive yield. As Eberhard puts it, '[t]he EPR argument backfired. It was invented to demonstrate the shortcomings of the orthodox Copenhagen interpretation of quantum theory. It only ended up by showing an additional difficulty that an alternative theory describing reality would have to face. It would have to include faster-than-light influences'.[35] However, Bohm himself was prepared to grasp this particular nettle and admit nonlocality as a feature (albeit a puzzling feature) of the quantum domain just so long as the 'no-first-signal' rule applied; that is to say, so long as these remote correlations could not be used to convey messages at superluminal velocity. Since this was indeed the case – since there was no possible means of controlling or decoding the measurements obtained at either end – he took a fairly relaxed view of the EPR paradox and what Einstein called 'spooky action-at-a-distance'. In Bohm's words, '[i]f the price of avoiding nonlocality is to make an intuitive explanation impossible, one has to ask whether the cost is not too great'.[36]

Thus, for Cushing, there is no question of applying the strong sociological approach in such a way as to make these problems appear just a product of short-term cultural pressures among certain sections of the physics community at a certain time and place. On the other hand, he *does* argue that the readiness of so many physicists to endorse the Copenhagen view cannot be explained entirely on the basis of internal (i.e. strictly scientific, observational, or theoretical) criteria. After all, Bohm's alternative theory has a great many signal advantages, not least its avoidance of the 'Schrödinger's cat' problem (that is, the problem of defining any point where quantum indeterminacy must have an end), and its managing to incorporate far more of the principles, methods, and established results of classical physics from Galileo and Newton to Einstein. (It will seem less strange to call Einstein a 'classical' physicist if one considers his postulate of the speed of light as an absolute value within relativity theory, and, directly resulting from this, his

steadfast opposition to quantum mechanics on the orthodox interpretation.)
The Copenhagen theory entailed a whole range of anti-realist and counter-
intuitive beliefs, among them 'complementarity (the wave/particle duality),
inherent indeterminism at the most fundamental level of quantum
phenomena, and the impossibility of an event-by-event causal representation
in a continuous space–time background'.[37] It thus contrasted strongly with
Bohm's causal-realist account, which – to repeat – was observationally
equivalent to the Copenhagen theory while requiring nothing like such a
wholesale suspension of hitherto basic scientific laws. As Cushing describes
it, '[t]his interpretation, which in its nonrelativistic form represents a
microentity as a particle guided by a quantum potential (not as a wave *or* a
particle as does Copenhagen), lends itself readily to a realist construal of
even fundamental physical processes that develop completely
deterministically in a continuous space–time background'.[38]

What is basically at issue between these theories is the question whether
Bohr and Heisenberg were right in supposing any such realist ontology to be
strictly ruled out by the need to conserve the quantum formalisms and by
the fact that these could be construed only in statistical or probabilistic terms.
Cushing's argument (following Bohm) is that they were *not* so justified and
that the main obstacles to acceptance of a causal-realist account had to do
not so much with its intrinsic problems as with the strength of certain opposed
doctrinal attachments and other – strictly speaking – extraneous factors.
Thus a typical pronouncement in favour of the orthodox view (by an ex-student
and colleague of Heisenberg) rejects Bohm's theory on the grounds that it is
(1) 'observationally indistinguishable' from Copenhagen, and (2) laden with
'excess baggage' since it involves 'ontological assumptions which go, as far as
quantum theory can say today, beyond the realm of human knowledge, being
neither provable nor refutable with our means'.[39] But of course this takes for
granted the case against Bohm and indeed against any scientific theory –
like most of those advanced throughout the history of atomic and subatomic
physics – which have typically worked by inference to the best explanation as
regards the existence of as-yet unobservable entities.[40] That is to say, it adopts
the instrumentalist premise that such theories can never in principle be
justified, and of course ends up by confirming that premise in a purely circular
fashion. From which point it is a short step to the Copenhagen view that
quantum mechanics is *by its very nature* recalcitrant to any Bohm-type hidden-
variables theory which seeks to go beyond the existing evidence (i.e., the
established quantum formalisms) and offer a more satisfactory (causal-realist
or depth-ontological) account.

III

What quantum mechanics thus required was a readiness to abandon those
tenets of the classical worldview which Einstein had sought to preserve by
adopting the Lorentz transformations, i.e. the method for establishing

spacetime coordinate frameworks consistent with the speed of light taken as an absolute value. In Bohr's view – endorsed by Heisenberg and other proponents of orthodox QM – this system was clearly incapable of coping with phenomena such as quantum nonlocality or remote simultaneous anti-correlation. Yet we had no choice but to carry on describing and theorizing those phenomena in the language of 'classical' physics; that is to say, in a language whose conceptual resources – whose syntax, semantics, logical grammar, predicative structures, and so forth – had evolved as a part of our equipment for coping with objects and events in the non-quantum (macrophysical) domain. Thus, in Bohr's words:

> however far the phenomena transcend the scope of classical physical explanation, the account of all evidence must be expressed in classical terms. The argument is simply that by the word 'experiment' we refer to a situation where we can tell others what we have done and what we have learned and that, therefore, the account of the experimental arrangement and of the results must be expressed in unambiguous language with suitable application of the terminology of classical physics.[41]

The only way around this problem, Bohr thought, was to adopt a 'complementarity-principle' which acknowledged these ultimate limits on our power of univocal accurate description.[42] Thus, in cases like the wave–particle dualism, we should just have to use two different languages – along with their associated frameworks or ontologies – and avoid any conflict between them by accepting that each had its proper application relative to the experimental set-up or the kind of measurement performed. As for the reality of quantum phenomena – whatever lay 'behind' their observed manifestation – this should not be an issue for practising physicists since the theory's truth was measured by its power as a matter of confirmed statistical support and strong predictive yield.

John Honner has described Bohr's philosophical approach as a complex, elusive mixture of 'commonsense' pragmatism and Kantian ideas about a noumenal quantum 'reality' that must be thought to exist quite apart from our current best theories concerning it, but whose nature is so mysterious – so far beyond our powers of adequate representation – as to make those theories at best a matter of proven descriptive convenience.[43] All the same, Bohr clearly believed that the results so far obtained (whether through observation or through speculative thought experiments of the kind conducted in his series of dialogues with Einstein) were sufficient to force some large-scale revisions to our basic concepts of physical reality. That is to say, Bohr was enough of a realist – in this respect at least – to take it for granted (1) that replications of the two-slit experiment bore witness to the *actual repeated occurrence* of results confirming the wave–particle dualism, no matter how strange by hitherto accepted standards of scientific realism; and (2) that even in the case of those ingenious counterfactual thought experiments what counted was the rigour of consequential reasoning from (orthodox) QM premises to certain strictly unavoidable conclusions concerning events in the real-world microphysical domain.[44]

However, this is not the lesson drawn by some philosophers, Quine among them, who take quantum theory as a powerful illustration of the fact that changes of thinking in the physical sciences can force radical changes at every point in the fabric of accredited beliefs.[45] These changes – so it is argued – might extend all the way from high-level theories (which in the quantum context are famously 'underdetermined' by the evidence) to the ground rules of classical bivalent logic (which may perhaps need revising so as to admit paradoxical conclusions like the wave–particle dualism or the impossibility of assigning precise simultaneous values of particle location and momentum).[46] Thus quantum mechanics would appear to offer strong support for Quine's ontological-relativist thesis that there is no 'law of thought' or item of belief so firmly entrenched that it might not be subject to radical revision under pressure from conflicting evidence.

However, there are reasons to reject this argument, or at any rate to think it decidedly premature. First, it is based on just one interpretation of quantum mechanics – the 'orthodox' Copenhagen account – which even its advocates (Bohr and Heisenberg among them) admit to be fraught with unresolved problems and paradoxes. Second, that account has itself been challenged by alternative (Bohm-type 'hidden-variable') theories which entail nothing like such a drastic affront to our basic conceptions of scientific truth and method. And third, whatever the puzzles about quantum mechanics, they cannot be viewed as lending support to a doctrine of full-scale ontological relativity – or radical meaning-holism – where everything is thought of as somehow simultaneously up for grabs, from observation data to the ground rules of logical reasoning. For in that case it is hard to explain why those puzzles have continued to vex the minds of so many physicists and philosophers of science, from the year 1900 (when Planck first enounced the basic principles of quantum mechanics in connection with the phenomenon of black-body radiation), through Einstein's well-known series of debates with Bohr about quantum nonlocality and the wave–particle dualism, to more recent discussions in the wake of Bell's theorem and its sharpened re-statement of the issues. No doubt the standard (Copenhagen) view is one that consorts well enough with Quine's approach since it holds – in pragmatic-instrumentalist fashion – that the quantum theory has achieved a high measure of predictive success, which is all that should properly be expected of it, considering the kinds of problem that arise when one seeks to interpret those predictions in realist or causal-explanatory terms. But again there is an obvious problem here, namely the fact that every major development in quantum mechanics from its inception down has been spurred by just such problems concerning its conceptual foundations, its empirical warrant, its status *vis-à-vis* the 'laws' (or conventions) of classical two-valued logic, and so forth.

Thus, for instance, Einstein's disagreements with Bohr concerned precisely the question whether quantum mechanics could be thought of as a 'complete' physical theory, given the drastic choice that it seemed to impose between abandoning local realism – an option which Einstein found wholly unacceptable – or maintaining the possibility of some alternative account that would conserve the quantum formalisms but also the tenets of special relativity and the ground-rules of scientific reason. To Einstein the choice seemed clear: *against* any kind of 'spooky action-at-a-distance' (such as that entailed by quantum nonlocality) and *for* the maintenance of bivalent logic plus the speed of light as an absolute constant forbidding such impossible phenomena. Something had to give, both parties agreed, since the quantum theory despite its impressive degree of confirmation as a matter of statistical-predictive warrant nevertheless turned out to decree such highly paradoxical or counter-intuitive results. Where Einstein opted for retaining as much as possible of the standard (post-relativity) framework, Bohr took the view that any ultimate 'reality' subtending these quantum phenomena might lie forever beyond reach of an adequate descriptive-explanatory account. That is to say, we had no choice but to operate with the concepts and categories of 'classical' physics, even though there existed a body of evidence – from empirical observation and thought experiments – which pointed to their not holding good for processes or events at the subatomic level.

Bohr's philosophy thus worked out as a kind of extreme instrumentalism – 'never mind what it is or how it works so long as the formalisms match the results!' – combined with a version of the Kantian argument for a realm of noumenal reality to which we can never gain access, confined as we are to the phenomenal realm of humanly possible knowledge where intuitions must be brought under adequate ('classical') concepts. To this extent it might seem perfectly in keeping with the Quine–Kuhn doctrine of ontological relativity, or the notion that entities may be said to 'exist' just in so far as they play some role in this or that theory, paradigm, conceptual scheme, etc.[47] Even so, it may be argued that Bohr arrived at these conclusions only through a process of consequential reasoning on the evidence – or extrapolating logically from it through a series of ingenious thought-experiments – which retained a great many of those same classical concepts. Thus, in order for his arguments *contra* Einstein to possess probative force, they required (1) the resources of classical (two-valued) logic without which they could prove nothing either way, and (2) the supposition that any evidence thereby obtained – whether through empirical observation or through conjectural testing in 'the laboratory of the mind' – must have reference to processes or events in the quantum-physical domain.[48] For otherwise those arguments would belong to a realm of purely abstract hypotheses, a realm (that is to say) where the operative truth conditions were those of mathematics and formal logic rather than those of the physical sciences. There could then be no difference in point of ontological status between, say, the consequences of Gödel's undecidability theorem with respect to mathematical proofs and the consequences of quantum mechanics

– e.g. Heisenberg's uncertainty principle – as applied to our knowledge of what goes on at the subatomic level. That Gödel espoused a strictly Platonist view of mathematical truth is all the more reason for not running these arguments together as if they amounted to much the same thing.[49] Such is at any rate the pyrrhic upshot of Bohr's instrumentalist approach: that the whole apparatus of quantum mechanics – its concepts, descriptions, predictive hypotheses, probability functions, etc. – should be treated as a framework inescapably imposed by our own cognitive limits, and hence as affording no possible access to the putative 'reality' of quantum-physical events.

Of course it is still open for defenders of Bohr to protest that this was exactly his point. Thus we *must* continue to deploy such 'classical' concepts – from the ground-rules of logic to the framing of causal explanations and the idea that there exists a salient distinction between observer and observed – since they are built into the very structure of human understanding. And this despite our knowledge (somehow achieved from within that conceptual prison-house) that their deployment is no longer valid once the threshold is crossed from the macro- to the microphysical realm. Thus it is not hard to see why Quine – and others of a kindred persuasion – have used the example of quantum mechanics as a prime exhibit in their generalized case for the framework-relative character of even our most basic, firmly entrenched items of belief. What is not so clear is the justification firstly for adopting one particular (Bohr-derived) construal of the quantum-physical evidence, and secondly for assuming its lessons to apply outside and beyond the quantum domain. For it is a major problem with this interpretation – pointed out by Schrödinger through his thought experiment concerning the cat in the box – that it fails to explain how and where any cut-off point can be drawn between (supposedly) observer-induced microphysical events, like the collapse of the wavepacket, and (presumably) observer-independent events, like that of the cat's having died or not before the box was opened for inspection.[50]

I shall not here attempt to summarize the range of views – some of them mind-boggling in the extreme – which have grown up around this particular topic.[51] My main point is that none of these issues could ever have arisen – or these problems even registered as such – had physicists adopted the Quinean approach and counted everything in principle open to revision, from the logical 'laws of thought' to the theory-laden terms that figured in their various observation statements. The same applies to more recent arguments for and against the 'hidden-variables' theory, among them those of J.S. Bell concerning quantum nonlocality and the existence (as predicted by the standard theory) of superluminal anti-correlation effects.[52] For in this case it is a matter of showing that one is constrained to *make a choice* between alternative ways of resolving the issue, each of which requires some specified revision to accepted ('classical') concepts and categories, but none of which involves an outlook of wholesale revisionist licence such as that recommended by Quine.

Thus it follows from Bell's theorem that any causal-realist interpretation which adopts a Bohm-type 'hidden-variables' postulate while conserving the well-established quantum-statistical formalisms will also *necessarily* be constrained to admit the existence of superluminal interaction between widely separated particles. As so often, this result was first obtained through a thought experiment similar to those conducted by Einstein and Bohr, and only later – with the advent of more sophisticated measuring devices – borne out by a series of ingenious laboratory tests. But in neither case would the experiment have served its purpose (or narrowed the range of compossible options) had it been carried out in the Quinean belief that any 'recalcitrant' data could always be conjured away, whether by redistributing predicates, reinterpreting the observational evidence, or revising the logical ground-rules so as to accommodate any number of otherwise contradictory findings.[53] Indeed, the main reason why Bell's theorem has assumed such prominence in recent debate is the clarity with which it sets out this issue as between the rival (Copenhagen versus Bohm-type) theories and their various logical entailments. Thus Bohm for one accepted – following Bell – that any future defence of a realist or hidden-variables account would have to make terms with the idea of quantum-mechanical 'action at a distance', at least in so far as it wished to conserve the basic quantum formalisms. (As I have said, he saw no problem with this idea just so long as the phenomenon could not be used to transmit information over vast distances at superluminal velocity, a consequence which is anyway safely ruled out on other practico-theoretical grounds.) However – to repeat – these issues would never have arisen or the alternatives been posed so sharply had Bell, Bohm, and other physicists elected to adopt Quine's principle of wholesale ontological relativity.

That doctrine finds its closest parallel in Bohr's interpretation of quantum mechanics, or rather his agnostic refusal to offer any such interpretation, given what he sees as the unbridgeable gulf between quantum 'reality' and the descriptive-conceptual-explanatory resources available to human enquirers. But in Quine's case the problems are even more acute since he extrapolates directly from the micro- to the macrophysical domain and thus raises the issue of Schrödinger's perhaps ill-fated cat in a peculiarly trenchant (if typically insouciant) way. That is, Quine takes it pretty much for granted that (1) quantum mechanics may indeed force revisions at any point in the total 'fabric' of accredited beliefs; (2) this fabric extends all the way from analytic (logical) 'truths of reason' to synthetic or empirical observation-statements; and (3) we can therefore justifiably conclude that ontological relativity affects every item of belief, whether concerning such issues as particle location and momentum or the life-or-death predicament of macroscopic items like Schrödinger's cat.[54] However, one could turn these arguments around point for point and mount a case against Quine's general doctrine as well as his analogy with quantum physics. Thus (1) there are reasons – some of which I have instanced above – to reject that full-scale holistic view of the sorts of revision that may be forced upon us by developments in the quantum-theoretical domain; (2) these reasons have to do with the necessary role of logical thought (and of 'classical' two-valued logic at that) as a means of defining the relevant issues in quantum mechanics as elsewhere; and (3) those issues are hopelessly blurred if one fails to distinguish different *ontological* levels of enquiry and the different *epistemological* lessons that may properly be drawn concerning them.

IV

Wesley Salmon, one of the most resourceful defenders of a causal-realist approach in philosophy of science, may well be right when he echoes David Mermin's remark that anyone of a like persuasion who is not worried about quantum mechanics 'has rocks in their head'.[55] All the same, I would suggest that some of the interpretative problems with orthodox QM result more from its own deep strain of anti-realist prejudice than from anything entailed by consequent reasoning on the observational evidence. Indeed, those criteria – of rationality, consistency, evidential warrant – would be simply incapable of conjoint application if one took Bohr's pronouncements and those of his disciples at anything like face value. Of course there are no adequate grounds, as yet, for asserting the *truth* of Bohm's hidden-variables theory or for fully endorsing any such alternative account that assumes the incompleteness of orthodox QM and which meets the conditions for a realist interpretation. To do so would be to invite the standard instrumentalist charge, i.e. that realism goes 'beyond the evidence' and thus carries an unnecessary weight of surplus metaphysical baggage. Besides, as we have seen, it still has to cope with the Bell-type (EPR-derived) paradoxes, and moreover involves some added complications when it comes to describing how the particle is 'guided' by its associated wave-like probability function.[56] So clearly the realist will be ill-advised to claim that Bohm has got it right – as against the orthodox view – since his theory captures the truth (the 'deep further fact' or ultimate reality) of quantum physics. But this is just what the realist should *never* want to say, given the non-finality of even the best-supported scientific theories and the overwhelming likelihood, on past evidence, that any such claim will at length be subject to replacement or qualification. Rather, it is the realist's strongest suit that truth in such matters is 'verification-transcendent' *precisely in so far* as it does not depend on our current best knowledge or the limits of our present-day theories, observational means, investigative methods, etc. What is so odd about the orthodox QM line is that it makes a veritable dogma of the opposite argument, i.e. that those limits are somehow intrinsic to the very nature or structure of quantum 'reality'.

For some (mainly followers of Bohr), this issue is effectively closed since there is no way beyond the recalcitrance of quantum phenomena to any kind of realist construal. For others – Bohm and Cushing among them – realism is a basic requirement of any adequate physical theory, and orthodox QM must therefore be considered 'incomplete' in so far as it fails to meet that requirement. Such was of course Einstein's view in his series of debates with Bohr, a view that he maintained until the end of his life and which finds its most eloquent expression in the following passage from a tribute to James Clark Maxwell.

> The latest and most successful creation of theoretical physics, namely Quantum Mechanics, is fundamentally different in its principles from the two programmes which we will briefly call Newton's and Maxwell's .... Yet I incline to the belief that physicists will not be permanently satisfied with such an indirect description of Reality, even if the theory can be fitted successfully to the General Relativity postulates. They would then be brought back into the attempt to realize that programme which may suitably be called Maxwell's: the description of physical reality by fields which satisfy without singularity a set of partial differential equations.[57]

Others again take full stock of the problems confronted by any realist interpretation in the wake of Bell's results and the Aspect experiments, but hope that some solution may yet be found – perhaps in accord with Bohm's hidden-variables theory – that manages to resolve those problems within a realist (if not a local-realist) framework.[58] This is also the position of philosophers (among them Ian Hacking) who are not so much engaged with the finer points of QM dispute but who adopt a kind of midway, instrumentalist-realist approach. On this view particles are assumed to exist – and moreover to possess certain properties, locations, space-time trajectories, etc. – in so far as they play an explanatory role in our current best scientific theories or can be shown to have definite (observable) effects under given laboratory conditions. Thus for Hacking, in the case of positrons, 'if you can spray them, they are real!', as opposed to the standard instrumentalist line which in principle refuses to allow their reality on such merely inferential grounds.[59]

The last group includes thinkers like J.S. Bell who are the hardest to classify in terms of this conventional 'realist versus instrumentalist' line-up. As we have seen, Bell's theorem was – and remains – the major source of arguments against any realist construal since it showed that no hidden-variables theory could satisfy *both* the well-established quantum formalisms *and* the requirements of local realism. Yet it is clear from various statements in his work that Bell undertook this programme of research not so much with a view to undermining the realist case as in order to specify with maximum precision just what criteria would have to be met if that case were to hold up under pressure from orthodox QM. Indeed he repeatedly expressed a conviction that the orthodox theory *must* be in some way 'incomplete' and, moreover, that a realist construal of quantum phenomena was the only approach that held out any prospect of improved scientific understanding since it alone offered a genuine trial of substantive (ontologically committed) truth-claims or hypotheses.[60] Thus Bell, like Bohm, espoused an ontology wherein it still made sense to talk of really existing physical quantities –

'beables' as distinct from 'observables' – which are taken to possess precise simultaneous values for ever parameter, for example position or momentum, whatever the associated measurement problem and (as he found himself compelled to admit) the resultant paradox of nonlocal simultaneous interaction.

Of course this commitment has struck many commentators as the weak point in Bell's philosophical armoury and the reason for his being self-confessedly perplexed at the outcome of his own elaboration on the EPR experiment. Moreover, it is the main point at issue between defenders of a realist-objectivist approach (among them most notably Einstein, Bell and Bohm) and others – such as Bohr but also the proponents of quantum field theory – who would urge that we abandon that entire metaphysics as merely a relic of old (pre-quantum) thinking about substances, attributes, properties, and so forth. In Chapter 2 we saw how Putnam invoked a quantum field-theoretical approach as against what he considered the naive belief of Hacking that particles could be reidentified or treated as in any way numerically distinct.[61] Other writers – including Paul Teller in what is probably the most accessible introduction to this topic – have likewise argued that we need to unlearn that whole ontology of 'primitive thisness' (Teller's phrase) if we want to take the measure of quantum theory and its implications for our knowledge of the physical world.

Still, there is a question of whether this proposal can be carried through without at some point falling back on just the kinds of 'primitive' substance-attribute thinking which it claims to leave behind. Thus:

> things with primitive thisness can be *counted*; that is, we can think of the particles as being counted out, the first one, the second one, the third, and so on, with there being a difference in principle in the order in which they are counted, a difference that does not depend on which particle has which properties. By way of contrast quanta can be *aggregated*; that is, we can only heap them up in different quantities with a total measure of one, or two, or three, and so on, but in these aggregations there is no difference in principle about which one has which properties. The difference between countability and susceptibility to being merely aggregated is like the difference between pennies in a piggy-bank and money in a modern bank account.[62]

From this homely analogy Teller goes on to develop the case for quantum field theory as an approach that requires some considerable effort of revision to our basic ('commonsense') world picture but which repays that effort – at least for quantum physicists and philosophers – by resolving some of the deepest problems in current scientific debate. All the same, one has to note that his own terminology and way of framing these issues constantly has recourse to just the kinds of language – along with its 'primitive' ontological assumptions – that he blames for causing all the trouble. For the most part, these lapses from his own strict requirement are allowed to pass without mention, or treated as a matter of descriptive convenience which the reader should be able to dispense with once she has climbed far enough to achieve the new field-theoretical perspective and to kick the ladder away. However, there are passages where Teller is more explicitly (and perhaps uncomfortably) aware of his need to continue talking in the old manner, despite its misleading implications and the massive obstacle it poses to any genuine grasp of these issues in the quantum domain. Thus for instance:

> I will use the words 'field' and 'quantum' in the relatively precise senses I have introduced .... I follow what I take to be at least one traditional usage and use 'waves' to talk about field configurations that may be superimposed. I will use the word 'particle' with its prequantum meaning, with at least the suggestion of exact trajectories and primitive thisness, although the word is to some extent vague and fails to have clearly settled criteria of application. In particular, I will use the word 'particle' when attempting to dissect the felt conflict between the ideas of fields and particles. The strategy is to move to concepts whose mutual fit can be more clearly judged because the concepts are more clearly delineated.[63]

It is hard to know what to make of this recourse to a whole range of terms, meanings, and stipulative (however 'vague') definitions which are meant to be taken – one assumes – as a mere expository device on the way to better understanding, but which Teller is unable to expunge from his language even in other, less overtly concessive or recidivist passages. At very least it would suggest that quantum field theory has problems of its own, ontologically speaking, and that in cases such as that of Putnam versus Hacking – or the orthodox QM theorists versus Bohm and Bell – any verdict is at this stage decidedly premature.

The same applies to Bell's much-criticized but (in my view) strongly argued case for preserving the distinction between 'beables' and 'observables' for quantum-theoretical purposes. Here again I must quote at some length since Bell's position is often made out to be more dogmatic than it actually is in the face of those complicating factors that arose in consequence of his own EPR-derived theorem. 'It would be foolish', he writes,

> to expect that the next basic development in theoretical physics will yield an accurate and final theory. But it is interesting to speculate on the possibility that a future theory will not be *intrinsically* ambiguous and approximate. Such a theory could not be fundamentally about 'measurements', for that would again imply incompleteness of the system and unanalyzed interventions from outside. Rather it should again become possible to say of a system not that such and such may be *observed* to be so but that such and such *be* so. The theory would not be about '*observ*ables' but about '*beables*'. These beables need not resemble those

of, say, classical electron theory; but at least they should, on the macroscopic level, yield an image of the everyday classical world, for 'it is decisive to recognise that, however far the phenomena transcend the scope of classical physical explanation, the account of all evidence must be expressed in classical terms'.[64]

There are two chief points to note about this passage, aside from its markedly tentative character and Bell's perhaps unfortunate choice of the homespun term 'beable' to make his ontological point. One is the fact that it clearly comes out on the side of Einstein and Bohm, that is to say, in favour of an alethic-realist or objectivist interpretation that rejects the opposing (instrumentalist) view according to which observation, prediction, and measurement are the jointly sufficient criteria of 'completeness' for any candidate physical theory. Thus, according to Bell, it may yet turn out – *and should indeed be the case* if quantum theory is to make genuine progress – that its present 'ambiguous and approximate' language will yield to a better, more accurate means of describing and explaining quantum phenomena. In order for such progress to occur, he thinks, it will need to renounce the empiricist veto on realist ('beables') talk and conceive its purpose as that of providing an objectively truthful and valid account. That orthodox QM falls short in this respect is Bell's main reason (like Einstein's before him) for judging it necessarily 'incomplete'. But there is a further aspect of his argument – something of a sting in the tail – when Bell cites that well-known passage from Bohr about the need for all descriptions of quantum phenomena to be 'expressed in classical terms', however far they must be thought to 'transcend the scope of classical physical explanation'. For his point in so doing, I take it, is to stress that *even by its own criteria* the Copenhagen doctrine points to a reality beyond phenomenal appearances or beyond the limits of a purely instrumentalist (observational-predictive) approach.

The standard objection – here as with Bohm – is that this argument tends to presuppose what it sets out to prove, i.e. the existence of particles that possess precise simultaneous objective values of position and momentum despite the impossibility (as stated by Heisenberg's uncertainty principle) of obtaining measurements for such conjugate variables. Moreover, it assumes that particle position, rather than momentum, is the prime concern of any realist account that would save the objectivity of quantum physics and counter the arguments routinely brought against it by proponents of orthodox QM. Tim Maudlin makes this point (though he doesn't endorse it) in a passage that states the issue with exceptional clarity. 'In Bohmian mechanics', he writes,

particles with determinate positions are added to the Scientific Image, with the particle trajectories being determined by the wave function. A common complaint against the theory is that the choice of particle positions as the 'beables' of the theory is *arbitrary*: the wave function, after all, can be represented in momentum space as easily as in position space. So why not choose particle *momenta* as the beables of the theory and produce a Bohmian dynamics of how momentum changes in time? Or why not choose any of the other myriad 'observables' of standard quantum theory as the beables added to round out the Scientific Image?[65]

This would appear to be a strong argument against any realist ontology premised (like Bohm's and implicitly Bell's) on particle position as the chief objective value that needs to be maintained against the rival, i.e. orthodox-empiricist or Copenhagen view. However, Maudlin then goes on to question that argument on grounds that derive from Wilfrid Sellars' distinction between the 'Manifest' and 'Scientific' images of the world, i.e. an image 'shorn of theoretical posits' and one that 'advances with the postulation of new imperceptible entities and the laws which govern their behavior'.[66] What is required of a realist theory on this account is that it offer some means of bringing those images into an isomorphic relation, or a scheme that allows for the ready translation from one descriptive framework to the other. In which case, he argues,

> [t]he justification for Bohm's choice of beables is simple and powerful. It is relatively easy to discover an isomorphism between the Manifest Image and a Scientific Image which contains particles with determinate positions. It is not a hard task to construct a passable *doppelgänger* for the world revealed by experience using particles in motion. Cats in the Manifest Image correspond to cat-shaped collections of particles in the Scientific .... But try, in contrast, to describe your immediate environment in terms of particles which have *only* momentum and not position. It is hard to know even where to begin. The manifest momentum of the objects around me is almost uniformly zero. And although one does notice motion, it is always the motion of located objects, never momentum neat. There is simply no obvious way to sketch any isomorphism between a world of particles which have only momentum and the world as we experience it.[67]

From the Copenhagen viewpoint this would seem just a typical case of what happens when a realist (or objectivist) approach is extended to events in the quantum domain, i.e. to a realm where 'classical descriptions' no longer have any but an ambiguous, approximative, or *faute de mieux* application. However, as we have seen, that argument is itself highly problematic and at times self-subverting, whether it takes the orthodox (Bohr-derived) form of an empiricist veto on realist talk or – as with Teller – invokes quantum field theory as a means of avoiding such 'naive' objectivist conceptions. At any rate, Maudlin has a strong point – as against these opposing views – when he takes it to be a signal virtue of Bohm's theory that it comes out far more closely in accord with the manifest image of the physical world as delivered by pre-quantum scientific theories and their various stages of advance beyond the level of naive (commonsense-intuitive) belief.

This helps to explain Bell's steady conviction that orthodox QM must be somehow 'incomplete' and that there could not be any reason in principle – or in the quantum-physical nature of things – why a fuller and more adequate objective account should be ruled a priori unattainable. The following passage describes his reaction when he belatedly came across Bohm's hidden-variables theory and de Broglie's earlier pilot-wave hypothesis on which that theory was based.

> Why then had Born not told me of the 'pilot-wave'? If only to point out what was wrong with it? Why did von Neumann not consider it? More extraordinarily, why did people go on producing 'impossibility' proofs, after 1952, and as recently as 1978? When even Pauli, Rosenfeld, and Heisenberg could produce no more devastating criticism of Bohm's version than to brand it as 'metaphysical' and 'ideological'? Why is the pilot-wave picture ignored in text books? Should it not be taught, not as the only way, but as an antidote to the prevailing complacency? To show that vagueness, subjectivity, and indeterminism are not forced on us by experimental facts, but by deliberate theoretical choice?[68]

Thus Bell can scarcely be enlisted on the side of orthodox QM, even though his most famous result – the violation of Bell's inequality for anti-correlated spin measurements – is standardly taken as placing him in that camp. As Cushing succinctly puts it, 'Bell inequalities are the necessary and sufficient conditions on the joint distributions for a common cause explanation of the actually observed outcome of the experiments. Since these inequalities are violated, then, at least for this *one* experiment (and there are others), no common-cause explanation is possible'.[69] What is here meant by 'common cause' is basically the local-realist (EPR-type) assumption, i.e. that the anti-correlated values for particles A and B are explainable in terms of their originating in a singlet-state pair of combined spin-value zero, in which case – by the conservation law respecting angular momentum – they will always yield a sum-zero value when measured at any time thereafter. However, it emerged from Bell's calculations that any causal-realist account along these lines was contravened – or statistically 'violated' – by the QM prediction that any measurements obtained for either particle must somehow have their outcome momentarily determined by measurements carried out on the other. (This finding can also be generalized to instances where there is no strict anti-correlation as with the singlet-state case but rather an exceptionally high statistical preponderance of such results which violates the laws of probability.) In other words, Bell established the absolute in-principle incompatibility of quantum mechanics with the tenets of EPR-type local realism. Nevertheless, he left it an open question whether realism actually *required* the locality condition or whether there might be some way of lifting that condition so as to reconcile QM with a modified (i.e. non-local) realist ontology.

Of course this approach would be altogether vacuous – or collapse straight back into the orthodox (instrumentalist) line – if it entailed redefining 'realism' to the point where that term simply changed meaning in accordance with its new context. For, as Cushing remarks, '[i]f we say that quantum realism is that realism required by quantum mechanics, we have not thereby helped anyone to comprehend just what realism *is* as a representation of the world'.[70] Thus a 'realist' (in this sense) with regard to quantum phenomena

> can take seriously an 'ontic blurring' of many of the variables that can be observed in a quantum system. The notion of 'quantum particle' is introduced for an object that exhibits wave–particle duality to distinguish it from the traditional 'classical particle'. The non-separability characteristic of quantum systems is taken to indicate the 'holistic character' of such systems. In a sense, what has been done is to take some of the unique aspects of the quantum formalism and then assign names and ontic status to them. This in itself does not produce any sense of understanding of the physical phenomena.[71]

This may remind us of the problems with Quine's version of the argument from and to meaning-holism, i.e. his ontological-relativist thesis that it is only in the context of some overall scientific worldview – some total 'fabric' or 'web' of interconnected beliefs – that we are able to interpret particular terms, predicates, observation statements, explanatory hypotheses, and so forth. Indeed, Quine may well have been influenced in adopting this position not only by Duhem's cognate ideas concerning the theory-laden character of observation statements and the underdetermination of theory by evidence, but also by the doctrine of relational holism as applied to (or derived from) quantum mechanics.[72] Moreover, as I have argued, both are open to a similar objection: that if consistently applied they would just explain away – rather than explain or even seek to explain – any conflicts that arose between empirical evidence and standing theoretical beliefs. Cushing makes the point with regard to orthodox QM that it simply adjusts the 'ontic status' of terms such as *particle* by redefining them in accordance with the quantum theory, itself taken to redefine 'reality' as that which cannot be described or explained except by reference to the wavefunction likewise holistically construed. Thus here – as in Quine – the upshot is a form of ontological-relativist doctrine which effectively decrees that quantum-physical reality *cannot be other* than the way it is represented as being under some given (however paradoxical) description.

V

'Is it possible in such circumstances', Cushing asks, 'to produce any explanation that allows us to understand or comprehend how the observed phenomena come about?'[73] By 'understand' and 'comprehend' Cushing clearly means something more than just 'construe in accordance with the quantum formalism' or 'describe in terms that respect the limits on our knowledge of quantum phenomena as laid down by orthodox QM'. Rather, it involves the claim to provide a better, more detailed and depth-ontological account, one that goes beyond the observational evidence – and likewise beyond the instrumentalist appeal to statistical-predictive warrant – but which does so precisely in virtue of its greater explanatory powers. Thus Bohm's hidden-variables theory, although perfectly compatible at every point with the established QM results, is also (to its credit) able to show 'that more microstructure is consistent with it than had previously been appreciated'.[74] In this respect it gains strong support from the earlier history of particle physics which witnessed a steady advancement in the knowledge of ever more recondite structures from the molecular to the atomic and thence to the subatomic and subnuclear levels.[75] At the very least, it must be counted as a point in favour of Bohm's realist approach that it avoids the imposition of a stipulative limit – such as that entailed by orthodox QM – beyond which no further progress can be hoped for in this hitherto promising direction. After all, as Cushing very reasonably comments, '[o]ne ought not to accept, as logically required, constraints that are more restricting than nature actually dictates'.[76]

Of course this goes against the orthodox QM argument that nature does in some sense 'dictate' those constraints, as for instance by making it '*in principle* impossible that position (and, hence, a trajectory in time and space) is a possessed property of a microsystem'.[77] However, it is precisely Bohm's point that such arguments beg all the main questions as against his own hidden-variables-based account. For this latter is (1) observationally and predictively equivalent to orthodox QM, (2) likewise able to incorporate nonlocality on the no-first-signal condition, and (3) – most important – unburdened with conceptual problems (such as the observer-induced collapse of the wavepacket or the Schrödinger's cat paradox) which remain unresolved on the Copenhagen version. As concerns item (2), Bohm is obliged to grasp the nettle more firmly since for the realist nonlocality becomes an objective feature of the world rather than a 'logically' requisite feature of the theory describing it. But this is in any case a direct consequence of the quantum theory – on whatever interpretation – when construed in accordance with Bell's theorem. And it can scarcely be considered a virtue in the orthodox model that it takes the instrumentalist line of least resistance by refusing to admit the existence of 'properties [objectively] possessed by the microsystem' and thus keeps an escape route open whenever such problems loom. For Bohm, on the contrary, '[i]f the price of avoiding nonlocality is to make an intuitive explanation impossible, one has to ask whether the cost is not too great'.[78] Just as an adequate physical theory must be taken as applying to individual particles and their associated pilot-wave systems, rather than statistical ensembles, so likewise nonlocality should be taken as a property of the world rather than a matter of descriptive convenience or deduction from QM principles. Besides, 'the actual nonlocality demanded by nature turns out to be of a fairly benign variety; we cannot signal with it and it does not so entangle the world as to prevent us from doing science as we have traditionally known it'.[79] Thus Bohm's interpretation has the great advantage of asking what the theory actually entails if true of physical reality rather than what implications it has for quantum-theoretical debate.

A further advantage is the fact that, unlike orthodox QM, it avoids positing a total break with previous physical theories and hence does not face the problem of explaining just where the boundary should be taken to fall between events at the subatomic (quantum) level and events in the macrophysical domain. On the usual account, as Cushing succinctly describes it, '[a] theory that supersedes a previous one must reduce to the old one in a suitable limit'.[80] That is to say, the 'old' theory will still be valid for certain well-defined scientific purposes or within a certain restricted observational domain, but will yield to the new theory whenever it is a question of accounting for phenomena that previously lacked any adequate explanation. Thus, according to special relativity, there is a means of calculating the degree to which – for velocities less than the speed of light – its own equations will approximate to those of classical (Newtonian) mechanics and thereby produce results that can be taken as practically equivalent. In general relativity, likewise, 'the limit of weak gravitational fields (or small space–time curvature) leads to Newtonian gravitational theory'.[81] Moreover, it is often the case that the later theory is able to explain just why – by reason of what specific doctrinal attachments, conceptual limitations, observational shortcomings, etc. – the predecessor theory failed to attain a more adequate or comprehensive grasp.

Such is at any rate the usual account of how physics progressed from Galileo to Newton, from Newton to Einstein, and then – but here the story encounters certain problems – from relativity theory to quantum mechanics. Where those problems arise is with the attempt to explain how QM could either be reconciled with Einsteinian physics or held to provide a more powerful explanatory theory that converges with it 'in a suitable limit'. This is not just a matter of the conflict between Einstein's precept of local realism (relative to the speed of light as an absolute constant) and the QM requirement of remote simultaneous correlation between widely separated particles. After all, one purpose of relativistic quantum field theory was to produce a set of equations that were consistent both with special relativity and with basic quantum postulates. Rather the problem had to do with the fact that this solution was arrived at by applying the standard QM formalism *twice over*, that is to say, by adopting the 'double quantization' technique which abstracts yet further from anything conceivable in classical (realist) terms. As Teller describes this process:

> In field quantization we start with something we were thinking of as a

classical field …. In second quantization we proceed in exactly the same way except that instead of starting with the description of a classical field, we start with the state function resulting from ordinary first quantization on the one-quantum theory: that is, we proceed exactly as in field quantization except that we treat the first-quantized state function exactly as if it were itself a description of a physical field![82]

Teller thinks of this as a definite advance since it represents a further, more decisive break with those ideas of 'primitive thisness' – of particles as possessing numerical identity and definite space–time locations or coordinates – which create such problems for the 'classical' view. Other writers tend to treat this procedure of second quantization as a problem in itself since it involves more reliance on complex mathematical techniques and thus places further obstacles in the way of any plausible realist interpretation. Thus, in Euan Squires' words, '[i]f, in our previous, non-relativistic discussion, we regarded the wavefunction as a part of reality, we now have to replace this by the wavefunctional, which is even further removed from the things we actually observe'.[83] So the problem is not so much the technical issue of QM nonlocality – 'technical' so long as it entails no violation of the no-first-signal requirement – but rather the way that quantum theory responds to any realist challenge by retreating to yet more abstract realms of hypothetical or probabilistic conjecture. It is in this respect chiefly that Einstein's realism comes into conflict with orthodox QM, even when the latter is 'adjusted' to accord with the equations of special relativity. For there is still – so to speak – all the difference in the world between a theory (Einstein's) that deploys those equations in the quest to achieve a more adequate, encompassing knowledge of physical reality and a theory (derived 'logically' enough from orthodox QM precepts) which rules such knowledge to be strictly and forever beyond hope of attainment.

One can see this prejudice very clearly at work in Heisenberg's firm belief that 'nature works only in such a way as not to violate the quantum mechanical formalism'.[84] Thus, for Heisenberg, there is no longer any question as to the 'completeness' of quantum mechanics and its claim to fix an absolute limit not only on the measure of precision attainable in our knowledge of physical reality but also on the workings of 'nature' itself as dictated by QM principles. It was this aspect of the orthodox theory that Einstein found so objectionable and that seems to have prompted his own very marked shift of allegiance – deplored by many of his colleagues – from a Mach-derived instrumentalist or positivist philosophy to the realist position adopted so tenaciously in his series of debates with Bohr.[85] What those debates brought home to Einstein, quite apart from the various technical problems of nonlocal causality, wave–particle superposition, and so forth, was the strict *impossibility* of ever achieving a match between special relativity (realistically construed) and *any version* of quantum mechanics that adopted the standard instrumentalist line. 'I am therefore inclined to believe', he wrote, 'that the description of quantum mechanics ... has to be regarded as an incomplete and indirect description of reality, to be replaced at some later date by a more complete and direct one.'[86] For on the orthodox account there could be no means of explaining how the QM formalism and its derivative equations might approximate those of relativity theory 'in the limit', i.e. at the point of transition from the micro- to the macrophysical domains.

Hence the predicament of Schrödinger's 'superposed' cat, somehow (inconceivably) caught in a state of suspension between life and death, and thus providing an apt emblem of the quandary imposed by orthodox QM with its precept that the 'collapse of the wave-packet' could occur only through the act of observation. Schrödinger (like Einstein) took this as sufficient proof that the theory *must* be in some way defective – or 'incomplete' – since it led to results that were plainly in conflict with experience, observation, and our everyday knowledge of the world, as well as with the more specialized findings of special and general relativity. For Bohr and Heisenberg, conversely, truth resided in the quantum formalism and the various *strictly inescapable* conclusions drawn from its rigorous deployment as a means of assigning probability values to the wavefunction. Thus, in their view, any notion of a reality 'beyond' or 'behind' quantum appearances was just a sign that its defender had lapsed into old-style ontological-realist assumptions that were no longer tenable given the immense predictive success of QM on the standard (Copenhagen) interpretation. So long as the formalism is self-consistent and continues to produce correct predictive results through use of the appropriate equations, then there is simply no need for a physical interpretation in accord with such 'metaphysical' requirements.

This is why Einstein perceived such a drastic conflict between orthodox QM and the entire previous development of the physical sciences from Galileo and Newton to relativity theory. In Cushing's words:

> Heisenberg believed that a successful mathematical formalism of a physical theory, such as classical mechanics, was of a piece or whole and that it could not be modified in any essential way without destroying the entire structure .... When such a formalism encounters difficulties (as classical mechanics did with quantum phenomena), it is not possible to modify that formalism successfully. A radically new formalism must be found to accommodate these new features of the physical world.[87]

However – and this was Einstein's chief objection, like Bohm's after him – it is not so much the 'physical world' that dictates how the new formalism is developed, refined and applied but rather the formalism which thenceforth decrees what must be the case with respect to those 'new features' of physical reality. In which case, again, there can be no prospect of achieving convergence 'in a suitable limit' between theories (such as general relativity and orthodox QM) which find themselves deeply at odds with respect to certain basic ontological issues. Rather, those theories should be seen as strictly incommensurable to the extent that they involve two quite different formalisms, each setting its own terms for what shall count as an admissible feature of 'reality'. Moreover – according to Heisenberg – the formalism (along with its derived structure of equations, predictions, ontological commitments, etc.) must in each case be construed as standing or falling 'of a piece and whole' and thus, *ex hypothesi*, as incapable of being modified in any way 'without destroying the entire structure'.[88]

So it is not hard to see why orthodox QM became a standard *topos* for philosophers, such as Quine and Kuhn, who made it central to their generalized case for ontological relativity or for the history of science as a series of paradigm shifts from one such structure to the next.[89] Indeed, there is a 1963 interview between Kuhn and Heisenberg where the latter recollects being 'very much afraid' that Schrödinger's version of wave mechanics might produce a 'new interpretation of the thing' (i.e. of the quantum formalism) that enabled an alternative construal of QM phenomena along realist or observer-independent lines. What emerges very clearly from this interview is Heisenberg's defensive attitude in the face of any threat to the orthodox account. Thus, 'that was a disappointment with Schrödinger .... I felt: now Schrödinger puts us back into a state of mind which we have already overcome, and which certainly has to be forgotten'.[90] At any rate there is good reason to believe that one major factor in the widespread acceptance of Kuhn's ideas about scientific theory change and paradigm incommensurability is the existence of a doctrine (orthodox QM) which raises such ideas to a high point of scientific principle.

It was just this aspect of the orthodox approach that Einstein rejected as involving a full-scale obscurantist retreat from the proper aims and objectives of any adequate physical science. More specifically, it failed the basic test of describing and explaining physical reality in a form that was not so abstrusely mathematical as to lose all touch with our powers of conceptual-intuitive grasp. For 'in a sense', as Cushing points out,

> Einstein's general theory of relativity provided an understandable (picturable) causal explanation in terms of a curved space–time background (whose specific structure is determined by the distribution of masses) through which gravitons (or gravitational waves) propagate to transmit physical influences of one mass upon another.[91]

Orthodox QM possessed nothing like this intuitively 'picturable' character, and indeed became all the more remote from physical reality with each new refinement (like the double quantization technique) introduced in order to square its results with those of relativity theory. At the same time this sharpened the demarcation problem, that is to say, the problem of specifying a limit in which the various QM requirements – uncertainty, probability, statistical causality – were somehow to be rendered consistent with those of macrophysically observable objects and events. Bohr's response was effectively to push the problem out of sight by refusing to draw any such line since quantum mechanics was in some sense true of 'reality' on every physical scale. Instead, he proposed the idea of 'complementary' descriptions as a means of avoiding any awkward conflict, whether between particle and wave ontologies, standard and deviant (quantum) logics, or the 'classical' world picture endorsed by physicists from Galileo to Einstein and the non-picturable quantum 'world' where an altogether different reality obtained. Thus, in the words of one commentator, '[c]omplementarity suggests developing an ontological conception of an independent reality ... not describable by the terms of experience. Beyond such a generalization, little else can be said regarding the positive nature of such an ontology'.[92] Still, one may sympathize with Cushing's response: that if this is the best that Bohr's defenders can do by way of elucidating comment then 'complementarity' would seem to create more problems than it offers constructive solutions.

Bohm agreed with Einstein that there had to be something wrong or demonstrably incomplete about a theory that concerned the ultimate nature of physical reality, and which claimed an overwhelming measure of predictive success, yet was driven to adopt such evasive strategies rather than explain or interpret its own results. In his view, the only promising way forward from this conceptual impasse was to offer an alternative realist account of events at the quantum level in terms that would be fully consistent with (and not merely 'complementary to') those that applied in the macrophysical domain. Thus Bohm's interpretation has the following features which set it decisively apart from all versions of orthodox QM theory. First, it resolves the inaugural problem of quantum mechanics – that of the wave–particle dualism – by adopting the suggestion that de Broglie put forward at the 1927 Solway conference, namely his 'principle of the double solution' whereby particles were assumed to possess definite locations and trajectories but also to be 'guided' by a phase wave whose properties were those derived from the standard QM equations.[93] De Broglie had arrived at this conjecture partly in consequence of his own early work on the close mathematical correlation between wave optics and classical mechanics, and partly by elaborating Einstein's theory of photons (or light quanta) which likewise appeared to manifest a form of wave–particle duality. Moreover, as Cushing remarks, Einstein's support for the pilot-wave hypothesis was hardly surprising 'since [he] had previously, in the context of general relativity, attempted to treat "particles" as the singularities in an underlying field'.[94]

Hence the second main advantage of Bohm's theory: that it offers a solution to the QM measurement problem, or that of explaining just how and when the wavepacket should be taken to 'collapse' and thus yield the kinds of determinate (non-probabilistic) result commonly observed for macrophysical objects, processes, and events. Here of course we are back with Schrödinger's cat and also with the closely related issue as to why orthodox QM fails to reduce to previous theories (such as Newton's and Einstein's) 'in a suitable limit', that is, at the point of maximal convergence where its equations approximate to theirs for all practical purposes. On Bohm's account the measurement problem simply disappears since the particle is considered to have definite (objective) values of location and momentum *at every point in its trajectory*, rather than yielding such values only as and when a measurement is performed on this or that chosen parameter. At any rate – contrary to widespread report – his theory offers strong grounds for rejecting any premature verdict as to the 'completeness' of orthodox QM and the consequent demise of causal realism as a viable scientific worldview.

Endnotes

1 This distinction is common to empiricist philosophers, such as Hume, who argue on largely naturalistic grounds and others, like Kant, who frame it in different (a priori or transcendental) terms. More recently it was taken up by the logical positivists and logical empiricists for whom the only meaningful statements were those that concerned either matters of verifiable fact – e.g. observation statements in the physical sciences – or self-evident (hence tautologous) logical truths. In this latter connection see especially A.J. Ayer (ed.), *Logical Positivism* (New York: Free Press, 1959); Richard B. Braithwaite, *Scientific Explanation* (Cambridge: Cambridge University Press, 1953); Rudolf Carnap, *The Logical Structure of the World and Pseudoproblems in Philosophy*, trans. R. George (Berkeley & Los Angeles: University of California Press, 1969); Carl Gustav Hempel, *Fundamentals of Concept Formation in Empirical Science* (Chicago: University of Chicago Press, 1972); Hans Reichenbach, *Experience and Prediction* (Chicago: University of Chicago Press, 1938). The distinction came increasingly under strain as critics pointed out various problems with the logical-empiricist programme, among them its failure to formulate the verification principle in a manner that satisfied its own strict requirements of empirical and/or logical warrant. For a root-and-branch attack on this entire way of thinking, see W.V.O. Quine, 'Two Dogmas of Empiricism', in *From a Logical Point of View*, 2nd edn (Cambridge, MA: Harvard University Press, 1961), pp. 20–46.

2 Kant, *Critique of Pure Reason*, trans. N. Kemp Smith (London: Macmillan, 1964).

3 David Hume, *An Enquiry Concerning Human Understanding*, ed. L.A. Selby-Bigge, rev. P.H. Nidditch, 3rd edn (Oxford: Clarendon Press, 1975).

4 For some useful recent discussions, see Paul Guyer (ed.), *The Cambridge Companion to Kant* (Cambridge: Cambridge University Press, 1992).

5 Johann Gottlieb Fichte, *Foundations of Transcendental Philosophy (Wissenschaftslehre 1796–99)*, trans. Daniel Breazale (Ithaca, NY: Cornell University Press, 1992); also Frederick C. Beiser, *The Fate of Reason: German philosophy from Kant to Fichte* (Cambridge, MA: Harvard University Press, 1987).

6 See for instance Hilary Putnam, *The Many Faces of Realism* (La Salle, IL: Open Court, 1987); *Representation and Reality* (Cambridge: Cambridge University Press, 1988); *Realism With a Human Face* (Cambridge, MA: Harvard University Press, 1990); *Renewing Philosophy* (Harvard University Press, 1992). For a critique of Putnam's position and some illuminating commentary on these current debates, see William P. Alston, *A Realist Conception of Truth* (Ithaca, NY: Cornell University Press, 1996).

7 See especially J. Alberto Coffa, *The Semantic Tradition from Kant to Carnap: to the Vienna Station* (Cambridge: Cambridge University Press, 1991).

8 In this connection, see Kant, *Metaphysical Foundations of Natural Science*, trans. J. Ellington (Indianapolis: Bobbs-Merrill, 1970); also *Opus Postumum*, ed. E. Förster, trans. E. Förster and M. Rosen (Cambridge: Cambridge University Press, 1993); Gordon G. Brittan, *Kant's Theory of Science* (Princeton, NJ: Princeton University

Press, 1978); Michael Friedman, *Kant and the Exact Sciences* (Brighton: Harvester, 1992).

9   P.F. Strawson, *Individuals: an essay in descriptive metaphysics* (London: Methuen, 1959) and *The Bounds of Sense: an essay on Kant's Critique of Pure Reason* (Methuen, 1966).

10 Strawson, *Scepticism and Naturalism: some varieties* (London: Methuen, 1985).

11 For further discussion, see David Z. Albert, *Quantum Mechanics and Experience* (Cambridge, MA: Harvard University Press, 1993); Bernard D'Espagnat, *Veiled Reality: an analysis of present-day quantum-mechanical concepts* (Reading, MA: Addison-Wesley, 1995); Peter Forrest, *Quantum Metaphysics* (Oxford: Blackwell, 1988); John Honner, *The Description of Nature: Niels Bohr and the philosophy of quantum physics* (Oxford: Clarendon Press, 1987); Max Jammer, *Philosophy of Quantum Mechanics* (New York: Wiley, 1974); Josef M. Jauch, *Are Quanta Real? a Galilean dialogue* (Bloomington, IN: Indiana University Press, 1973); Alasdair I.M. Rae, *Quantum Physics: illusion or reality?* (Cambridge University Press, 1986); A. Sudbury, *Quantum Mechanics and the Particles of Nature* (Cambridge University Press, 1986).

12 See for instance Niels Bohr, *Atomic Theory and the Description of Nature* (Cambridge: Cambridge University Press, 1934); *Atomic Physics and Human Knowledge* (New York: Wiley, 1958); also Dugald Murdoch, *Niels Bohr's Philosophy of Physics* (Cambridge University Press, 1987); Henry J. Folse, *The Philosophy of Niels Bohr: the framework of complementarity* (Amsterdam: North-Holland, 1985); Honner, *The Description of Nature* (op. cit.).

13 Albert Einstein, B. Podolsky and N. Rosen, 'Can Quantum-Mechanical Description of Reality be Considered Complete?', *Physical Review*, series 2, Vol. 47 (1935), pp. 777–80: Niels Bohr, article in response under the same title, *Physical Review*, Vol. 48 (1935), pp. 696–702; also Bohr, 'Conversation with Einstein on Epistemological Problems in Atomic Physics', in P.A. Schilpp (ed.), *Albert Einstein: philosopher–scientist* (La Salle, IL: Open Court, 1969), pp. 199–241.

14 Peter Gibbins, *Particles and Paradoxes: the limits of quantum logic* (Cambridge: Cambridge University Press, 1987), p. 11.

15 See J.S. Bell, *Speakable and Unspeakable in Quantum Mechanics: collected papers on quantum philosophy* (Cambridge: Cambridge University Press, 1987); James T. Cushing and Ernan McMullin (eds.), *Philosophical Consequences of Quantum Theory: reflections on Bell's Theorem* (Notre Dame, IN: University of Notre Dame Press, 1989); Arthur Fine, *The Shaky Game: Einstein, realism, and quantum theory* (Chicago: University of Chicago Press, 1936); Don Howard, 'Einstein on Locality and Separability', *Studies in the History and Philosophy of Science*, Vol. 16 (1985), pp. 171–201; Tim Maudlin, *Quantum Nonlocality and Relativity: metaphysical intimations of modern science* (Oxford: Blackwell, 1993); Michael Redhead, *Incompleteness, Nonlocality and Realism: a prolegomenon to the philosophy of quantum mechanics* (Oxford: Clarendon, 1987).

16 James T. Cushing, *Quantum Mechanics: historical contingency and the Copenhagen hegemony* (Chicago: University of Chicago Press, 1994).

17 See David Bohm, *Causality and Chance in Modern Physics* (London: Routledge & Kegan Paul, 1957); David Bohm and B.J. Hiley, *The Undivided Universe: an ontological interpretation of quantum theory* (London: Routledge, 1993); also Evadro Agazzi (ed.), *Realism and Quantum Mechanics* (Amsterdam and Atlanta: Rodopi, 1997); David Z. Albert, 'Bohm's Alternative to Quantum Mechanics', *Scientific American*, No. 270 (May 1994), pp. 58–63; F.J. Belinfante, *A Survey of Hidden Variable Theories* (Oxford: Pergamon Press, 1973); S.V. Bhave, 'Separable Hidden Variables Theory to Explain the Einstein–Podolsky–Rosen Paradox', *British Journal for the Philosophy of Science*, Vol. 37 (1986), pp. 467–75; Peter Holland, *The Quantum Theory of Motion* (Cambridge: Cambridge University Press, 1993).

18 David Bohm, 'A Suggested Interpretation of the Quantum Theory in Terms of "Hidden" Variables, I and II', *Physical Review*, Vol. 85 (1952), pp. 166–79 and 180–93; cited in Cushing (op. cit.), pp. 77–8.
19 Erwin Schrödinger, *Letters on Wave Mechanics* (New York: Philosophical Library, 1967); also John Gribbin, *In Search of Schrödinger's Cat: quantum physics and reality* (New York: Bantam Books, 1984).
20 Cushing (op. cit.), p. 39.
21 See for instance J.A. Wheeler and W.H. Zurek (eds.), *Quantum Theory and Measurement* (Princeton, NJ: Princeton University Press, 1983).
22 Albert Einstein, *Relativity: the special and the general theories* (London: Methuen, 1954); also J.R. Lucas and P.E. Hodgson, *Spacetime and Electro-Magnetism* (Oxford: Clarendon Press, 1990).
23 Cushing (op. cit.), p. 39.
24 P.H. Eberhard, 'The EPR Paradox: roots and ramifications', in W. Schommers (ed.), *Quantum Theory and Pictures of Reality* (Berlin: Springer Verlag, 1989), pp. 49–88; p. 51.
25 Cushing (op. cit.), p. 108.
26 See for instance Barry Barnes, *About Science* (Oxford: Blackwell, 1985); David Bloor, *Knowledge and Social Imagery* (London: Routledge & Kegan Paul, 1976); Steve Fuller, *Philosophy of Science and its Discontents* (Boulder, Colorado: Westview Press, 1989); Andrew Pickering (ed.), *Science as Practice and Culture* (Chicago: University of Chicago Press, 1992); Steven Shapin, 'History of Science and its Sociological Reconstruction', *History of Science*, Vol. 20 (1982), pp. 157–210; Steve Woolgar, *Science: the very idea* (London: Tavistock, 1988).
27 Christopher Norris, *Against Relativism: philosophy of science, deconstruction and critical theory* (Oxford: Blackwell, 1987); also J. Aronson, R. Harré and E. Way, *Realism Rescued: how scientific progress is possible* (London: Duckworth, 1994); W.H. Newton-Smith, *The Rationality of Science* (London: Routledge & Kegan Paul, 1981); John Passmore, *Science and its Critics* (London: Duckworth, 1978); Karl Popper, *Realism and the Aim of Science* (London: Hutchinson, 1983); Peter J. Smith, *Realism and the Progress of Science* (Cambridge: Cambridge University Press, 1981).
28 Paul Forman, 'Weimar Culture, Causality, and Quantum Theory, 1918–1927: adaptation by German Physicists and Mathematicians to a hostile intellectual environment', in *Historical Studies in the Physical Sciences*, Vol. 3 (1971), pp. 1–115 and 'The Reception of an Acausal Quantum Mechanics in Germany and Britain', in S.H. Mauskopf (ed.), *The Reception of Unconventional Science*, AAAS Selected Symposium, No. 25 (1979), pp. 11–50; also Cushing, *Quantum Mechanics: historical contingency and the Copenhagen hegemony* (op. cit.).
29 Cited by Cushing (op. cit.), p. 98.
30 Ibid, p.99.
31 Max Jammer, *Philosophy of Quantum Mechanics* (op. cit.); also *The Conceptual Development of Quantum Mechanics* (New York: McGraw-Hill, 1966).
32 Cushing (op. cit.), p. 100.
33 Cushing (op. cit.), p. 25.
34 See entries under Notes 13 and 15, above; also A. Aspect, P. Graingier and C. Roger, 'Experimental Realization of the E–P–R–B Paradox', *Physical Review*, Vol. 48 (1982), pp. 91–4.
35 Eberhard, 'The EPR Paradox' (op. cit.), p. 57.
36 Cited in Cushing, p. 22.
37 Cushing, p. 24.
38 Ibid, p. 204.
39 Cited in Cushing, p. 156.
40 See for instance Martin Gardner, 'Realism and Instrumentalism in Nineteenth-Century Atomism', *Philosophy of Science*, Vol. 46, No. 1 (1979), pp. 1–34.

41 Bohr, cited in Gibbins, *Particles and Paradoxes* (op. cit.), p. 54.

42 See entries under Note 12, above.

43 Honner, *The Description of Nature* (op. cit.).

44 See Notes 13 and 15, above.

45 Quine, 'Two Dogmas of Empiricism' (Note 1, above).

46 For further discussion, see G. Birkhoff and J. von Neumann, 'The Logic of Quantum Mechanics', *Annals of Mathematics*, Vol. 37 (1936), pp. 823–43; E.G. Beltrametti and Bas C. van Fraassen, *Current Issues in Quantum Logic* (New York: Plenum, 1981); Martin Gardner, 'Is Quantum Logic Really Logic?', *Philosophy of Science*, Vol. 38 (1971), pp. 508–29; Peter Gibbins, *Particles and Paradoxes* (op. cit.); Susan Haack, *Deviant Logic: some philosophical issues* (Cambridge University Press, 1974); Peter Mittelstaedt, *Quantum Logic* (Princeton, NJ: Princeton University Press, 1994); Hilary Putnam, 'How to Think Quantum-Logically', *Synthèse*, Vol. 74 (1974), pp. 55–61.

47 Quine, 'Two Dogmas of Empiricism' (op. cit.); Thomas S. Kuhn, *The Structure of Scientific Revolutions*, 2nd edn (Chicago: Chicago University Press, 1970).

48 See James Robert Brown, *The Laboratory of the Mind: thought experiments in the natural sciences* (London: Routledge, 1991) and *Smoke and Mirrors: how science reflects reality* (Routledge, 1994); also Paul Davies, 'The Thought that Counts: thought-experiments in physics', *New Scientist*, May 6th 1995, pp. 26–31 and Roy Sorensen, *Thought Experiments* (New York: Oxford University Press, 1992).

49 See S.G. Shanker (ed.), *Gödel's Theorem in Focus* (London: Routledge, 1987).

50 See Note 19, above.

51 For some useful surveys of the field – including the 'many-minds' and 'many-worlds' theories – see Paul Davies, *Other Worlds* (London: Dent, 1980); Paul Davies and J.R. Brown (eds.), *The Ghost in the Atom* (Cambridge: Cambridge University Press, 1986); David Deutsch, *The Fabric of Reality* (Harmondsworth: Penguin, 1997); Bryce S. DeWitt and Neill Graham (eds.), *The Many-Worlds Interpretation of Quantum Mechanics* (Princeton, NJ: Princeton University Press, 1973); J.A. Wheeler and W.H. Zurek (eds.), *Quantum Theory and Measurement* (Princeton University Press, 1983).

52 See Note 15, above.

53 See Quine, 'Two Dogmas of Empiricism' (op. cit.); also Sandra G. Harding (ed.), *Can Theories Be Refuted? essays on the Duhem-Quine thesis* (Dordrecht & Boston: D. Reidel, 1976).

54 For further discussion see Quine, *Theories and Things* (Cambridge, MA: Harvard University Press, 1981); Quine and J.S. Ullian, *The Web of Belief* (New York: Random House, 1970); also Robert Barrett and Roger Gibson (eds.), *Perspectives on Quine* (Oxford: Blackwell, 1989); Christopher Hookway, *Quine: language, experience and reality* (Cambridge: Polity Press, 1987).

55 Wesley C. Salmon, *Four Decades of Scientific Explanation* (Minneapolis: University of Minnesota Press, 1989), p. 186.

56 See Notes 17 and 18, above.

57 Cited by Eberhard, 'The EPR Paradox' (op. cit.), p. 80.

58 See for instance Rae, *Quantum Physics: illusion or reality?* (op. cit.) and Euan Squires, *The Mystery of the Quantum World*, 2nd edn (Bristol and Philadelphia: Institute of Physics Publishing, 1994).

59 Ian Hacking, *Representing and Intervening: introductory topics in the philosophy of natural science* (Cambridge: Cambridge University Press, 1983), p. 23.

60 See Bell, *Speakable and Unspeakable in Quantum Mechnics* (op. cit.).

61 Hacking, *Representing and Intervening* (op. cit.), p. 23; Hilary Putnam, *Pragmatism: an open question* (Oxford: Blackwell, 1995), p. 59.

62 Paul Teller, *An Interpretive Introduction to Quantum Field Theory* (Princeton, NJ: Princeton University Press, 1995), p. 12.

63 Ibid, pp. 103–4.
64 Bell, *Speakable and Unspeakable in Quantum Mechanics* (op. cit.), p. 41.
65 Tim Maudlin, 'Descrying the World in the Wave Function', *The Monist*, Vol. 80, No. 1 (January 1997), pp. 3–23; p. 15.
66 Ibid, pp. 5 and 6.
67 Ibid, p. 16.
68 Cited by Cushing, *Quantum Mechanics* (op. cit.), p. 192.
69 Cushing, p. 16.
70 Cushing, p. 17.
71 Ibid, pp. 16–17.
72 See Pierre Duhem, *The Aims and Structure of Physical Theory*, trans. Philip Wiener (Princeton, NJ: Princeton University Press, 1954); also Harding (ed.), *Can Theories Be Refuted?* (op. cit.).
73 Cushing (op. cit.), p. 10.
74 Ibid, p. 42.
75 See for instance Gardner, 'Realism and Instrumentalism in Nineteenth-Century Atomism' (op. cit.); also Mary Jo Nye, *Molecular Reality* (London: MacDonald, 1972); J. Perrin, *Atoms*, trans. D.L. Hammick (New York: van Nostrand, 1923); Squires, *The Mystery of the Quantum World* (op. cit.); A. Sudbury, *Quantum Physics and the Particles of Nature* (op. cit.).
76 Cushing (op. cit.), p. 26.
77 Ibid, p. 26.
78 Cited by Cushing (op. cit.), p. 22.
79 Ibid, p. 57.
80 Ibid, p. 39.
81 Ibid, p. 39.
82 Teller, *An Interpretive Introduction to Quantum Field Theory* (op. cit.), p. 81.
83 Squires, *The Mystery of the Quantum World* (op. cit.), p. 111.
84 Cushing (op. cit.), p. 111.
85 See especially P.A. Schilpp (ed.), *Albert Einstein: philosopher–physicist* (La Salle, IL: Open Court, 1994), especially Einstein's 'Autobiographical Notes', pp. 1–95; also Fine, *The Shaky Game* (op. cit.) and other entries under Note 15, above.
86 Cited in Bell, *Speakable and Unspeakable in Quantum Mechanics* (op. cit.), p. 145.
87 Cushing (op. cit.), p. 114.
88 See especially Werner Heisenberg, *The Physical Principles of the Quantum Theory* (New York: Dover, 1949) and *Physics and Philosophy* (New York: Harper & Row, 1958); also Patrick A. Heelan, *Quantum Mechanics and Objectivity: a study of the physical philosophy of Werner Heisenberg* (The Hague: Nijhoff, 1965).
89 See Quine, 'Two Dogmas of Empiricism', and Kuhn, *The Structure of Scientific Revolutions* and Notes 1 and 47, above.
90 Cited by Cushing, p. 116.
91 Ibid, p. 13.
92 Folse, *The Philosophy of Niels Bohr* (op. cit.), p. 257.
93 See Louis de Broglie, *An Introduction to the Study of Wave Mechanics*, trans. H.T. Flint (London: Methuen, 1930) and *Physics and Microphysics*, trans. M. Davidson (New York: Pantheon Books, 1955).
94 Cushing (op. cit.), p. 105.

4 Quantum worlds without end

The multiverse according to Deutsch

I

Popularizing books on quantum mechanics tend often to dwell on the kinds of far-out speculative theory embraced by advocates of the rival 'many-worlds' or 'many-minds' interpretations.[1] Such theories mostly take rise from the various well-known paradoxes of QM, such as the wave–particle dualism and the so-called 'collapse of the wave-packet' brought about – so it is thought – by the act of observation.[2] They are invoked in order to explain how all possible outcomes of every measurement can be somehow simultaneously 'realized', whether in the minds of different individual observers or in different worlds which branch off when the wavepacket collapses, and thereafter coexist as a multitude of parallel universes with just occasional quantum interference-effects to signal their shadowy 'reality'.[3] Since the observer likewise splits off at every point into a series of multiple selves, each with a continuous lifeline through one such proliferating world series only, it follows that they can have no direct awareness of this omnipresent but intangible quantum 'multiverse' and may therefore be tempted to find the whole idea quite fantastic. Such is at any rate how David Deutsch – currently its most vigorous champion – explains both the absolute *necessity* of adopting 'many worlds' as an answer to the QM paradoxes and also the strong resistance it encounters from a commonsense-intuitive standpoint. Thus he glosses 'quantum theory' quite simply as 'the theory of the physics of the multiverse', since in his view there is just no other means of accounting for such observed QM phenomena as photon interference or deflection in Bell-type delayed-choice or multiple-path experiments.[4] Indeed, 'if the best theory available to physics did not refer to parallel universes, it would mean that we needed a better theory, one that did refer to parallel universes, in order to explain what we see' (Deutsch, p. 51).

Deutsch is an out-and-out realist with regard to these multiple coexisting parallel worlds and spends a good deal of time chastising instrumentalists for their abject evasion of the issue. In this respect he is fully in accord with Einstein, contending that it must be the aim of any adequate physical theory to describe and explain the way things stand in reality, rather than merely to 'save the phenomena' by proving them predictively and observationally

consistent with some given formalism or set of results.[5] The following passage
– with reference to the classic two-slit experiment and its various later
refinements – presents his argument in typically forthright terms.

> The key fact is that a real, tangible particle *behaves differently* according to
> what paths are open, elsewhere in the apparatus, for something to travel
> along and eventually intercept the tangible photon. Something does travel
> along those paths, and to refuse to call it 'real' is merely to play with
> words. 'The possible' cannot interact with the real: non-existent entities
> cannot deflect real ones from their paths. If a photon is deflected, it
> must have been deflected by something, and I have called that thing a
> 'shadow photon'. Giving it a name does not make it real, but it cannot be
> true that an actual event, such as the arrival and detection of a tangible
> photon, is caused by an imaginary event such as what that photon 'could
> have done' but did not do. It is only what really happens that can cause
> other things really to happen. If the complex motions of the shadow
> photons in an interference experiment were mere possibilities that did
> not in fact take place, then the interference phenomena we see would
> not, in fact, take place.
>
> (Deutsch, pp. 48–9)

I have quoted Deutsch at some length since this passage brings out very
clearly just how many and various are the meanings attached to the word
'realism' in the context of QM debate. Even Niels Bohr could profess to be a
realist in this sense at least: that he acknowledged the existence of a noumenal
quantum 'reality' behind or beyond phenomenal appearances, albeit one that
must remain inaccessible to creatures equipped with our particular
('classical') framework of concepts, categories, perceptual powers, epistemic
modalities, etc.[6] Thus he subscribed to something very like Kant's doctrine
of 'empirical realism' conjoined with 'transcendental idealism', a doctrine
whose ultimate effect was to make things easy for the sceptic by driving an
insurmountable wedge between 'reality' and our knowledge of it.[7] Moreover,
this makes things additionally hard for the realist since she must likewise be
committed to what seems an identical precept, i.e. that any truths concerning
that reality are 'verification-transcendent' in so far as they hold good
objectively and quite aside from our current best theories or beliefs. Such is
at any rate the standard (sceptical) line of counter-argument: that realism
must always, by this self-subverting logic, give rise to an outlook of
thoroughgoing scepticism as concerns our knowledge of so-called 'external
reality'. Hence the otherwise remarkable ease with which a Bohr-type 'realist'
QM philosophy flips over into the kind of dogmatic instrumentalist doctrine
that on principle renounces any prospect of attaining a knowledge of quantum
'reality'.

As I have said, Deutsch is implacably opposed to this or any other
interpretation that avoids taking sides on the realism issue or coming clean

as regards its own ontological commitments. Thus he remarks – very much to the point – that orthodox (Copenhagen) QM theory emerged 'during the heyday of positivism in philosophy of science', and fell in with this wider movement of retreat from any form of realist or depth-explanatory approach (Deutsch, p. 329). What is more, those who rejected the orthodox account but could envisage no alternative theory consistent with the established QM observations and predictions were themselves driven to adopt the same outlook in all but name. As Deutsch puts it:

> [r]ejection (or incomprehension) of the Copenhagen interpretation, coupled with what might be called *pragmatic instrumentalism*, became (and remains) the typical physicist's attitude to the deepest known theory of reality. If instrumentalism is the doctrine that explanations are pointless because a theory is only an 'instrument' for making predictions, pragmatic instrumentalism is the practice of using scientific theories without knowing or caring what they mean.
>
> (Deutsch, p. 329)

One could scarcely wish for a stronger statement of the anti-instrumentalist case or indeed – as it might appear – a statement more closely and emphatically in line with realist views on the interpretation issue. However, this appearance is deceptive, as should already have emerged in the long passage from Deutsch's book that I cited one paragraph above. For what he means by quantum 'reality' is something very different from the arguments put forward by Einstein, David Bohm, and other defenders of a realist approach to the quantum measurement issue.[8] On their account, this interpretation would (1) perfectly match the observational–predictive results of orthodox QM, while also (2) finding room within a suitable limit for the best-established theories of earlier physics from Newton to Maxwell and Einstein, and (3) remaining sufficiently in touch with experiential reality. As regards item (1) they converge with Deutsch in so far as both interpretations – 'hidden variables' and 'many worlds' – are fully consistent with all the evidence produced in support of orthodox QM. With respect to (2), there is no direct conflict between them except that Deutsch would presumably view those earlier theories – along with Bohm's classically based QM interpretation – as valid only within (and not across) each of the various proliferating worlds that constitute the quantum 'multiverse'. For it is his contention that these worlds can be known to interact only through certain very slight interference effects – such as those produced in the two-slit or multiple-path experiments – when a particle is momentarily deflected by one of its ghostly other-world counterparts.

Thus, '[t]he only thing in the universe that a shadow photon can be observed to affect is the tangible photon that it accompanies .... Shadow photons would go entirely unnoticed were it not for this phenomenon and the strange pattern of shadows by which we observe it' (Deutsch, p. 44). On this view 'reality' extends far beyond anything conceivable on Bohm's interpretation, confined as the latter is – like all of classical physics – to just one of the manifold alternative worlds that branch off from every multiple-outcome event at the quantum level. In which case (3) is the chief point at issue between Deutsch and Bohm since the latter is committed to a form of realism which involves no such drastic revision to our basic ideas of what can plausibly count as a 'realist' ontology or worldview. For Deutsch, on the other hand, such revision is justified – indeed absolutely required – if we accept the manifest *reality* of quantum interference or multipath 'shadow' phenomena, along with the established QM formalisms and predictive-observational results. In short: 'the quantum theory of parallel universes is not the problem, it is the solution. It is not some troublesome, optional interpretation emerging from arcane theoretical considerations. It is the explanation ... of a remarkable and counter-intuitive reality' (Deutsch, p. 51).

Deutsch's book contains many such passages, all of them asserting the truth of his theory as a matter of straightforward inference to the best (indeed the only possible) realist account of interference phenomena along QM-compatible lines. In fact he sometimes appears to be making a yet stronger claim to the effect that those phenomena demand a many-worlds explanation *in and of themselves*, and can hence be taken to rule out any other theory quite aside from the detailed technicalities of quantum debate. Thus:

> [p]erhaps because the debate began among theoretical physicists, the traditional starting-point has been quantum theory itself. One states the theory as carefully as possible, and then one tries to understand what it tells us about reality .... But as regards the issue of whether reality consists of one universe or many, it is an unnecessarily complicated approach. This is why I have not followed it in this chapter. I have not even stated any of the postulates of quantum theory. I have merely described some quantum phenomena and drawn some inescapable conclusions.
>
> (Deutsch, p. 50)

To which might be added – in support of Deutsch's claim – that some philosophers (among them David Lewis) have argued on other than QM grounds for the reality of coexisting parallel worlds, and likewise refused to accept any compromise version of the doctrine which treats them as merely 'virtual', 'possible', or products of theoretical convenience.[9] Lewis himself arrives at this conclusion by way of modal logic and the argument that necessary truths are those that hold good across all possible worlds rather than obtaining only in a certain limited subset of worlds which happen to resemble our own in respect of various contingent features.[10]

In this form the theory goes back to Leibniz and involves the essentially rationalist belief that thinking can indeed deliver such real-world applicable truths through a priori reflection on the scope and limits of human knowledge in general. Thus, for Leibniz, the difference between matters of logical necessity and matters of contingent (e.g. historical or natural-scientific) truth is that the former can be known *to us* through a process of rigorous deductive reasoning from first principles while the latter, although still necessary in some ultimate sense, involve such a lengthy and complex chain of concatenated causes and effects that they can only be known by a God-like intelligence that surveys every relevant link in the chain and is subject to none of our creaturely limitations.[11] So, for instance, when we state it as a fact borne out by our best (though none the less fallible) sources of evidence that 'Caesar crossed the Rubicon', we thereby concede the limits placed upon certain kinds of human knowledge and – contrasted with this – the possibility of a knowledge that would so far transcend those limits as to render that statement a necessary truth.

My point is that Deutsch's argument in support of the 'many-worlds' hypothesis shares certain features with Leibniz's doctrine of logical necessity and also with the recent revival of that doctrine by metaphysically minded modal logicians such as Lewis. That is to say, it works on the strong rationalist principle that one can derive certain necessary truths about the quantum 'multiverse' – truths that hold good across all possible worlds or 'universes' – by a process of purely deductive reasoning from self-evident premises. No doubt Deutsch would say that I have misrepresented his position since the case for many-worlds rests crucially on the evidence of QM interference and multipath phenomena, and only then makes appeal to the mode of a priori deductive inference under logical closure that typifies rationalist metaphysics in the Leibniz–Lewis style. Thus:

> [w]e do not need deep theories to tell us that parallel universes exist - single-particle interference phenomena tell us that. What we need deep theories for is to explain and predict such phenomena: to tell us what the other universes are like, what laws they obey, how they affect one another, and how all this fits in with the theoretical foundations of other subjects.
>
> (Deutsch, p. 51)

In which case we should have even less need of metaphysical arguments – such as those of Lewis – by way of support for what should be self-evident to anyone capable of drawing the appropriate conclusion, i.e. that the many-worlds theory is the only one fully and logically consistent with observed QM phenomena.

Still, there is room for doubt whether Deutsch's argument really proceeds on this basis of a straightforward appeal to the evidence quite apart from any prior commitment to 'deep theories' or to anything so suspect as metaphysical justification. After all, to repeat, he lays it down as a matter of undoubted (rationally inescapable) truth that 'if the best theory available to physics did not refer to parallel universes, it would merely mean that we needed a better theory ... in order to explain what we see' (Deutsch, p. 51). And this despite his own repeated insistence that 'what we see' is never just a matter of what is plainly and objectively *there* to be seen, but is always theory laden in so far as it involves a whole complex and interactive range of perceptual modalities, cognitive frameworks, ontological commitments, pre-existing beliefs, conceptual-explanatory schemes, and so forth. Indeed, Deutsch has a chapter on 'Virtual Reality', where he goes a long way – further than can easily be squared with the above-cited statement – towards dissolving the objectivist distinction between actual and conceivable or real and 'virtual' worlds. On the one hand, 'we realists take the view that reality is out there: objective, physical, and independent of what we believe about it'. On the other hand, 'we never experience that reality directly', since '[e]very last scrap of our external experience is of virtual reality' (p. 121). Moreover,

> every last scrap of our knowledge – including our knowledge of the non-physical worlds of logic, mathematics and philosophy, and of imagination, fiction, art and fantasy – is encoded in the form of programs for the rendering of those worlds on our brain's own virtual-reality generator.
>
> (p. 121)

Now there is nothing here – at least on the face of it – that any realist should find objectionable. After all, Deutsch appears to come out very firmly on the realist's side as concerns the existence of objective (verification-transcendent) truths and the mistake of supposing that they are any less objective – or reality any less real – for the fact that we can access them only by way of our various perceptual, cognitive, or reality-generating 'programs'. Yet there is also a sense in which Deutsch's whole argument inverts the realist order of priority which holds such truths to obtain irrespective of our present-best beliefs or state of knowledge concerning them. For on his account the truth of the many-worlds theory is taken to follow necessarily from certain (as he thinks) likewise indubitable premises, among them the truth of orthodox QM theory, at least in so far as it excludes any possible alternative realist explanation.

This is not to say that Deutsch endorses every aspect of the orthodox model. Indeed, as we have seen, he rejects that theory on account of its instrumentalist approach, its acceptance of the doctrine that 'explanations are pointless' since 'a theory is only an "instrument" for making predictions', and again (worse still) its practice of 'using scientific theories without knowing or caring what they mean' (p. 329). However, Deutsch shares at least one major tenet of orthodox QM: the belief that any viable theory – one that accepts the observational evidence while conserving the quantum predictions and formalisms – will necessarily entail a decisive break with what counted as a realist ontology or worldview before the advent of quantum mechanics. That is, he follows Bohr in accepting the 'completeness' of the orthodox theory in so far as it is taken to specify the requirements that *any* interpretation must meet if it is not to fall back upon naive (pre-quantum) ideas about objective or observer-independent physical 'reality'.[12] Thus the multiverse hypothesis may seem utterly 'bizarre' and 'counter-intuitive' when judged by hitherto prevailing (i.e. 'classical') standards of realism and rational argument in the physical sciences. Indeed, it involves a baroque proliferation of 'worlds' that will surely strike the classical realist as a piece of sheer ontological extravagance of the kind best left to speculative metaphysicians. All the same – Deutsch argues – it is the only theory that can reasonably be upheld by anyone who accepts the *reality* of quantum phenomena (photon deflection, multipath interference, etc.) and who seeks a genuine explanation for them rather than a handy instrumentalist escape route.

Thus the great virtue of the multiverse hypothesis, from his point of view, is that it brings about a radical change in our conception of the physical world – a change commensurate with the inherent strangeness of quantum phenomena – while none the less leaving that conception pretty much intact as concerns our everyday or commonsense-intuitive modes of knowledge and experience. For it is just Deutsch's point that interference effects of the sort on which the many-worlds argument rests are 'usually so weak and hard to detect' that they have escaped recognition until very recently and even now show up only under certain very special conditions. The reasons for this are, first, that any given particle is deflected (or interfered with) only by its conspecific counterparts in other quantum universes; second, that such interference can be observed to occur only when their paths 'separate and then reconverge' at exactly the right spatio-temporal point in their trajectories; and third, that 'the detection of interference between any two universes requires an interaction to take place between *all the particles whose positions and other attributes are not identical* in the two universes' (Deutsch, p. 49; italics in the original). What this means, in effect, is that the universes have to be very like each other – their similarity resulting from a vast range of convergent interactive subatomic events – in order for us to observe or detect such interference phenomena.

In which case it is hardly surprising – and no argument against the many-worlds hypothesis – that they show up only under highly controlled laboratory conditions and not as a feature of our everyday awareness of physical reality. After all, as Deutsch remarks,

> in all the experiments I have described, the interfering universes differ only in the position of one photon. If a photon affects other particles in its travels, and in particular if it is observed, then those particles or the observer will also become differentiated in different universes. If so, subsequent interference involving that photon will be undetectable in practice because the requisite interaction between *all* the affected particles is too complicated to arrange.
>
> (Deutsch, pp. 49–50)

Thus any experiment set up to detect these quantum-interference phenomena will involve the observation of just one particle whose effect upon others or theirs upon it (should they interact) will result in the observer herself being 'split' into as many different worlds as there are possible outcomes to the experiment in question. Since each of those worlds constitutes a separate 'reality' (a universe along with its observer), and since the observed interaction has been on such a limited (single-photon) scale, therefore it follows – according to Deutsch – that we can never have more than momentary or transient glimpses of the quantum multiverse wherein those outcomes are all equally real. For in order to be aware of that reality we should need to devise an inconceivably complex experiment that allowed us to observe the full range of interactions between every particle and its host of shadowy counterparts. Only then could we command a panoptic view of the various divergent worlds (or realities) that must otherwise make it strictly impossible for any one observer to know what is occurring in a world other than her own, or to grasp the truth of the multiverse theory as a matter of intuitive self-evidence rather than a powerful explanatory hypothesis arrived at by deduction from QM phenomena.

In short, this theory has a full explanation – one that necessarily presupposes the truth of many-worlds – for the fact that its results will appear so 'bizarre' and 'counter-intuitive' to anyone who is not thus convinced. The explanation is that we have no means of epistemic access from one such world to another except through the fleeting glimpses offered by localized interference effects, themselves unnoticed except under special conditions (i.e. experiments of the single-photon type) that paradoxically render such access impossible. For their upshot, to repeat, is a situation in which the entire system – particle + interacting shadow counterparts + observer – splits off into a multiverse of parallel 'worlds' that could be bridged only through a subsequent interaction 'between *all* the affected particles' which is far 'too complicated to arrange'. So the many-worlds theory cannot be refuted by experience, even though Deutsch approvingly cites Dr Johnson as having delivered an effective riposte to Berkeleian idealism by kicking the stone and commonsensically declaring: 'Sir, I refute him thus!'. What he (Deutsch) derives from this anecdote is yet further support for the many-worlds theory of quantum phenomena. In short: '[s]hadow photons kick back by interfering with the photons that we see, therefore shadow photons exist' (Deutsch, p. 88). However, this tone of sturdy commonsense realism should not persuade us to forget the extent to which Deutsch's theory conflicts with – or extrapolates massively beyond – any 'evidence' concerning the nature of physical reality except in so far as that evidence is construed *always* with reference to QM theory and *always* as supporting his own preferred (i.e. 'many-worlds') interpretation of it. That is to say, there is a strong a priori commitment to the truth of that interpretation, and a resulting tendency to treat any others – Bohm's 'hidden-variables' theory among them – as non-starters simply because they fail to acknowledge that truth. For it can scarcely be said of the many-worlds hypothesis (as it can of Bohm's theory) that this is the interpretation of QM phenomena that is most in agreement with 'classical' physics, from Newton to Einstein, and which also involves least departure from our suitably adjusted realist intuitions. On the contrary, it is Deutsch's constant assertion of the *real* (not merely 'virtual' or 'possible') existence of those multiplex parallel worlds that places his theory at the furthest remove from any version of realism previously known to the physical sciences.

II

Indeed, one could argue that his 'kick-back' criterion for assessing reality claims is one that kicks back against the many-worlds theory by showing it to involve ontological claims wildly in excess of the evidence. There are two passages in Deutsch's book where this irony can be felt most keenly. One has to do with the physicist Hugh Everett, who was first to propose the many-worlds theory and who encountered widespread scepticism from fellow researchers, among them the quantum theorist Bryce DeWitt.[13] After a series of detailed technical criticisms, DeWitt ended up 'on an informal note, pointing out that he just couldn't feel himself "split" into multiple, distinct copies every time a decision was made' (Deutsch, p. 328). Deutsch interprets this as an object lesson in the need to stand back from such commonsense-intuitive grounds of rational assurance, and to treat them as always revisable under pressure from recalcitrant evidence such as that forced upon us by quantum interference effects. Thus, 'Everett's reply echoed the dispute between Galileo and the Inquisition. "Do you feel the Earth move?" he asked – the point being that quantum theory *explains* why one does not feel such splits, just as Galileo's theory of inertia explains why one does not feel the earth move' (pp. 328–9). Apparently DeWitt conceded the point, thus earning recognition from Deutsch as a tardy but none the less welcome convert to the cause. However, one may doubt that there is really such strong support to be had for the many-worlds theory from this suggestive and morally resonant parallel with the case of Galileo *contra* the Inquisition. The main difference is that Galileo's theory was subject to a range of experimental tests, the result of which was to close the gap between commonsense belief and scientific knowledge to the point where commonsense adjusted to the theory and it ceased thereafter to pose any such problem. But this scarcely applies to the many-worlds hypothesis since here – as Deutsch himself concedes – there is no prospect of the theory ever acquiring that degree of intuitive acceptance owing to the extreme rarity and weakness of quantum interference effects and the consequent limits upon epistemic access from one world to another.

Of course the theory might be true all the same, and our problems with it merely the result of our not being suitably equipped – in respect of our various perceptual, cognitive, or intellectual capacities – to grasp the deepest principles of quantum reality. However, it strains credibility to the limit when Deutsch claims Galilean warrant for a thesis that involves so radical a break not only with commonsense-intuitive 'knowledge' but also with just about every major tenet of the previous physical sciences, Galileo's not least among them. This leads on to the second of those passages in Deutsch's book where – as I suggested above – his realist avowals have the ironic effect of undercutting his case for the many-worlds theory as the one and only possible realist account of what transpires in the quantum multiverse. Again it concerns Galileo and has to do with the vacuity of certain overcomplicated pseudo-explanations, such as the Catholic Church's doctrine that the Sun and planets must be thought to revolve around a central stationary Earth, but that the whole affair is set up in such a way – with so many complex intersecting orbits – that terrestrial observation may support the hypothesis of a heliocentric solar system.[14] This example is important to Deutsch because he views the resistance to his many-worlds theory as stemming partly from doctrinal adherence within the scientific community and partly from the kind of commonsense realism that responded to Galileo's argument by remarking that it didn't *feel* as if the Earth were moving at a great speed under our feet. Moreover, the orthodox (theologically approved) version had just as good a claim to fit the cosmological data and explain why the heavens appeared as they did to terrestrial observers. 'And yet it moves!' as Galileo famously (perhaps apocryphally) murmured, since his own hypothesis entailed far less in the way of needless complication. In the same way, Deutsch suggests, the multiverse theory is sure to strike most people as a wildly extravagant hypothesis since – for reasons summarized above – they lack any means of epistemic access to worlds other than their own. And yet that theory is the only one capable of explaining quantum interference phenomena without producing all manner of unwanted complications or eventually retreating to the standard instrumentalist line.

Thus Deutsch casts himself very much in the role of a latter-day Galileo, maintaining a strong realist position *vis-à-vis* the quantum multiverse but up against the kind of obdurate resistance that has often attended such major challenges to received notions of reality and truth. After all, he invites us, 'consider what it would feel like if we did exist in multiple copies, interacting only through the imperceptibly slight effects of quantum interference' (Deutsch, p. 89). In conducting this thought experiment, we are following Galileo's example when he asked what sensations would actually be experienced by earth-dwellers if their planet were orbiting around the Sun, and deduced – in accordance with his theory – that the effects of that motion would be imperceptible. In both cases they can show up only through the use of sophisticated measuring equipment, whether in the interference patterns created by controlled photon emission or in the gradually rotating arcs described by a Foucault pendulum. But this is quite enough – so Deutsch maintains – to establish their objective or real-world status according to the standard 'kick-back' criterion that he borrows from Dr Johnson. For '[i]t is only an accident of evolution ... that the senses we are born with are not adapted to feel such things "directly"' (p. 89). It is a leading point in his argument (as I have said already) that *all* human knowledge is 'indirect' in the sense that it depends upon perceptual inputs and modes of higher level cognitive processing which necessarily place us at a multiple remove from real-world objects and events.

This is why Deutsch thinks of 'virtual reality' as extending far beyond the realm of computer simulation to even the most (seemingly) objective or mind-independent experiential data. It is also a main plank in his argument for regarding the many-worlds theory *not* as some abstruse metaphysical doctrine based on just one interpretation of some highly debatable QM phenomena but rather as the sole adequate theory for explaining those phenomena and much else besides. That is to say, every item of human knowledge is 'verification-transcendent' if one assumes (in positivist, instrumentalist, or naive realist terms) that the limits of 'verifiability' are those laid down by a straightforward appeal to the as-yet untheorized observational evidence. However, quite simply, there is no such evidence that is not already an outcome or product of our various cognitive modes of engagement with an otherwise inscrutable reality. Thus we are always perceiving or interpreting the world by way of some 'virtual-reality generator', whether as concerns the most basic components of everyday knowledge and experience or with respect to scientific theories – such as quantum mechanics on the many-worlds interpretation – which seem to involve a far greater degree of speculative licence. In which case, Deutsch suggests, we should take a lesson from Galileo and acknowledge that reality transcends the limits of our commonsense-intuitive grasp yet does so in ways that can still be understood by application of our best (scientifically informed) explanatory hypotheses.

This argument is crucial to Deutsch's defence of the many-worlds theory, so it worth getting clear about just what he means by the claim that all knowledge comes to us via such forms of 'virtual reality'. First, and most important, they are *not* to be thought of as 'falling into the same philosophical category as illusions, false trails, and coincidences', that is, 'phenomena which seem to show us something real but mislead us' (pp. 102–3). If these had any part in 'virtual reality', then Deutsch's whole argument would amount to just an update on Berkeleian idealist themes, a 'token of the coarseness of human faculties', or a reminder of certain 'inherent limitations on the capacity of human beings to understand the physical world' (p. 102). It would then be open to just the kind of knockdown commonsense-realist riposte that Dr Johnson famously delivered when Boswell acquainted him with Berkeley's doctrine. In fact, it is precisely the opposite lesson that Deutsch wishes us to draw and which provides – as he sees it – the strongest argument in support of the many-worlds theory. No doubt the physical sciences spend a lot of time avoiding various kinds of perceptual distortion and other such sources of illusory 'commonsense' belief. However, 'virtual reality is not in that category', and this for reasons that Deutsch thinks decisive in the quest for a rational and realist interpretation of quantum phenomena. In short:

the existence of virtual reality does not indicate that the human capacity to understand the world is inherently limited but, on the contrary, that it is inherently unlimited. It is no anomaly brought about by the accidental properties of human sense organs, but is a fundamental property of the multiverse at large. And the fact that the multiverse has this property, far from being a minor embarrassment for realism and science, is essential for both – it is the very property that makes science possible.

(Deutsch, p. 103.

This is a crucial passage for Deutsch's argument and one that would bear a good deal of ampliative commentary. Suffice it to say that he rests the plausibility of his many-worlds case on the three major theses: (1) that it is consistent with the as-yet uninterpreted findings of orthodox instrumentalist QM; (2) that it falls square with 'commonsense' realism except under special (experimentally induced or laboratory-specific) conditions; and (3) that it involves no greater excursion into the realm of 'virtual reality' than even the most basic or everyday forms of perceptual-cognitive enquiry. For we could not get to understand anything – so his argument runs – were it not for our capacity to interpret incoming sensory data in a way that progressively abstracts from naive sense certainty and is thus able to correct or make allowance for the numerous sources of error and illusion built into our naive 'commonsense' view of the world. Chief among these – according to Deutsch – is the error that persuades us (as it likewise persuaded Galileo's opponents) that reality extends only so far as the limits of our direct perceptual acquaintance or, in this case, our direct knowledge of just one among the multitude of universes that contain all our various duplicate selves.

Thus the first step toward accepting Deutsch's theory is to acknowledge that 'virtual reality' is the nearest we can get to any understanding of the physical world beyond our private sensorium. The second is to grasp how quantum phenomena (interference effects, multipath deflections, and the like) require a many-worlds explanation that is no more 'virtual' or further removed from the data of straightforward perceptual acquaintance than any other theory in the physical sciences or indeed any item of everyday commonsense knowledge. The third is to take Deutsch's point that many-worlds is *uniquely* successful in resolving the various problems – of observer-induced wavepacket collapse, etc. – which dog other variants of QM theory such as the 'many-minds' interpretation. And the fourth is to recognize (*contra* instrumentalists and upholders of the orthodox Copenhagen view) that we are thus fully justified in taking Dr Johnson's line against Berkeley and declaring those worlds to be as real as anything that 'kicks back' or offers resistance like the stone when struck by Johnson's boot. For the only thing that stands in the way of our accepting this massively expanded ontology is our natural prejudice in favour of established (commonsense-intuitive) worldviews and the fact that quantum interference phenomena are mostly so transient and hard to detect. Nevertheless – Deutsch argues – it is now the one theory that can honestly claim to respect and apply the standards for valid scientific reasoning that were first enounced by Galileo and have since given rise to every major advance in the physical sciences.

This stance, however, although rhetorically effective, will appear less plausible if one considers that Deutsch's case for the many-worlds theory rests entirely on a number of premises drawn from orthodox quantum mechanics, itself a theory that is highly problematic as regards its own premises or the interpretation that can best (most plausibly) be placed upon them. After all, the most striking difference between his and Galileo's hypothesis is that Galileo succeeded in explaining a whole range of physical phenomena in terms that were *both* theoretically well-supported *and* consistent with the evidence in a way that required no conjuring up of speculative 'worlds' or 'universes' other than that which he and his fellow observers actually inhabited. Indeed, this is just Deutsch's point when he sides with Galileo against the Church authorities and their resort to an overly complex instrumentalist theory in order to defend the traditional view. In this mode he wields Occam's Razor with great relish and briskly rejects any theory or hypothesis that complicates matters beyond the simplest possible (or ontologically least extravagant) interpretation. Thus, for instance, 'observation alone can never rule out the theory that the Earth is enclosed in a giant planetarium showing us a simulation of a heliocentric solar system, and that outside the planetarium there is anything you like, or nothing at all' (Deutsch, p. 77). And again, there is nothing to exclude the theory that the planets follow their appointed course owing to the pressure exerted by angels whose existence is clearly proven by the laws of celestial mechanics.

Deutsch takes these – justifiably enough – as test case illustrations of the fact that there exist any number of possible theories (some of them wildly extravagant) which 'fit the evidence' or 'save appearances' while possessing not the least claim to genuine scientific or explanatory merit. They are properly ruled out on the basic principle that no theory is scientifically valid if it takes for granted some simpler, more direct explanation but then proceeds to deny that explanation by mounting a different, more complex, or roundabout explanation of its own. In the case of the Church versus Galileo,

> [t]he Inquisition's explanation is that the planets are seen to move in complicated loops because they really are moving in complicated loops in space; *but* ... this complicated motion is governed by a simple underlying principle: namely, that the planets move in such a way that, when viewed from earth, they appear just as they would if the Earth were in simple orbits around the Sun.
>
> (Deutsch, p. 79)

In short, this is a bad (scientifically disreputable) theory because it drags in a whole range of superfluous – doctrinally imposed – complications in order to discredit an alternative account which it none the less accepts as operationally valid for all practical purposes. Thus, '[o]ne cannot understand the world through the Inquisition's theory unless one understands the heliocentric theory first' (p. 79). The same applies to those other, more extravagant specimens – the planetarium hypothesis and the theory of angel-driven planetary motion – which likewise involve an unwarranted leap to abstruse 'explanations' that are wholly redundant *even by their own professed observational criteria*. The angel theory has certain limited merits in so far as it explains (or at any rate allows for) the fact that observed planetary motions are at variance with those prescribed or entailed by the doctrine of celestial spheres. However, as Deutsch points out, 'it does not explain why the angels should push the planets along one set of orbits rather than another, or, in particular, why they should push them as if their motion were determined by a curvature of space and time, as specified in every detail by the universal laws of the general theory of relativity' (pp. 88–9). His point in all this is that causal realism and inference to the best explanation are jointly sufficient to exclude any theory which involves an appeal to duplicate entities – such as angels or the giant planetarium – that do no genuine explanatory work since they merely go proxy for elements in another, more adequate physical theory.

Still, as I have said, there is a sense in which this whole line of argument may be seen to rebound against the many-worlds (or multiverse) hypothesis and the kinds of reasoning that Deutsch himself offers in support of that hypothesis. It emerges most strikingly when Deutsch goes on to translate the 'angel' theory from the realm of cosmology to that of quantum mechanics. Thus he asks how this theory would fare if applied to the famous two-slit experiment, usually interpreted (on orthodox QM) as establishing the wave–particle dualism by reason of the interference effects that continue to appear on the photosensitive screen even at the lowest rate of emission when light is detected as passing through the slits in the form of discrete quanta, i.e. as 'individual' photons.[15] According to Deutsch, these effects can be explained – realistically and rationally explained – only by positing the existence of shadow photons which belong to some other parallel quantum universe and which sometimes (under special conditions) interact with our own so as to produce such observable results. Now one might apply the angel hypothesis here, just as in the case of celestial mechanics, and come up with a yet more elaborate alternative theory that matched the observations, predictions, formalisms, etc., right down to their last detail. And indeed, 'to postulate that angels come through the other slits and deflect our photons would be better than nothing' (p. 89). That is to say, it would at any rate make some attempt (albeit an exorbitantly far-fetched attempt) at explaining the observed phenomena, rather than retreating to the standard instrumentalist line of least resistance. However, as Deutsch commonsensically remarks,

> we can do better than that. We know exactly how those angels would have to behave: very much like photons. So we have a choice between an explanation in terms of invisible angels pretending to be photons, and

one in terms of invisible photons. In the absence of an independent explanation for why angels should pretend to be photons, that latter explanation is superior.

(Deutsch, p. 89)

One may readily agree that if this were the *only* available choice – if interference effects could be accounted for only on one or other of these two hypotheses – then the many-worlds theory minus the angels would have a far stronger claim to scientific credibility. But there is no compelling reason to accept Deutsch's claim that this is indeed how the choice works out, since (as he believes) any other account of QM phenomena will either be driven to invoke the instrumentalist opt-out clause or end up by embracing some profligate ontology pretty much on a par with the angels hypothesis in point of explanatory merit.

What is particularly striking in this regard is the minimal attention that Deutsch accords to David Bohm's hidden-variables theory.[16] (See Chapter 3 of this book for a fuller account of that theory and its strong realist credentials in contrast to the orthodox, i.e. the Bohr-derived instrumentalist or empiricist approach.) He does cite it briefly as the one main exception to an otherwise general rule, namely that 'opponents of the multiverse theory *as an explanation* have seldom advanced rival explanations' (p. 335). Elsewhere there is a paragraph-long summary of Bohm's argument but one that effectively treats it – like the angels hypothesis – as a needlessly complex and roundabout way of endorsing the many-worlds account. Thus Deutsch thinks it a mistake, applying Dr Johnson's realist (or 'kick-back') criterion, to accept Bohm's theory at face value when it claims to avoid the ontological problems of orthodox QM by providing a more 'realistic' account of phenomena such as wave–particle dualism or multipath interference. Rather, it is 'a theory with predictions identical to those of [orthodox] quantum theory, in which a sort of wave accompanies every photon, washes over the entire barrier, passes through the slits, and interferes with the photon that we see' (p. 93). Moreover, it offers not so much a genuine alternative to the many-worlds interpretation as a means of avoiding any *overt* reliance on it while implicitly taking its truth for granted and merely coming up with a different (more complicated) set of equations in order to accommodate the wave-guided particle hypothesis.

On this view, Bohm's hidden-variables theory can best be compared with those other ontologically profligate doctrines – such as the planetarium hypothesis or the angel-driven theories of celestial mechanics and photon interference – which break the first rule of scientific explanation by pointlessly multiplying entities. Thus:

Bohm's theory is often presented as a single-universe variant of quantum theory. But, according to Dr. Johnson's criterion, that is a mistake. Working out what Bohm's invisible wave will do requires the same computations as working out what trillions of shadow photons will do.

Some parts of the wave describe us, the observers, detecting and reacting to the photons; other parts of the wave describe other versions of us, reacting to photons in different positions. Bohm's modest nomenclature – referring to most of reality as a 'wave' – does not change the fact that in his theory reality consists of large sets of complex entities, each of which can perceive other entities in its own set, but can only indirectly perceive entities in other sets. These sets of entities are, in other words, parallel universes.

(Deutsch, pp. 93–4)

Now if this were an accurate or non-prejudicial description of Bohm's theory, then one would have to agree with Deutsch that it offers none of the advantages claimed on its behalf by Bohm and other more recent advocates.[17] Those advantages – to state them very briefly – are (1) that it preserves a realist ontology wherein particles possess determinate values of space–time location and momentum; (2) that they continue to possess such values *between* various acts of observation–measurement, rather than acquiring them only in consequence of being measured with respect to this or that parameter; and (3) that this allows for greater continuity with certain components of classical (pre-quantum) physics such as the conservation laws respecting matter–energy and angular momentum. Moreover (4), Bohm's hypothesis produces results in perfect accordance with those obtained in orthodox QM by means of the Schrödinger-derived wave probability function, while (5) avoiding any recourse to mysterious ideas of the wavepacket collapse as somehow brought about by observer intervention or only on the instant – in Schrödinger's parable – when the box is opened up for inspection and the cat thus released from its 'superposed' (dead-and-alive) quantum predicament.[18] Lastly (6), Bohm's theory also seeks to explain quantum effects such as photon deflection or multipath interference without proposing a massively expanded ontology of parallel worlds, shadow universes, multiple intersecting realities, etc.

Such is at any rate the challenge Bohm's theory poses to orthodox QM and also, in virtue of item (6), to Deutsch's many-worlds interpretation. For if Bohm's theory is true – that is to say, if those effects can indeed be explained by adopting a hidden-variables account of the established QM formalisms – then of course there is no need for any such excursions into the realm of quantum hyperreality. As we have seen, Deutsch makes a point of stressing that his interpretation is likewise thoroughly realist in so far as it refuses to treat those shadow particles (or shadow worlds) as any less real than the world which we 'actually' inhabit and in which we conduct our observation–measurement of 'actually' existing particles. On the contrary, they are all equally real and may therefore be quantified over with the same degree of ontological assurance. And this despite the fact that some of them – the vast majority – belong to universes that have branched off from our own through some previous multiple-outcome event, and are hence epistemically accessible to us only on the evidence of certain transient and scarcely perceptible quantum interference effects. Thus, 'a real, tangible photon *behaves differently* according to what paths are open, elsewhere in the apparatus, for something to travel along and eventually intercept the tangible photon. Something does travel along those paths, and to refuse to call it real is merely to play with words' (Deutsch, pp. 48–9). And again, with a similar implied appeal to Dr Johnson and the realist 'kick-back' criterion: '[i]f the complex motions of the shadow photons … were mere possibilities that did not in fact take place, then the interference phenomena we see would not, in fact, take place' (p. 49).

Deutsch's argument may appear to lose something of its knock-down force, however, if we note how far he has to reinterpret Bohm – or twist Bohm's theory into line with his own fixed preconceptions – in order to clear the field of this particular (on the face of it more realistic and more intuitively plausible) account. Thus it may well be the case as Deutsch claims – and as Bohm would not deny – that 'working out what Bohm's invisible wave will do requires the same computations as working out what trillions of shadow photons will do' (p. 93). Certainly it is no part of Bohm's case or that mounted by advocates such as Cushing that the pilot-wave theory in any way avoids the kinds of mathematical complexity entailed by orthodox QM or by other interpretations (including many-worlds) that accept the necessity of continuing to work with the basic QM formalism and equations. However, Deutsch misses the point – or perhaps chooses not to see it – when he takes this as a pretext for maintaining that Bohm's theory, although 'often presented as a single-universe variant', in fact amounts to just an overly complicated version of many-worlds. It is the same sort of argument that is frequently encountered among defenders of orthodox QM when they reject Bohm's theory on the instrumentalist grounds that it is better – more sensible or less risky – to stick with the empirical evidence plus the well-tried quantum formalisms and avoid taking on any surplus ontological baggage such as that entailed by a hidden-variables or realist account. What is thus ruled out is the motivating premise of Bohm's entire argument (like Einstein's before him): namely, that an adequate physical theory should describe and explain those features of objective reality that exist and exert their causal powers quite apart from our current best theories concerning them. That is to say, it should not be content merely to save phenomenal appearances by producing an abstract formalism and set of equations which successfully predict observational results. In the case of orthodox (Copenhagen) QM, this outlook went along with the positivist belief that empirical adequacy plus a formalized proof procedure was the best that any theory could properly aspire to, and hence that science should have no truck with realist or causal-explanatory hypotheses.

Such ideas still have their following among some philosophers of science, very often – as with Bas van Fraassen's 'constructive empiricist' approach – extending all the way from quantum mechanics to issues of knowledge and theory construction in the macrophysical domain.[19] Thus it is hardly surprising that Bohm's theory should have run up against strong resistance from those of an instrumentalist persuasion (philosophers and physicists alike) who have viewed it as a needless 'metaphysical' extravagance or as carrying too much in the way of non-entailed ontological content. On this point at least he is fully in agreement with Deutsch: that instrumentalism manages to protect itself against such a charge only by avoiding realist commitments of the sort that must play a central role in any genuine or worthwhile physical theory. Thus Deutsch defines Copenhagen QM – in terms that would surely win approval from Bohm – as 'an idea for making it easier to evade the implications of quantum theory for the nature of reality' (p. 342). And again, on a more positive note:

> [f]ortunately, our best theories embody deep explanations as well as accurate predictions. For example, the general theory of relativity explains gravity in terms of a new, four-dimensional geometry of curved space and time. It explains precisely how this geometry affects and is affected by matter .... What makes [this theory] so important is not that it can predict planetary motions a shade more accurately than Newton's theory can, but that it reveals and explains previously unsuspected aspects of reality, such as the curvature of space and time.
>
> (Deutsch, pp. 2–3)

He then goes on to take issue with Steven Weinberg over the latter's instrumentalist claim that the 'important thing' from a scientific standpoint is 'to be able to make predictions about images on the astronomers' photographic plates, frequencies of spectral lines, and so on', and that it should be a matter of no concern 'whether we ascribe these predictions to the physical effects of gravitational fields on the motion of planets and photons (as in pre-Einsteinian physics) or to a curvature of space and time'.[20] Here again Deutsch comes out very much in sympathy with Bohm regarding the inadequacy of any such approach and its failure to offer a plausible account of the growth of scientific knowledge.

All the same, this broad, in-principle convergence on the issues of realism and causal explanation does not prevent him from radically mistaking the explanatory import of Bohm's hidden-variables theory. As I have said, Deutsch misses the point when he objects that the hidden-variables account is mathematically just as complex as its various rivals (including many-worlds) and therefore might just as well be abandoned in favour of his own interpretation. For this ignores the whole series of arguments that I have listed four paragraphs above in support of Bohm's claim that his theory provides a more detailed, realistic, intuitively plausible, and explanatorily adequate account of observed quantum phenomena. Deutsch thinks that any attempt to deny the many-worlds theory – for instance, by rejecting the reality of shadow particles and instead (like Bohm) construing interference effects as brought about by the wave-like distribution of forces as it acts upon particles with definite (this-world) location and momentum – is pretty much on a par with the 'angel' or 'planetarium' theories of celestial motion. The mathematics involved would 'require the same computational effort as working out the history of large numbers of shadow photons', and would moreover 'have to work its way through a story of what each shadow photon does: it bounces off this, is stopped by that, and so on' (p. 93). In which case, he concludes, 'just as with Dr Johnson's rock, and just as with Galileo's planets, a story that is in effect about shadow photons necessarily appears in any explanation of the observed effects' (ibid). An alternative account – such as Bohm's – which rejects the truth of many-worlds or the reality of shadow photons is thereby automatically disqualified as (1) lacking the presumed virtue of greater mathematical simplicity, (2) failing to explain anything that is not explained by many-worlds, and (3) in effect merely offering its own more evasive, equivocal, or roundabout version of the shadow-particle story. For it is just the 'irreducible complexity of that story' which 'makes it philosophically untenable to deny that the objects [i.e. shadow photons] exist' (p. 93).

It seems to me, on the contrary, that any comparison between the two 'stories' that does not start out from a strong a priori commitment to the truth of many-worlds will surely conclude in favour of Bohm on all the counts listed above. Chief among them are the sheer ontological extravagance entailed by Deutsch's version of quantum 'realism' and – sharply contrasted with that – the fact that Bohm's theory conserves far more of the 'classical' (Galileo-to-Einstein) conception of physical reality as well as far more of our basic, commonsense-intuitive understanding of the world. Indeed, it is a remarkable feature of Deutsch's argument that he contrives to turn this argument around so as to represent Bohm's as a far-fetched metaphysical or speculative construct and many-worlds as a matter of straightforward inference from known quantum phenomena. This he achieves partly by redescribing Bohm's theory in a way that burdens it with many of the problems which plague both orthodox (Copenhagen) QM and also, as I have argued, his own interpretation when construed from a different – dare one say less dogmatic and metaphysically loaded – standpoint. Thus, according to Deutsch, Bohm's hypothesis 'does not change the fact that in his theory reality consists of large sets of complex entities, each of which can perceive other entities in its own set, but can only indirectly perceive entities in other sets' (pp. 93–4). However, there is something decidedly odd – and decidedly at odds with Bohm's interpretation – in this talk about entities 'perceiving' other entities as if the state of the quantum system at any given time were somehow dependent on the photons themselves having mutual epistemic access and thus producing interference effects. That is to say, Deutsch is here translating Bohm's theory into his own preferential idiom, taking it for granted that the photons in question (along with their multiple shadow counterparts) exist across a range of quantum worlds which only interact on those rare occasions when such effects are perceived to occur. So it is that he can go straight on to draw the 'obvious' conclusion, i.e. that 'these sets of entities are, in other words, parallel universes' (p. 94). But this construal can be forced upon Bohm's theory – despite its professed ontological commitments – only if the multiverse interpretation is taken as self-evidently true and if all other theories which respect the QM evidence are therefore bound to acknowledge its truth, however obliquely or against their own professed intent. Otherwise, it must seem that Deutsch's redescription is really just a means of pre-empting the issue by smuggling in as an a priori doctrine what he claims to derive from observational data plus inference to the best explanation.

III

Heisenberg once described Bohr as one whose 'darkly metaphysical' ideas were put forward in the guise of a scientific theory concerning the ultimate nature of physical reality.[21] Bohr himself gave the argument a different spin – and invited an ironic rejoinder – by commenting that 'no man who is called a philosopher really understands what is meant by complementary descriptions'.[22] I think that these two comments are not unconnected and point towards some of the most striking aspects of quantum-theoretical debate. One is the fact that much of this discussion is indeed 'metaphysical' in so far as it involves a strong a priori commitment to the existence of various entities (such as shadow photons or multiple quantum universes) concerning which – *pace* Deutsch – the available evidence clearly cannot serve to adjudicate the issue. Another is the fact that arguments put forward by quantum physicists and philosophers are often 'verification-transcendent' in a sense sharply opposed to the scientific realist desire for causal-explanatory theories which transcend the limits of observational warrant or straightforward inductive inference. Deutsch is particularly tough on the inductivist approach, declaring that 'if we want to understand the true nature of knowledge, and its place in the fabric of reality, we must face up to the fact that inductivism is false, root and branch' (p. 61). In this he takes a lead from Popper's hypothetical-deductive account of how scientific theories are first devised with a view to maximal explanatory content and then tested against the evidence in order to falsify (rather than confirm) their purported veridical status.[23] However, there seems little virtue in applying such a theory if its truth conditions – or the criteria for what should count as a falsifying instance – are as widely contested or ill-defined as they are in the case of Deutsch's many-worlds hypothesis.

Of course, this is a general difficulty with Popper's philosophy of science and not just when that doctrine is applied to issues in quantum mechanics. It is the problem that arises as soon as one asks how *any* theory could *ever* be falsified unless by a process of inductively supported reasoning on the evidence that is likewise employed in verificationist procedures of the sort that Popper roundly rejects. It becomes much more of a problem, however, in the context of QM debate, especially when taken – as it is by Deutsch – to licence the kinds of far-reaching speculative theory (such as the many-worlds hypothesis) that are simply not subject to normal standards of probative or evidential warrant. That is to say, such hypotheses cannot be falsified – let alone verified – except by the appeal to a range of transient and hard-to-detect quantum interference phenomena which already come laden with various ontological commitments or preconceptions, according to the particular version of QM theory one happens to adopt. Thus, as Deutsch himself concedes, any claim regarding observational 'evidence' for the many-worlds theory has to take account of the extreme difficulty of setting up decisive experiments in order to elicit that evidence, as well as the rarity with which they occur outside such specially contrived situations and the scarcely perceptible effects that are produced even on those few occasions. Yet he still insists that the multiverse theory is true beyond reasonable doubt and, moreover, that the real (not merely 'virtual') existence of parallel universes is a truth whose denial cannot be sustained unless by either ignoring the evidence or failing to interpret it correctly. Thus, in Deutsch's disarmingly mock-modest words, 'I have merely described some physical phenomena and drawn inescapable conclusions' (p. 50).

Those conclusions may however seem somewhat less than inescapable if one examines some of the credibilizing strategies that Deutsch deploys in the course of his book. One is the technique of translating *all* reality into a species of 'virtual reality', or stressing the extent to which *all* human knowledge involves a construal of physical phenomena that goes far beyond the (so-called) observational 'data'. His main example here is Einstein's general theory of relativity, which explained gravitational attraction in terms of the curvature of space and time, and which thus came into conflict with Newton's theory as regards certain well-defined astronomical predictions, among them those of planetary motion on the classical (Newtonian) model. Thus, for instance, 'it [general relativity] correctly predicted that every year the planet Mercury would drift by about one ten-thousandth of a degree away from where Newton's theory said it should be'. And again, '[i]t also implied that starlight passing close to the Sun would be deflected twice as much by gravity as Newton's theory would predict' (Deutsch, p. 56). This latter claim was famously the subject of Eddington's 1919 expedition to test its truth under optimal conditions, a trial whose results were widely accepted as bearing out Einstein's hypothesis to a high degree of observational precision. That certain doubts have since been raised as to just *how* precise the results actually were is less important than the fact that repeated experiments – sometimes of a quite different nature – have continued to offer strong support for the theory.[24]

Now these successes have usually been marked down by philosophers and historians of science as a triumphant vindication of general relativity and also as a striking example of the way that certain, at the outset highly speculative, scientific claims can at length become subject to empirical testing through the deployment of new and more refined experimental techniques. Deutsch is himself quite content to draw this lesson but only in conjunction with his further point: that the results thus obtained – and their impact on our scientific worldview – were ever more closely bound up with minute differences of technologically assisted measurement and observation. In other words, they involved a law of simultaneously increasing and diminishing returns whereby the slightest change in some observed parameter could entail the most significant or far-reaching change in our entire cosmological picture. 'It may therefore seem', Deutsch writes,

> that we are inferring ever grander conclusions from ever scantier evidence. What justifies these inferences? Can we be sure that just because a star appeared millimetrically displaced on Eddington's photographic plate, space and time must be curved; or that because a photodetector at a certain position does not register a 'hit' in weak light, there must be parallel universes?
>
> (Deutsch, p. 57)

These are clearly not rhetorical questions in the sense that they invite a negative response or that Deutsch wishes to persuade us to scepticism concerning either of the cases in question. Thus he is no more in doubt that Eddington's results offered strong evidence for Einstein's theory than he is with respect to his own (purportedly equivalent) claim that the existence of certain well-nigh undetectable quantum phenomena should be taken as proof that 'there must be parallel universes'. But the logic of this argument is somewhat curious since it works not so much by investing his theory with an added measure of probity or rational support derived from the Einstein–Eddington case but rather by claiming that the many-worlds hypothesis is *no less* dependent on 'virtual reality' in the form of large inferences drawn from minute (scarcely perceptible) observation data.

What Deutsch seeks to bring home by means of this comparison is 'the fragility and the indirectness of all experimental evidence'. After all, he continues, 'we do not directly perceive the stars, spots on photographic plates, or any other external objects or events' (p. 57). We interpret them always as the end-result of an inferential chain that begins with the 'basic' perceptual processing of sensory data and which ascends through increasingly complex stages of conceptual, theoretical, or causal-explanatory grasp. But at no point in this chain – even the perception of incoming sensory stimuli – is the link between one and the next stage so completely fixed or determinate as to put us directly in touch with reality and leave no room for alternative possible outcomes. For, as Deutsch observes,

> the physical evidence that sways us, and causes us to adopt one theory or world-view rather than another, is less than millimetric: it is measured in thousands of a millimetre (the separation of nerve fibres in the optic

nerve), and in hundredths of a volt (the change in electric potential in our nerves that makes the difference between our perceiving one thing and perceiving another).

(p. 57)

From which he concludes that 'virtual reality' is the condition of all human knowledge and experience, from the most (seemingly) self-evident facts of our everyday dealing with the world to the most advanced theories of the physical sciences, quantum mechanics in particular. Yet so far from counting this a pessimistic conclusion – or one that gives rise to scepticism concerning the very possibility of scientific knowledge – Deutsch considers it the crucial insight which alone permits us to explain how such knowledge could conceivably occur. Thus – to repeat – the ubiquitous character of 'virtual reality' indicates not so much the limited scope of human understanding but, on the contrary, its 'inherently unlimited' character. More than that, 'it is no anomaly brought about by the accidental properties of human sense organs, but is a fundamental property of the multiverse at large' (p. 103). For it is in virtue of just that property – so Deutsch maintains – that human beings are enabled to understand the world through the kinds of interactive commerce with it (such as those revealed by quantum phenomena) that would otherwise not be possible.

It seems to me that Heisenberg's quip about Bohr applies with equal force to Deutsch's elaboration of the multiverse hypothesis. That is to say, he is here embarked upon an exercise of speculative metaphysics in the high rationalist tradition descending from Leibniz and lately revived by thinkers like David Lewis.[25] Deutsch accords Lewis only a passing mention – with a wrong page reference in the index – as one who has 'postulated the existence of a multiverse for philosophical reasons alone' (p. 340, not p. 349). This accords with his generally low estimate of philosophy *vis-à-vis* the physical sciences, a view that extends to all philosophers of science with the signal exception of Popper, and then only in so far as the case for many-worlds can be made out plausibly in keeping with Popper's hypothetico-deductive prescription. Still, one could suggest that Deutsch might have found reason to reconsider some of his far-out speculative claims had he taken more account of the various philosophical objections that rise against them. Peter Gibbins strikes a very different note at the close of his book on issues in quantum logic. 'Both philosophers and physicists', he writes,

have their own special contributions to make to the philosophy of physics in general, and to the philosophy of quantum mechanics in particular. Physicists are not only better at the physics, they also have a truer feel for what is physically reasonable and for what isn't. The philosopher's lack of the physicist's *feel* for the physics can be a strength, freeing him to explore the unreasonable, which is after all what quantum mechanics and quantum logic are.[26]

Deutsch, however, has no time for such goodwilled ecumenical gestures. Thus he takes it pretty much for granted that one sure sign of progress in knowledge is the tendency of certain problems to start out as matter for philosophical debate and then – through better understanding – to acquire the status of scientific theories, hypotheses, or research programmes. So it was, for example, that 'Einstein's general relativity swallowed geometry, and rescued both cosmology and the theory of time from their former purely philosophical status, making them into fully integrated branches of physics' (Deutsch, p. 344). From this point of view Lewis's many-worlds version of modal logic can only strike Deutsch as a similar conclusion arrived at on merely 'philosophical' grounds, or a speculative theory which cannot lay claim to the kinds of decisive evidence provided by quantum interference phenomena. And this despite Lewis's strong attestation to the truth of that theory not only as a matter of speculative reasoning from modal-logical premises but also as regards the existence of those worlds as a matter of plain ontological-realist commitment. For Deutsch, the main interest of Lewis's agreement on the reality of multiple worlds is the fact that it shows how quantum theory – rightly interpreted – can resolve a whole range of erstwhile philosophical issues that now belong squarely within its domain. Thus, '[t]he fruitfulness of the multiverse theory in contributing to the solution of long-standing philosophical problems is so great that it would be worth adopting even if there were no physical evidence for it at all' (p. 339).

This last statement seems to me quite remarkable given Deutsch's declarations elsewhere to the effect that many-worlds is a theory whose truth follows directly from the physical evidence and depends far less – if at all – on QM principles and premises. ('I have not even stated any of the postulates of quantum theory. I have merely described some physical phenomena and drawn inescapable conclusions.' [p. 50]) But in that case it is difficult to see why Deutsch's hypothesis should be taken to exert such a strong claim to truth compared with its various rivals, Bohm's hidden-variables interpretation among them. For if the many-worlds argument ultimately rests *neither* on the physical evidence *nor* on the postulates of quantum theory, then the question arises what could be the rational motivation for anyone to propound it in the first place or to accept its ontologically extravagant claims. Of course, these particular remarks by Deutsch should not be taken at face value since his argument does in fact presuppose *both* the validity of basic QM theory – its formalisms, equations, methods for calculating the wavefunction, etc. – *and* the 'physical evidence' as Deutsch construes it, i.e. the evidence of various quantum phenomena which the theory was designed to accommodate. All the same, they are not isolated statements but find numerous echoes elsewhere in his book. Thus, to repeat, 'we do not need deep theories to tell us that parallel universes exist', since 'single-particle interference phenomena tell us that' (p. 51).[27] And again, as if to quell any thought that there might be some viable alternative account: 'if the best theory available to physics did not refer to parallel universes, it would mean that we needed a better theory ... in order to explain what we see' (p. 51). In short, Deutsch takes it as a matter of rational self-evidence that many-worlds is the sole adequate account and will either defeat those rival theories or effectively subsume them by showing – as we have seen him to argue – that they presuppose the truth of many-worlds while purporting to offer a genuine alternative. For this is not – we recall – 'some troublesome, optional interpretation emerging from arcane theoretical considerations'. Rather, quite simply, 'it is the explanation – the only one tenable – of a remarkable and counter-intuitive reality' (p. 51).

This passage pinpoints the tension in Deutsch's argument between a rhetoric of bluff, no-nonsense appeal to the evidential facts of the case and a nagging recognition that his theory, if true, involves a massive and unprecedented break with every tenet of the scientific-realist worldview as hitherto conceived. Of course there is no question of rebutting that argument by adopting an equally dogmatic appeal to the commonsense-intuitive truths of experience or our knowledge of the world as it presents itself plainly and straightforwardly to human understanding. Deutsch himself rehearses all the standard (post-Kantian) objections to any such direct realist view, even if he pushes them so far towards the realm of 'virtual reality' as to make it very hard – at certain points in his book – to distinguish between the various orders of empirical, theoretical, hypothetico-deductive, metaphysical, and (sometimes) purely speculative or fictive claims concerning the many-worlds hypothesis. At any rate, it is clearly a line of counter-argument that the realist would be ill-advised to adopt after so many signal developments over the past two centuries – in mathematics, geometry, and the physical sciences as well as in logic and epistemology – which have left little if any room for confident appeals to a priori intuition as a sure means of access to truths about the world or even truths about the scope and limits of human understanding.[28]

To this extent Deutsch is fully justified in claiming (see the passage cited above) that the 'counter-intuitive' character of his multiverse hypothesis cannot be invoked against it by anyone possessing an informed grasp of those developments. Yet it is also apparent that Deutsch's reasons for endorsing many-worlds are presented in a manner which can only be construed as aprioristic and indeed as 'metaphysical' in just the sense of that term that Kant applied to rationalist thinkers like Leibniz and Wolff.[29] That is to say, Deutsch thinks to derive certain necessary truths about the way things stand in reality from his own interpretation of a set of first principles – those of orthodox QM theory – whose validity is placed beyond question since they form the very framework of operative concepts within which he elaborates his multiverse theory. For there is otherwise no accounting for his claim that the truth of many-worlds would be forced upon us as a matter of rational self-evidence *quite aside* from any 'deep theory' in physical science and *even were it not* for those decisive phenomena (such as multipath photon deflection) which supposedly indicate the momentary convergence of multiple quantum universes.

This would tend to suggest that Deutsch – like Bohr in Heisenberg's description – is practising a kind of quantum-based metaphysics that erects a large structure of speculative thought on its own favoured version of ultimate reality. Of course there is a sense of the term 'metaphysical' that applies to all scientific theories and which cannot (or should not) be run together with its usage as a vaguely pejorative shorthand for whatever theories one happens to find unacceptable. This usage goes back to the heyday of logical positivism, finds a prominent source text in E.A. Burtt's 1932 book *The Metaphysical Foundations of Modern Science*, and is also to be found in some of the more reductionist present-day approaches to the history and sociology of knowledge.[30] Beyond that – on the wilder postmodernist fringes of literary academe – it figures as a yet more offhand and all-purpose term of abuse for the kind of thinking that has not caught up with the truth that all truth claims are a species of naive illusion brought about by our attachment to old 'metaphysical' ideas of a reality outside the realm of discursive or textual representation.[31] Against this currently fashionable *doxa* the point needs making that scientific theories are by no means discredited by arguments which show them to entail or incorporate certain strictly metaphysical commitments, that is to say, certain presuppositions which cannot be justified on straightforward evidential grounds but which still have a claim to rational warrant as a matter of inference to the best (most adequate) causal explanation. However, it is hard to see how this could possibly apply to a theory – such as that of many-worlds QM – which ventures so far into the realm of purely abstract or (in a different sense) 'metaphysical' thought that it loses all touch with normative criteria of reality and truth.

Endnotes

1 See among others Paul Davies and James R. Brown (eds.), *The Ghost in the Atom* (Cambridge: Cambridge University Press, 1986); George Gamow, *Thirty Years that Shook Physics* (New York: Doubleday, 1966); John Gribbin, *In Search of Schrödinger's Cat: quantum physics and reality* (New York: Bantam Books, 1984); Fred Alan Wolf, *Taking the Quantum Leap* (New York: Harper & Row, 1989); also – more advanced but readable and informative – Alastair I.M. Rae, *Quantum Physics: illusion or reality?* (Cambridge University Press, 1986) and Euan Squires, *The Mystery of the Quantum World*, 2nd edn (Bristol and Philadelphia: Institute of Physics Publishing, 1994).

2 On these topics see for instance Michael Audi, *The Interpretation of Quantum Mechanics* (Chicago: University of Chicago Press, 1973); Robert G. Colodny (ed.), *Paradigms and Paradoxes: the philosophical challenge of the quantum domain* (Pittsburgh, Pa.: University of Pittsburgh Press, 1972); Bernard D'Espagnat, *Veiled Reality: an analysis of present-day quantum-mechanical concepts* (Reading, MA: Addison-Wesley, 1995); Peter Forrest, *Quantum Metaphysics* (Oxford: Blackwell, 1988); Max Jammer, *Philosophy of Quantum Mechanics* (New York: Wiley, 1974); Josef M. Jauch, *Are Quanta Real? a Galilean dialogue* (Bloomington, IN: Indiana University Press, 1973); A. Sudbury, *Quantum Mechanics and the Particles of Nature* (Cambridge: Cambridge University Press, 1986).

3 See David Deutsch, *The Fabric of Reality* (Harmondsworth: Penguin, 1997); also Bryce S. DeWitt and Neill Graham (eds.), *The Many-Worlds Interpretation of Quantum Mechanics* (Princeton, NJ: Princeton University Press, 1973); Roger Penrose, *The Emperor's New Mind: concerning computers, minds, and the laws of physics* (Oxford: Oxford University Press, 1989); J.A. Wheeler and W.H. Zurek (eds.), *Quantum Theory and Measurement* (Princeton University Press, 1983).

4 Deutsch, *The Fabric of Reality* (op. cit.), p. 53; henceforth 'Deutsch' with page number references in the text.

5 See especially Albert Einstein, B. Podolsky and N. Rosen, 'Can Quantum-Mechanical Description of Reality be Considered Complete?', *Physical Review*, series 2, Vol. 47 (1935), pp. 777–80: Niels Bohr, article in response under the same title, *Physical Review*, Vol. 48 (1935), pp. 696–702; also Bohr, 'Conversation with Einstein on Epistemological Problems in Atomic Physics', in P.A. Schilpp (ed.), *Albert Einstein: philosopher-scientist* (La Salle, IL: Open Court, 1969), pp. 199–241; Arthur Fine, *The Shaky Game: Einstein, realism, and quantum theory* (Chicago: University of Chicago Press, 1986); Don Howard, 'Einstein on Locality and Separability', *Studies in the History and Philosophy of Science*, Vol. 16 (1985), pp. 171–201; Tim Maudlin, *Quantum Nonlocality and Relativity: metaphysical intimations of modern science* (Oxford: Blackwell, 1993); Michael Redhead, *Incompleteness, Nonlocality and Realism: a prolegomenon to the philosophy of quantum mechanics* (Oxford: Clarendon Press, 1987).

6 See entries under Note 2, above; also Niels Bohr, *Atomic Theory and the Description of Nature* (Cambridge: Cambridge University Press, 1934); *Atomic Physics and Human Knowledge* (New York: Wiley, 1958); Henry J. Folse, *The Philosophy of Niels Bohr: the framework of complementarity* (Amsterdam: North-Holland, 1985); John Honner, *The Description of Nature: Niels Bohr and the philosophy of quantum physics* (Oxford: Clarendon, 1987); Dugald Murdoch, *Niels Bohr's Philosophy of Physics* (Cambridge University Press, 1987).

7 Immanuel Kant, *Critique of Pure Reason*, trans. N. Kemp Smith (London: Macmillan, 1964). For further discussion see Christopher Hookway, *Scepticism* (London: Routledge, 1992); A.C. Grayling, *The Refutation of Scepticism* (London: Duckworth, 1985); Alan Musgrave, *Common Sense, Science and Scepticism: a historical introduction to the theory of knowledge* (Cambridge: Cambridge University Press, 1993); John Watkins, *Science and Scepticism* (Princeton, NJ: Princeton University Press, 1984); Michael Williams, *Unnatural Doubts: epistemological realism and the basis of scepticism* (Princeton, NJ: Princeton University Press, 1996).

8 See Note 2, above; also David Bohm, *Causality and Chance in Modern Physics* (London: Routledge & Kegan Paul, 1957); David Bohm and B.J. Hiley, *The Undivided Universe: an ontological interpretation of quantum theory* (London: Routledge, 1993); David Z. Albert, *Quantum Mechanics and Experience* (Cambridge, MA: Harvard University Press, 1993) and 'Bohm's Alternative to Quantum Mechanics', *Scientific American*, No. 270 (May 1994), pp. 58–63; F.J. Belinfante, *A Survey of Hidden Variable Theories* (Oxford: Pergamon Press, 1973); James T. Cushing, *Quantum Mechanics: historical contingency and the Copenhagen hegemony* (Chicago: University of Chicago Press, 1994); Peter Holland, *The Quantum Theory of Motion* (Cambridge: Cambridge University Press, 1993).

9 David Lewis, *On the Plurality of Worlds* (Oxford: Blackwell, 1986).

10 See also Saul Kripke, *Naming and Necessity* (Oxford: Blackwell, 1980); M. Loux (ed.), *The Possible and the Actual* (Ithaca, NY: Cornell University Press, 1979); Alvin Plantinga, *The Nature of Necessity* (Oxford: Oxford University Press, 1974); R.C. Stalnaker, *Inquiry* (Cambridge, MA: M.I.T. Press, 1987).

11 G.W. Leibniz, *Monadology*, trans. E. Latta (Oxford: Oxford University Press, 1972); *New Essays Concerning Human Understanding*, trans. A.G. Langley (La Salle, IL: Open Court, 1926); also Hidé Ishiguro, *Leibniz's Philosophy of Logic and Language* (London: Duckworth, 1972).

12 See Notes 5 and 6, above.

13 See DeWitt and Graham (eds.), *The Many-Worlds Interpretation of Quantum Mechanics* (op. cit.).

14 See Alexandre Koyré, *Galilean Studies* (Brighton: Harvester, 1978); also – from a very different (instrumentalist and arguably crypto-theological) standpoint – Pierre Duhem, *To Save the Phenomena: an essay on the idea of physical theory from Plato to Galileo*, trans. E. Dolan and C. Maschler (Chicago: University of Chicago Press, 1969).

15 For good brief accounts of this experiment in its various forms, see Rae, *Quantum Physics: illusion or reality?* and Squires, *The Mystery of the Quantum World* (Note 1, above).

16 See entries under Note 8, above.

17 See Cushing, *Quantum Mechanics: historical contingency and the Copenhagen hegemony* (op. cit.); also Holland, *The Quantum Theory of Motion* (op. cit.).

18 See Erwin Schrödinger, *Letters on Wave Mechanics* (New York: Philosophical Library, 1967); also John Gribbin, *In Search of Schrödinger's Cat* (op. cit.).

19 See Bas C. van Fraassen, *The Scientific Image* (Oxford: Clarendon Press, 1980); *Quantum Mechanics: an empiricist view* (Clarendon Press, 1992).

20 Steven Weinberg, *Gravitation and Cosmology* (New York: Wiley, 1972), p. 147; cited by Deutsch, p. 4.

21 Cited by Peter Gibbins, *Particles and Paradoxes: the limits of quantum logic* (Cambridge: Cambridge University Press, 1987), p. 9.

22 Cited by Cushing, *Quantum Mechanics* (op. cit.), p. 32.

23 See especially Karl Popper, *The Logic of Scientific Discovery*, 2nd edn (New York: Harper & Row, 1959); *Conjectures and Refutations* (Harper and Row, 1963); *Objective Knowledge: an evolutionary approach* (Oxford: Clarendon Press, 1972); *Realism and the Aim of Science* (London: Hutchinson, 1983).

24 See for instance R.B. Angel, *Relativity: the theory and its philosophy* (Oxford: Pergamon Press, 1980); also J.R. Lucas and P.E. Hodgson, *Spacetime and Electro-Magnetism* (Oxford: Clarendon Press, 1990).

25 See Notes 9 and 10, above.

26 Gibbins, *Particles and Paradoxes* (op. cit.), p. 167.

27 As my colleague Vivian Beedle remarked, 'this is quite fantastic, isn't it? Even to be able to *grasp* the notion of "single-particle interference phenomena", a pretty deep piece of theory is required' (private communication).

28 On these various developments, see especially J. Alberto Coffa, *The Semantic Tradition from Kant to Carnap: to the Vienna Station* (Cambridge: Cambridge University Press, 1991).

29 Kant, *Critique of Pure Reason* (op. cit.); also *Prolegomena to Any Future Metaphysics*, ed. L.W. Beck (Indianapolis: Bobbs-Merrill, 1950) and *Metaphysical Foundations of Natural Science*, trans. J. Ellington (Bobbs-Merrill, 1970).

30 E.A. Burtt, *The Metaphysical Foundations of Modern Physical Science: a historical and critical essay* (London: Routledge & Kegan Paul, 1932); also – prototypically – Rudolf Carnap, 'The Elimination of Metaphysics through Logical Analysis of Language', in A.J. Ayer (ed.), *Logical Positivism* (New York: Free Press, 1959), pp. 60–81.

31 See for instance Jean-François Lyotard, *The Postmodern Condition: a report on knowledge*, trans. Geoff Bennington and Brian Massumi (Manchester: Manchester University Press, 1984).

5   Should philosophers take lessons from quantum theory?

I

In this chapter I shall have more to say about Deutsch's many-worlds (or 'multiverse') theory of quantum mechanics and the kinds of philosophical problem to which it gives rise.[1] At the outset it is perhaps worth recalling what the theory entails and how it developed in response to various difficulties with the orthodox (Copenhagen) school of QM interpretation. Basically, many-worlds can be thought of as offering an answer – one of several – to the problem that Schrödinger posed with his famous 'cat-in-the-box' thought experiment, designed as a *reductio ad absurdum* of orthodox QM and as a means of effectively forcing the issue with regard to its more bizarre implications for our knowledge of physical reality.[2] Chief among these was its failure to explain at just what point the transition occurred from the realm of quantum uncertainty (where the cat must somehow be conceived as existing in a 'superposed' state between life and death) to that of macrophysical objects and events where it must surely be thought of as *either* alive *or* dead, regardless of whether the box has been opened up and the wavepacket thereby 'collapsed' into one or another definite state through an act of human observation.

According to orthodox QM, as theorized by Bohr and Heisenberg, such questions were ill-framed and merely reflected our inability to get our minds around the inherent strangeness – to our way of thinking – that typified quantum phenomena.[3] Much better to adopt an instrumentalist position and not be concerned about the reality 'behind' quantum appearances just so long as the theory continues to work as a matter of predictive–observational warrant. For others (Einstein pre-eminent among them, and latterly David Bohm) this approach seemed to beg all the main issues, not least the issue of whether QM was a 'complete' physical theory.[4] Thus there might exist some viable alternative – some 'hidden-variables' account – that would perfectly match the well-established quantum predictions while conserving the idea of an objective (mind-independent) microphysical reality which was not somehow conjured up in or through the very act of observation.[5] On this view particles would possess definite values of position and momentum quite aside from whether or not those values were subject to measurement at any given time and irrespective of the kind of measurement performed, e.g. the polarization setting of a spin-detector placed in their path. In other words, it would avoid the epistemic fallacy of conflating ontology with epistemology, or supposing the limits of our present-best knowledge to be somehow intrinsic to the very nature of quantum-mechanical phenomena.

Bohm's theory has lately received a good deal of attention after many years of rejection or neglect by advocates of the orthodox QM instrumentalist line.[6] However, it has attracted nothing like the degree of popular interest enjoyed by the many-worlds interpretation that was first proposed during the 1950s by Hugh Everett and Bryce S. DeWitt, and which Deutsch now defends – against all comers – as the sole candidate with a genuine claim to account for the quantum-physical evidence and provide an adequate realist ontology consistent with QM principles.[7] According to this theory – in brief – *every possible* outcome of *every* wavepacket collapse is *simultaneously and actually realized* through a constant branching of alternative quantum worlds which are all equally 'real', although only one is epistemically accessible to any individual observer at any particular time. For the observer must likewise be thought of as having previously split into a whole multitude of sentient selves, each of them consciously inhabiting a 'world' whose history is itself just one among the manifold world versions that have evolved up to that time through the proliferating series of wavepacket collapses. Thus he or she will have any number of counterpart 'selves' distributed across those worlds, and each possessing a lifeline which, if traced back far enough, will rejoin the others at just that point where their destinies first diverged. (Readers who have seen the film 'Sliding Doors' or who enjoy the fiction of Borges, Vonnegut and Italo Calvino will perhaps have the best imaginative feel for this kind of thing.) Only through certain transient and barely detectable phenomena – such as the interference patterns between 'real' and 'shadow' photons in QM multipath experiments – can we obtain evidence for the reality of these countless coexistent quantum worlds. Otherwise our sole means of access to them is by way of hypothetico-deductive reasoning to what Deutsch considers the necessary truth of his own 'multiverse' theory, as against (say) Bohm's hidden-variables account or the orthodox – in his view intellectually disreputable – Copenhagen instrumentalist approach.

Thus Deutsch, like Bohm, is most emphatically a realist in the sense that he rejects the evasive retreat to an idea of QM theory as having to do only with measurements, predictions, 'empirical adequacy', etc., and therefore as not requiring – indeed ruling out – any concern with deeper ontological issues.[8] However, his attitude of intransigent realism *vis-à-vis* those multiple worlds is one that inevitably comes into sharp conflict with other, more familiar varieties of realist thinking, whether in everyday (commonsense-intuitive) or specialized (philosophico-scientific) terms. Indeed, one could hardly envisage an outlook more drastically at odds with the realist worldview as defended by various philosophers and scientists from Galileo to Einstein and beyond. I had better make the point through some of Deutsch's more vivid illustrations of his many-worlds thesis lest the reader think that I have misrepresented his position. Thus for instance:

[w]hen I introduced tangible and shadow photons I apparently distinguished them by saying that we can see the former, but not the latter. But who are 'we'? While I was writing that, hosts of shadow Davids were writing it too. They too drew a distinction between tangible and shadow photons, but the photons they called 'shadow' include the ones I called 'tangible', and the photons they called 'tangible' are among the ones I called 'shadow'.[9]

Another example has to do with Michael Faraday and the counterfactual supposition that, if Faraday had died in 1830 (i.e. before making his 1831 discovery about electromagnetic induction), then the progress of scientific knowledge or technology would surely have been set back by many years. Such judgements play a large part in even the most 'factual' and avowedly non-speculative genres of historical discourse, as recent theorists have set out to prove by way of discrediting naive ('positivist') notions of enquiry in the natural and social sciences.[10] But in that case, Deutsch wonders, 'what does it mean, in the context of spacetime physics, to reason about the future of non-existent events?' (p. 274). After all, it can hardly make sense to speculate about the consequences for scientific progress of Faraday's death in 1830 if no such event can be thought of as having somehow 'really' occurred, i.e. as having happened in some parallel universe where its results are even now being worked through in a history no less real than that which we ourselves have chanced to inherit. Moreover, it is feasible – just about – that Faraday's earlier (counterfactual) death might have resulted in a speeding-up of scientific progress through somebody else's having produced a more advanced or a radically different theory concerning what we now think of, thanks to Faraday, as electromagnetic phenomena.

Perhaps we might keep these perplexities in check 'by imagining only spacetimes in which, though the event in question is different from that in actual spacetime, the laws of physics are the same' (p. 275). However, Deutsch sees no justification – this-world prejudice apart – for 'restricting our imagination in this way', or refusing to countenance the *real* possibility of things (the 'laws of physics' presumably among them) having turned out altogether different. At any rate if those laws are thought of as having necessarily remained the same, then 'the event in question *could not* have been different, since the laws determine it unambiguously from the previous history' (ibid). Yet clearly Faraday *might* have died earlier since we can easily imagine all sorts of counterfactual but perfectly plausible circumstances that would have led to his death in 1830. In which case, *ex hypothesi*, we shall have to acknowledge that the previous history must also have differed in respect of various details, the extent of that difference depending on the scope of our counterfactual imagining, or the restrictions applied by our use of such limiting clauses as 'other things being equal'. In some of these alternative scenarios, technological progress would presumably have been slowed down, whereas in others it might just conceivably have jumped a stage – and thus transformed the future of the physical sciences – through the advent of a more powerfully unified theory of electromagnetic phenomena, or indeed a theory that explained those phenomena (and others besides) in entirely different terms.

Thus the question remains as to *which* of these alternative possible worlds we are referring to when we reason counterfactually by constructing hypotheses in the form of an '*if ... then*' or subjunctive conditional statement. On Deutsch's account, we can make sense of them only by adopting the multiverse theory and assuming that *all* possibilities have been realized in one or another of the multiple worlds that diverge at every point where some particular world-specific event (such as Faraday's death) happens to occur. And of course this process continues thereafter, with each world splitting momentarily into all the various possible results of every such multiple-outcome event in its own future history. Moreover, as I have said, these alternative outcomes must each involve a retroactive effect on the entire chain of 'previous' occurrences that are taken to have led up to them. For of course that history will differ from world to world according to which of the various branching series happens to contain oneself (or some particular version of oneself) along with everything pertaining to one's own knowledge, experience, future possibilities, and so forth. All of which leads Deutsch to the conclusion that we can attach no definite meaning to the phrase 'other things being equal', since in the realm of counterfactual hypotheses any change of world specification, no matter how seemingly slight, can bring about world-transformative effects which reach far back into history and forward to everything that falls within the scope of subjunctive-conditional imagining.

At this stage the commonsense realist will no doubt respond that Deutsch has been betrayed into yet another bout of whacky metaphysical extravagance through his insistence that these other worlds are *not* to be thought of as speculative, fictive, or merely hypothetical, but must rather be conceived as entirely on a par with our own in point of reality or full-fledged ontological status. However, as might be expected, Deutsch will have none of this. 'Try as we may', he writes, 'we shall not succeed in resolving this ambiguity [i.e. that of 'other things being equal'] within spacetime physics. There is no avoiding the fact that in spacetime exactly one thing happens in reality, and everything else is fantasy' (p. 275). Yes indeed, the realist might well reply: 'fantasy' in so far as Deutsch can assert the reality of his multiple worlds only through a wholesale confusion of realms between the actual, the possible, the hypothetical, the speculative, the fictive, and the downright fantastic. Indeed, there is now a large body of work – by modal logicians and others – which seeks to clarify these various distinctions in terms of the criteria for transworld identity, epistemic access, kinds or degrees of deviation from the norms of this-world assertoric warrant, and so forth.[11]

Thus, for instance, in fictive possible worlds this deviation may extend well beyond the limits established for the use of counterfactual procedures in scientific hypothesis testing or in the conduct of historical enquiry.[12] That is, it may involve not only a suspension of contingent (might-have-been-otherwise) facts about the world but also a suspension of the laws of nature and even – at the limit – of necessary truths such as those of elementary logic. Within literary theory these distinctions have been subject to further refinement, as by ranging narrative genres on a scale from so-called 'faction' at the one extreme (involving various real-world historical characters, actions, and events), via the more traditional forms of novelistic realism, to postmodernist or experimental fictions at the opposite extreme where – in principle at least – there is no limit on the scope of counterfactual imagining.[13] This approach has also been used by historiographers in order to account for the different modes of historical discourse and the recent emergence of mixed-mode genres that exploit certain fictional narrative techniques with a view to shaking up the old conventions of (so-called) positivist scholarship.[14] What emerges from such analyses is a range of criteria for distinguishing between counterfactual-historical and literary-fictive discourse, even though certain authors – historians and novelists alike – may seek to confound any tidy parcelling-out of the two narrative domains.

Among modal logicians likewise there is widespread agreement on the need for a careful demarcation of bounds between actual and possible worlds, and, again, between the various kinds or orders of restricted possibility (physical, epistemic, metaphysical, logical, etc.) some of which hold only as a matter of contingent, this-world fact and some of which obtain for all conceivable worlds in virtue of their necessary (transworld applicable) truth.[15] What is most important – from this point of view – is to prevent the kind of massively swollen ontology that results from confusing those orders and attributing reality to even the wildest flights of counterfactual conjecture. Thus interpreted, modal logic is in danger of opening the way to all those problems that Russell famously encountered with Brentano's thesis concerning the 'reality' of inexistent or imaginary objects such as centaurs, golden mountains, square circles, and the like.[16] At any rate, few would go along with David Lewis – or presumably with Deutsch – in espousing the full-blown ontological parity principle, that is, the belief that every mode of counterfactual or subjunctive-conditional thinking is necessarily committed to the existence of each and every world wherein its hypotheses would hold true.[17]

However, once again, Deutsch will hear nothing of these merely 'philosophic' expedients for avoiding the literal truth of the multiverse theory. In his view philosophers and logicians have never come close to solving the problem of counterfactual statements, i.e. the problem of how such statements can possess any meaning, logic, or explanatory force if they are not taken as referring to what is actually the case with respect to those alternative worlds. Thus, '[w]e all know what such statements mean, yet as soon as we try to state their meaning clearly it seems to evaporate' (p. 275). However, the problem itself simply evaporates – so Deutsch would have us believe – if one looks elsewhere than to philosophy or logic in hope of an adequate solution.

For 'the source of this paradox is not in logic or linguistics, it is in physics – in the false physics of spacetime' (ibid). What is required in order to make sense of counterfactual statements and our reasoning upon them is a multiverse physics that allows those statements a sufficient range of real-world ontological domains over which their truth values can be held to apply. 'In the multiverse', as Deutsch describes it,

> there are almost certainly some universes in which Faraday dies in 1830, and it is a matter of fact (not observable fact, but objective fact none the less) whether technological progress in those universes was or was not delayed relative to our own. There is nothing arbitrary about which variants of our universe the counterfactual 'if Faraday had died in 1830 ...' refers to: it refers to *the variants which really occur* somewhere in the multiverse.
>
> (Deutsch, p. 276)

Thus Deutsch has no need of the refinements developed by modal logicians or epistemologists – let alone literary theorists – by way of distinguishing the various orders of logically necessary, factual-contingent, hypothetical, counterfactual, or subjunctive-conditional statement. On his account these all make reference to one or another of the multiple worlds wherein every such statement possesses a definite truth- or falsehood-value as determined by the way things actually stand with respect to that particular world. In which case clearly it is physics, not philosophy, to which we should look in quest of a new and more adequate worldview that promises to resolve these long-standing issues of mind, knowledge, and reality.

II

I shall press somewhat further with this discussion of Deutsch's book partly because it makes such (literally) exorbitant claims for the many-worlds theory, and partly because those claims are advanced – despite all his protests to the contrary – on the basis of certain highly speculative theses that invite philosophical comment and criticism. Indeed, it is one of the most striking features of quantum-theoretical debate since the early years of this century that it has involved deep-laid differences of view with regard to metaphysical, ontological, and epistemological questions that would seem to fall more within the province of philosophy than that of the natural sciences.[18] Of course this is not to suggest that there exists – or should properly exist – any absolute distinction between those realms, such as the logical positivists required when they laid down separate criteria for the conduct of empirical (observation-based) enquiry on the one hand and the procedures of rational inference (or scientific theory construction) on the other.[19] Nor is it by any means to imply – again like the logical positivists – that science can have no legitimate business with 'metaphysical' pseudo-statements, i.e. those to be found just about everywhere except in statements of empirically verifiable fact and in the strictly tautologous propositions of formal logic.[20] If one thing has emerged from the demise of old-style positivist thinking, it is the extent to which *all* scientific methods, theories, hypotheses, or modes of explanation involve certain strictly metaphysical premises which alone permit science to proceed beyond the stage of random observation or piecemeal data-gathering. However, the backlash against logical positivism has also had time to produce some cautionary lessons, among them the fact that philosophy of science needs to maintain a fairly robust sense of reality and truth if it is not to fall prey to 'metaphysical' illusions in the other (pejorative) sense of that term. What is remarkable about some interpretations of quantum mechanics – including Deutsch's many-worlds candidate – is the way that they derive an entire (in this sense) 'metaphysical' worldview from a process of largely a priori reasoning such as I have described above. All the more so, in Deutsch's case, when the interpretation is avowedly realist and thus entails the actual existence of as many parallel worlds or universes as there exist counterfactual hypotheses that quantify over their domain.

At this point it is worth dwelling for a moment on some of the old 'philosophical' perplexities which, according to Deutsch, have found their solution – or been successfully transferred from philosophy to physics – with the advent of QM multiverse theory. One is the (perennial!) problem of time and the difficulty of explaining how the subjective or phenomenological aspect of temporal consciousness can be reconciled with the relativistic concept of time as the fourth component of a spacetime coordinate system that strictly allows for no valid distinction between past, present, and future events. Now there is a certain limited sense, Deutsch concedes, in which those subjective intuitions are valid enough. 'Relative to an observer, the future is indeed open and the past fixed, and possibilities do indeed become actualities.' (p. 287) However, this view of things reduces to nonsense – or generates all the well-known aporias of reflective time-consciousness – because it '[tries] to express these true intuitions within the framework of a false classical physics' (ibid). That is to say, it excludes the saving possibility that different moments or temporal locations of a 'single' consciousness may exist in different parallel worlds with no point of contact (or epistemic access) between them except through fleeting interference-effects or other such recondite manifestations of the quantum multiverse. Thus:

> [w]e exist in multiple versions, in universes called 'moments'. Each version of us is not directly aware of the others, but has evidence of their existence because physical laws link the contents of different universes. It is tempting to suppose that the moment of which we are aware is the only real one, or is at least a little more real than the others. But that is just solipsism. All moments are physically real. The whole of the multiverse is physically real. Nothing else is.
>
> (p. 287)

Again, the only reason we have remained (up to now) in ignorance of this multiverse reality is that its effects are so rare and so nearly imperceptible from within our own universe that they went unrecognized until the advent of a physical theory – quantum mechanics – which produced both the necessary means of observation and the concepts whereby to explain them. However, Deutsch thinks that we are at last in a position to resolve all those old 'philosophical' problems, including the antinomies of time and change as viewed from that limited (single-universe) perspective.

Now it seems to me – speaking no doubt as an unreconstructed single-universe or classical space-time realist – that this leaves all the traditional aporias very firmly in place and does nothing to resolve the problems about time that have been raised by philosophers from Aristotle and Augustine to the present. All it does is translate the problems from one language into another, that is, from the subjective idiom of past, present and future (where time is experienced as flowing equably through a series of successive 'moments') into Deutsch's preferred multiverse idiom where those moments are thought of as stacked up in vertical formation like a pile of snapshots only some of which – those belonging to the specious present – we can access directly from within our own world. As Deutsch develops this metaphor:

> [t]he snapshots which we call 'other times in our universe' are distinguished from 'other universes' only from our perspective, and only in that they are especially closely related to ours by the laws of physics. They are therefore the ones of whose existence our own snapshot holds the most evidence. For that reason, we discovered them thousands of years before we discovered the rest of the multiverse, which impinges on us very weakly by comparison, through interference-effects. We evolved special language constructs (past and future forms of verbs) for talking about them. We also evolved other constructs (such as '*if ... then*' statements, and conditional and subjunctive forms of verbs) for talking about other types of snapshot, without even knowing that they exist. We have traditionally placed these two types of snapshot – other times, and other universes – in entirely different conceptual categories. Now we see that this distinction is unnecessary.
>
> (pp. 278–9)

However it is hard to see how this provides a better, simpler, more adequate or less paradoxical explanation of those traditional problems with the concept of time which Deutsch considers to be merely a result of our trying to resolve them within the framework of a single-universe space–time conception. After all, if it is the 'laws of physics' that relate us most closely to our own universe – and that have up to now kept us from discovering the rest of multiverse reality – then what could be the physical (as opposed to metaphysical or speculative) grounds for supposing that multiverse reality to exist and, moreover, to require so drastic a break with all the hitherto established laws of physical science? Deutsch may respond that this claim is warranted first by the otherwise insoluble paradoxes of quantum mechanics – problems unknown to earlier science – and second by the fact that multiverse QM is the only theory capable of resolving the kindred (since quantum-related) paradoxes encountered with classical philosophies of time. However, his argument will seem less convincing if it is pointed out (1) that the quantum dilemmas arise only on certain construals of orthodox QM and *not* on a Bohm-type realist or hidden-variables interpretation[21], and (2) that Deutsch's proffered solution escapes the traditional paradoxes about time only by way of an alternative ontology that multiplies parallel worlds or universes far beyond the limits of any possible evidence for them. Again this seems to be more an excursion into speculative hyperreality than a genuinely science-based answer to well-worn philosophic problems.

Thus the multiverse theory 'solves' the problem of time by invoking a massively counter-intuitive ontology and also by creating difficulties elsewhere. For how should we interpret the notion of multiple selves distributed over those various alternative worlds – which in most cases scarcely interact or communicate – yet still somehow meeting identity criteria that suffice for them to count as versions or tokens of the 'same' individual, rather than as transient states of consciousness that emerge and disappear from moment to moment without the least degree of perduring or continuous selfhood? Of course this idea has philosophical precedents, notably in Hume's conception of the self as a passing show of impressions, memories, desires, anticipations, and the like, devoid of any deep further fact – any underlying principle of personhood – that would somehow unite them over time as experiences pertaining to the same self-identical consciousness.[22] On Deutsch's account it has more to do with the strange ontology of quantum mechanics and the problems which that theory supposedly creates for any realist worldview entailing the premise that individuals (whether objects or persons) possess unique identity conditions whereby they can reliably be picked out from one moment or space–time location to the next. Such is the basic conceptual scheme proposed by Strawson in his scaled-down 'descriptivist' version of Kantian metaphysics, and such – its proponents would further claim – the commonsense ontology presupposed in our everyday experience of objects, persons, and events.[23] Undoubtedly, orthodox QM throws a paradox into this scheme of things by requiring us to think of particles (or particle-like phenomena) as turning up here and there in statistical accordance with the wave probability function and hence lacking identity criteria across or between measurements. So it is that Deutsch can propose many-worlds as a theory that may seem ontologically extravagant and at odds with our deepest intuitions about time, space, and selfhood but which in fact offers the only solution to these quantum dilemmas.

However, this solution will still appear extravagant and needlessly counter-intuitive if one recalls that Bohm's hidden-variables theory manages to avoid all those problems without resorting to any such far-gone flights of metaphysical conjecture. For on Bohm's account – to recapitulate briefly – the particle is taken to follow a continuous space–time trajectory and therefore to possess definite values of position and momentum not only as and when measured but also between measurements.[24] If this is the case, then there is simply no need to propound a theory (such as Deutsch's many-worlds interpretation) that claims to resolve the paradoxes of orthodox QM but only at the cost of a massively inflated ontology and a notion of 'reality' so far out of touch with the usual requirements of physical hypothesis testing. Least of all can it help with those 'classical' problems concerning the nature of time or the conflict between our intuitive experience of temporal flow and the current scientific (post-Einsteinian) concept of a four-dimensional space–time continuum where there is strictly no room for subjective intuitions of past, present, and future.[25] What the many-worlds theory offers is not so much an answer to these problems as a transposition into different terms whereby they are rendered yet more acute through the positing of multiple 'realities' to which, *ex hypothesi*, we can have only fleeting epistemic access and whose very existence must therefore be inferred as the upshot of some highly speculative reasoning on minimal evidence. Moreover, if Bohm's interpretation is correct, then it is able to account for that same evidence – of remote 'entanglement', multipath deflection, quantum interference effects, and so forth – in accordance with both a realist ontology and a physical worldview that necessitates no such radical break with the entire prehistory of modern scientific thought.

III

All the same, Deutsch's book has drawn high praise from many reviewers, among them Julian Brown – writing in *New Scientist* – whose comments are cited on the paperback jacket. 'Reading it', he writes, 'might just change your life .... The theory of everything, quantum mechanics, virtual reality, scientific method, evolution, the significance of life, quantum computation, the nature of mathematics, time travel, the end of our Universe: all these and much more find their place ... Crackles with originality ... This is an awesome book'. I quote his remarks not merely as an instance of hyperbolic journalese but in order to suggest at least one source of the widespread appeal that has long been exerted by construals of quantum mechanics – 'many-worlds' and 'many-minds' chief among them – which promise some far-reaching transformation of every last aspect of human knowledge and experience. Those construals clearly answer to some deeply felt need for a new conception of reality and our place within it, suggesting that science has at last come around to an acceptance of many ideas which had hitherto enjoyed no kind of scientific credibility. Of course there is no reason to reject those QM theories merely because they happen to chime with the wider drift toward forms of 'virtual-reality' thinking, or because they lend credence to speculative notions – such as time travel by means of quantum tunnelling – which have sparked interest partly through fictional or cinematic treatment and partly through popularizing accounts of the 'new physics'.[26] Nor can Deutsch's book fairly be regarded as a 'popularizing' work in the pejorative sense that it avoids the hard questions and panders to this kind of quasi- or pseudo-scientific interest. On the contrary, it makes few concessions to the reader in quest of byte-sized dramatic pronouncements like those of Brown in the above-cited passage. Nevertheless, as I have said, it does invite criticism on the grounds that it pre-emptively excludes any theory (such as Bohm's) that would seek to reduce the mystery or paradoxicality of quantum phenomena, and thus contrives to present many-worlds – along with all its own mysteries and paradoxes – as the one interpretation properly in line with our current best knowledge of physical reality.

At this point it is perhaps worth recalling what Cushing has to say about the reception history of orthodox (Copenhagen) QM and the extent to which that theory's widespread acceptance – along with the resulting marginalization of Bohm's hidden-variables alternative – had to do with certain clearly marked historical, ideological, and socio-cultural factors.[27] Thus, on Cushing's account, one major reason was the intellectual climate in Germany after the First World War, a climate wherein the physical sciences were widely viewed with suspicion or downright hostility since they seemed to embody that positivistic ethos – that domineering drive to subjugate everything, nature and humanity alike, to the purposes of a purely *Wertfrei* instrumental reason – which had led to the catastrophe of European civilization.[28] Such was Max Weber's famous diagnosis, repeated with various philosophical refinements in Husserl's *The Crisis of the European Sciences*, and taken up in more sombrely dramatic style by Adorno and Horkheimer in their book *Dialectic of Enlightenment*.[29] As these thinkers saw it, the enlightenment project of reason in the service of emancipatory values had turned into an 'iron-cage rationality' that excluded all reference to human ethical values and socio-political interests.[30] Their account thus jibed with the widespread conviction – copiously documented by Cushing – that science and technology were part symptom, part cause of that descent into the barbarism of World War One that already looked like repeating itself with the breakdown of social-democratic values in 1930s Weimar Germany. Hence – to repeat – the understandable appeal of a scientific theory (orthodox QM) that appeared to humanize science by making 'reality' observer-relative, by adducing Heisenberg's uncertainty principle as an absolute limit on its powers of objective description, by offering an escape route from the implacable laws of old-style scientific determinism, and also by implying the existence of a participatory universe where mind and world were somehow united in a manner that was wholly unthinkable to classical physics. In short, there were many features of orthodox QM which made it the likeliest contender for hegemonic status among a physics community responsive to wider social, political, and ideological pressures.

Perhaps the feature with strongest appeal – then and since – has been its promise to resolve the issue of freewill versus determinism, an issue that has preoccupied not only philosophers but reflective individuals of various persuasion, physicists among them. Deutsch is quite right when he remarks that the philosophers have not yet come up with anything like an adequate or convincing solution. What their proposals have often amounted to is a kind of double-aspect theory, involving on the one hand (as in Kant's version) a realm of phenomenal experience where causality reigns supreme and determinism is strictly inescapable, and on the other hand a 'noumenal' realm where human beings discover their autonomy as thinking, willing, and decision-making agents under no such causal compulsion.[31] This theory has been much refined and updated, nowadays usually in the linguistic mode derived from late Wittgenstein and other sources. Here the appeal is to different 'language-games' with different 'criteria', some of them (e.g. those of the physical sciences) making sense according to a largely determinist or causal-explanatory worldview, while others (those of ethics, aesthetics, law, religion, and the human or social sciences) presuppose the possibility of freely willed choices, actions, and commitments.[32] Still, one can sympathize with Deutsch's claim that all this philosophico-linguistic refinement has left the problem firmly in place and achieved little more than a range of handy expedients for talking one's way around it.

Some philosophers, including G.E.M. Anscombe in a well-known essay, have invoked quantum physics as pointing a way forward from the old clash between freedom of will and the doctrine of scientific determinism.[33] That conflict has arisen, she suggests, mainly through the high prestige attached to certain kinds of scientific theory – such as Newtonian celestial mechanics – which are taken to embody the ideal of science as a strictly demonstrative order of knowledge that deduces the necessary course of events (e.g. planetary motions) from an understanding of the laws of nature expressed in mathematical form. However, such cases are untypical, according to Anscombe, since they involve a degree of law-governed regularity that is seldom encountered even in the physical sciences, let alone in domains of human enquiry outside the scientific remit. They are also misleading in so far as they induce us to equate causality with necessity, or the fact that there exists some causal nexus between events A and B with the idea that B *must* therefore follow A with just the kind of absolute, implacable force that is taken to characterize Newtonian physics. It is this false idea, she argues, that has given rise to the illusory conflict between (so-called) 'scientific determinism' and philosophies which take freedom of will as a simply indispensable feature or attribute of human experience. Once it is accepted that the verb *to cause* is by no means synonymous with the verb *to necessitate* – that effects may have causes without those causes necessarily producing those effects – then the conflict will seem nothing more than a symptom of confusion in our grasp of the logical grammar of causal verbs.

Besides, physics has itself moved beyond that old determinist conception according to which – as Laplace notoriously claimed – one could in principle deduce every last detail of every past and future state of affairs from an adequate knowledge of physical conditions obtaining at the present moment. This was always an idle claim given that such knowledge is humanly unattainable even should the natural sciences advance to a stage of development or causal-explanatory grasp far beyond anything yet achieved. But with the advent of quantum mechanics – on Anscombe's account – there is no longer any need to take the claim seriously even as a matter of 'in-principle' argument from the viewpoint of a God-like Laplacean intelligence possessed of omniscient predictive and retrodictive powers. For it is now understood that determinism fails to hold at the most basic level of subatomic physics where it is ruled out by such strictly irreducible quantum phenomena as the uncertainty affecting acts of measurement and the probabilistic nature of various predictive or observational results.[34] If these are combined with the further (orthodox QM) claim concerning the observer-induced 'collapse of the wavepacket', then indeed it would appear that science can pose no threat to our deep-laid intuitive belief as regards the scope for freedom of will in matters of decision and choice.

Anscombe's fictive example to make this point is similar to the case of Schrödinger's cat, although couched in somewhat less offensive terms. In brief, it has to do with a lump of fissile material, a Geiger counter linked to an electromagnetic device, and a quantity of dynamite that will either explode or not, depending on whether the fissile material happens to decay and hence emit a particle which triggers the device and thus detonates the explosive. Since the outcome is necessarily a matter of probability – an aggregate figure based on averaging out over the half-life of all the atoms contained in the radioactive lump – there is of course no way that it can be predicted according to the classical (determinist) laws of pre-quantum physics. In which case – Anscombe concludes – we should give up the old Laplacean idea of science as a mode of knowledge whose advancement goes along with a steadily decreasing margin of freedom for future-directed actions, choices, and commitments. Rather, we should see that the intrinsic randomness of quantum phenomena is just what is required in order to break with this bugbear notion of strict scientific determinism.

Still there are problems with Anscombe's interpretation of the quantum-theoretical 'evidence', among them its endorsement of orthodox (Copenhagen) QM as the sole authorized version and its failure to reckon with alternative accounts – such as Bohm's hidden-variables theory – that would yield nothing like the wished-for result. Also, there is the issue that Schrödinger raised through his famous thought experiment, namely the sheer *impossibility* of supposing that quantum phenomena (such as wave–particle superposition or the observer-induced wavepacket collapse) can be carried across to the macrophysical domain without thereby producing all manner of insoluble paradoxes and aporias.[35] However, the chief objection to Anscombe's argument concerns the link that it takes for granted between *randomness* as an intrinsic feature of quantum 'reality' and *freedom of will* as a defining attribute of human thoughts and actions. Deutsch makes this point very effectively in one of his many passing remarks on the way that philosophers have missed the mark when talking about quantum physics. Thus:

> replacing deterministic laws of motion by indeterministic (random) ones would do nothing to solve the problem of free will, so long as the laws remained classical. Freedom has nothing to do with randomness. We value our free will as the ability to express, in our actions, who we as individuals are. Who would value being random? What we think of as our *free* actions are not those that are random or undetermined but those that are largely *determined* by who we are, and what we think, and what is at issue.
>
> (Deutsch, p. 338)

One could scarcely wish for a sharper rebuttal of the fallacy that equates freedom of choice with randomness of outcome, or a clearer statement of the problems that beset any such attempt to resolve the classic antinomies of knowledge and will by appealing to orthodox QM theory.

However, those problems are just as apparent when Deutsch puts forward his own answer to the freewill–determinism issue, an answer that is based – predictably enough – on the many-worlds interpretation. As he sees it, the difficulty is not so much with determinism but rather with the false and strictly incoherent worldview imposed by our habit of thinking in terms of classical space–time physics. For this perspective is confined to just one among the multiplicity of worlds that contain all our various possible alternative acts, choices and decisions. In classical space-time, he writes,

> *something happens* to me at each particular moment in my future. Even if what will happen is unpredictable, it is already there, on the appropriate cross-section of spacetime. It makes no sense to speak of my 'changing' what is on that cross-section. Spacetime does not change, therefore one cannot, within spacetime physics, conceive of causes, effects, the openness of the future or free will.
>
> (Deutsch, p. 338)

On the many-worlds theory, conversely, what will happen to 'me' at every moment in my future existence is determined in advance only for that version of myself who inhabits the singular spacetime universe that 'I' currently inhabit, and not for those other multiple versions whose reality will have branched off at various points along the way. From 'my' perspective, narrowly (or classically) construed, there is just no escaping the paradoxes of time and choice that have perplexed philosophers from Aristotle down and which continue to defy the best efforts of conceptual analysis even after the impact of Einstein's revolution on our current scientifically informed ways of thinking about space, time, and the relationship between them. For General Relativity is still 'classical' in so far as it works with a single-universe theory of spacetime reality as opposed to the quantum-multiverse theory which follows inescapably – so Deutsch argues – from those well-attested QM phenomena that Einstein was himself so loath to accept. In this respect his claim finds support from Paul Davies's recent book *About Time*, where he speaks of Einstein's 'unfinished revolution' and the failure – so far – of relativity theory to explain the various temporal paradoxes thrown up by present-day particle physics.[36]

Still there is the question whether Deutsch's alternative theory fares any better when it comes to explaining both the source of those problems and the means of resolving them by way of many-worlds QM. 'Consider', he invites us, 'this typical statement referring to free will: "After careful thought I chose to do X; I could have chosen otherwise; it was the right decision; I am good at making such decisions"' (p. 339). If construed in accordance with the classical worldview and its single-universe spacetime framework, then this statement is, he thinks, 'pure gibberish'. That is to say, it assumes that the speaker had freedom of choice with respect to certain past actions and their future consequences, only one of which has in fact been realized – the best outcome resulting from the right decision – but all of which somehow remain open for the speaker's reflective (if somewhat self-satisfied) process of appraisal. But how can this make sense, Deutsch seems to be asking, if we construe his statements on the single-universe theory which treats all events, 'past' and 'future', as belonging to a block-like space–time continuum wherein such distinctions can be conceived only as relative to human temporal perceptions that have no place in a properly scientific understanding of physics and cosmology? From this point of view – a view from nowhere, phenomenologically speaking – it must indeed be 'pure gibberish' to claim that one *chose* to do X (when every such 'choice' was a matter of timeless or tenseless necessity); that one 'could have chosen otherwise' (when no such alternatives existed); that 'it was the right decision' (when the fact of one's having done X is the sole relevant fact or event in question); and that 'I am good at making such decisions' (when nothing can intelligibly *count* as a 'decision', good or otherwise, given the classical spacetime framework and its absolute veto on talk of alternative – humanly decidable – outcomes or courses of event. Thus, according to Deutsch, all the efforts of philosophers to carve out a space for freedom of will have either proceeded in ignorance of the current best theories of physical science (up to and including Einstein), or come to grief by taking those theories into account and hence running up against intractable aporias such as those outlined above.

However – he claims – this situation is transformed with the advent of a quantum multiverse theory which alone makes it possible to explain how 'free will and related concepts are now compatible with physics' (p. 339). All that is required is a simple paraphrase of the above sentence into terms that render it consistent with many-worlds QM by adding a series of supplementary clauses which specify the new interpretation. 'After careful thought I chose to do X' becomes 'After careful thought some copies of me, including the one speaking, chose to do X' (ibid). 'I could have chosen otherwise' works out as the multiverse-equivalent claim that 'other copies of me chose otherwise', those copies of course having been just as real as the one who made 'my' particular choice, although now so remote from me and my particular (spacetime individuated) world that they seem to exist only in a realm of modal hypotheses or counterfactual possibility. 'It was the right decision' has a longer paraphrase which itself makes use of certain modal-epistemic refinements, although not – as Deutsch is at pains to insist – refinements that in any way compromise or qualify his multiverse reality principle. Thus, '[r]epresentations of the moral or aesthetic values that are reflected in my choice of option X are repeated much more widely in the multiverse than representations of rival values' (p. 339). That is, the 'rightness' of my choice and its outcome is a function of their obtaining across the largest possible range of other universes, worlds that may since have diverged from my own in all sorts of respects but where *this particular* choice holds fast amid all the shifting scenery. It is 'right' in the sense of being borne out to a high degree by the fact of its having actually occurred – along with all its consequences – in a far greater number of alternative worlds than those that were witness to other conceivable choices and possible outcomes. From here it is but a short step to the last sentence – 'I am good at making such decisions' – which similarly translates into: '[t]he copies of me who chose X, and who chose rightly in other such situations, greatly outnumber those who did not' (p. 339). For this is just a matter of extending the many-worlds rightness criterion from alternative acts and their consequences to the alternative selves who made such choices and are happy to live with the outcome. The more widely those selves are diffused across the quantum multiverse, the stronger 'my' claim (within 'this' universe) to have chosen according to the right principles and brought about the right – most ethically desirable or humanly beneficial – consequences.

Now it seems to me that this offers no help towards resolving the freewill–determinism issue and indeed merely complicates that issue by introducing a whole raft of other problems connected with the many-worlds hypothesis and its extravagant ontological claims. The very fact that the above sentences translate so directly – through a kind of canonical paraphrase – into their various multiverse equivalents is itself a sure sign that Deutsch's theory amounts to just a minor conceptual variation on familiar themes. After all, what possible *difference* can it make with respect to the problem concerning freedom of will if one replaces the sentence 'I chose to do X' with the sentence 'some copies of me, including the one speaking, chose to do X'? This problem is precisely with the word *chose* – with the claim that human beings make choices undetermined by the laws of physics or causal necessity – and can therefore scarcely be resolved by a theory that has lots of multiverse selves each making the same or some different choice across a range of alternative worlds. Of course, it might be said that what we call 'free will' is in fact the illusion – the subjective mirage – created by random quantum phenomena, such as the collapse of the wavepacket, whose effect is momentarily to produce multiple divergent world series whereby we are carried along from one moment to the next while remaining unaware of all the other self-versions that have branched off from our own. However, it is clear from Deutsch's criticism of the argument from (so-called) quantum 'randomness' to freedom of will that this is not a reading he would wish to endorse or a solution that he would find in the least degree convincing. Nor does it help if one substitutes 'probabilistic' for 'random', since this merely swings the argument back towards some kind of large-scale determinist outcome when those quantum collapses are averaged out over a great enough range of measured values for localized wave-like or particle-like phenomena. In short, Deutsch's proposed solution to the old philosophic dilemma is one that effectively re-states that dilemma in terms of his own favoured theory but which offers nothing like the radical alternative – the ultimate answer to these questions – that his book gives us to expect.

IV

There is a similar problem with the other three sentences that Deutsch translates into multiverse language in order to show how his theory removes their paradoxical or aporetic character. Two of them ('I could have chosen otherwise' and 'I am good at making such decisions') come out as simple transpositions of the 'other copy' idiom – myself doing differently or doing the same across a maximum range of worlds – which really amounts to just a verbal subterfuge for avoiding any direct confrontation with the freewill–determinism issue. The third ('It was the right decision') presents a somewhat different case since it involves – as I have said – an excursion into areas of modal-epistemic thought where Deutsch mostly prefers not to venture since he considers them the province of philosophic types with an underdeveloped sense of the reality (albeit the counter-intuitive reality) revealed by physical science. His paraphrase here – to repeat – takes the somewhat more complicated form, '[r]epresentations of the moral or aesthetic values that are reflected in my choice of option X are repeated much more widely in the multiverse than representations of rival values' (p. 339).

However, it is at just this point that Deutsch's argument comes closest to the long tradition of speculative rationalist metaphysics that runs from Leibniz to present-day exponents of the many-worlds theory such as David Lewis.[37] Thus for Leibniz also, this theory was a matter of rigorous deduction from certain self-evident premises, among them the distinction between necessary truths (i.e. truths of reason that were known a priori and which obtained across all logically possible worlds), and on the other hand contingent matters of fact which held good only with respect to one particular world and the course of events within it.[38] Nevertheless, he maintained, we should view this distinction as an artefact of our own restricted knowledge, or of the limits placed upon human understanding by our perceptual apparatus, conceptual powers, historical perspective, or again – most crucially – our confinement to a given spatio-temporal locale from which we are denied epistemic access to 'worlds' other than our own. Still, we can conceive of a God-like, omniscient intelligence subject to none of these creaturely limits and for whom the truth of contingent facts – such as Caesar's having crossed the Rubicon – is a matter of the strictest necessity when traced through the whole concatenated series of operative causes and effects.

For Leibniz this followed from another self-evident premise, namely the principle of sufficient reason, according to which no action or event was without some ultimate explanation could one but enquire deeply enough into the preordained rational scheme of things. To suppose otherwise – that certain events might be purely contingent or lacking such rational warrant – was to doubt God's omniscience and benevolence in having created the best of all possible worlds. Thus Leibniz's theodicy supplied the chief motive for his developing a modal (multiverse) conception of truth that sought to justify the ways of God to man by removing the paradoxes thought to arise from the existence of so much suffering and evil, or the apparent lack of just proportion between human acts and their consequences. Any answer to these problems – so Leibniz deduced – must entail the existence of numerous other worlds containing all the various events that might conceivably have come to pass, but only one of which worlds (our own) God had chosen to realize for reasons best known to himself and glimpsed by us only through an exercise of speculative transworld or counterfactual reason. Hence our failure to grasp the necessity of historical actions and events such as Caesar's having crossed the Rubicon, events that must strike *us* as contingent since from our – as likewise from Caesar's – restricted point of view they involved an element of human choice and might therefore have turned out differently, thus giving rise to a whole different series of subsequent real-world historical outcomes. However, Leibniz holds, we shall cease to think in this way – and cease to repine at the absurd disproportion between acts and their consequences or between divine and human orders of wisdom – if we can only learn to think of our particular world as just one (but the best conceivable) among the multitude that God might else have brought into existence. For we shall then understand that there is an order of rational necessity that explains why things *must* have turned out in this way despite all the manifest evil and suffering that will otherwise surely strike us – from our limited human perspective – as throwing a huge paradox into the doctrine of God's omniscience, omnipotence, and benevolence.

Hence Leibniz's famous (or notorious) conclusion that 'all is for the best in this best of all possible worlds'; a doctrine which he took to follow necessarily from this set of jointly theological, metaphysical, and modal-logical premises, but which hostile critics – Voltaire and Dr Johnson among them – found both philosophically absurd and morally repugnant. What aroused their moral indignation was the tone of complacent superior wisdom which claimed somehow to account for the facts of human pain and misery by reference to a higher providential scheme of things wherein those facts would assume their proper (subordinate) place as contributing to the ultimate working-out of God's rational purpose. I shall not here linger on the details of this well-known controversy save to remark that, in my view, the critics were right and that Leibniz's doctrine amounts to something more like a *reductio ad absurdum* of theological attempts to vindicate God's inscrutable wisdom. What interests me more in the present context is the striking similarity between Leibniz's version of the many-worlds thesis and that put forward by Deutsch as a matter of (purportedly) straightforward deduction from observed quantum phenomena. Indeed – it is worth recalling – Deutsch goes so far as to state that 'the fruitfulness of the multiverse theory in contributing to the solution of long-standing philosophical problems is so great that it would be worth adopting even if there were no physical evidence for it at all' (p. 339). In which case it seems fair to suggest that we are here dealing with something very like what Heisenberg noted in Bohr: a 'darkly metaphysical' doctrine presented in the guise of a scientific inference to the best (most rational) explanation.[39]

It is also worth recalling, in this connection, that Leibniz engaged in a spirited controversy with the English Newtonians – notably Samuel Clarke – concerning Newton's physics and what Leibniz considered its flagrant offence against the principle of sufficient reason.[40] That is to say, it signally failed to explain why anything in the world should be as it is or have turned out just the way it has given the Newtonian idea of absolute space as existing quite independently of the objects contained within it, and likewise the idea of time as flowing equably from one moment to the next unaffected by the various temporal events that marked different stages in its passage. For Leibniz this was a wholly untenable conception since it reduced those objects and events to an order of sheer random occurrence, and moreover treated space and time as belonging to a purely abstract dimension with no reason in the nature of things, or in God's providential wisdom, why they should not have been created otherwise. Thus the universe might just as well have sprung into being five minutes earlier, or a few feet to the left, since on Newton's account its spatio-temporal location cannot be explained – that is to say, justified – in any terms other than its merely having happened that way, which of course (for Leibniz) is no explanation or justification at all. In response he proposes an alternative cosmology based on the argument that space and time are not, as in Newton, abstract coordinate systems existing quite apart from objects and events but must rather be conceived in relational terms with reference always to those same objects and events.[41]

I should mention at this point that recent commentators are far from agreed in considering Newton (or the English Newtonians) to have gained the upper hand in this controversy regarding 'absolute' versus 'relativist' conceptions of space and time. Indeed, some historians of science have made out a case that Leibniz's theory 'anticipates' general relativity in so far as it allows for the reciprocal involvement of spatial and temporal parameters.

More specifically, it is able to embrace the idea – first proposed by Einstein and then borne out by the results achieved on Eddington's famous expedition – that time, since relative to the speed of light, may be contracted or dilated through the presence of large bodies with magnetic fields which deflect the passage of neighbouring lightwaves or photon streams.[42] So there is some room for doubt concerning the claim that Newton's apologists won the argument for physical science and that Leibniz was shown up as a speculative thinker wholly in the grip of his pet metaphysical system. All the same, it will surely be conceded that some of his reasons for holding this position (as likewise for maintaining the many-worlds hypothesis from which it derives) were indeed 'metaphysical' in the sense that they involved a priori principles and commitments which could not be verified or falsified by any method known to the physical sciences. Among them was the argument from God's capacity to have created any number of alternative possible worlds – for instance, one in which Adam would not have sinned – yet his having created this particular world (with Adam's sin and all its consequences) for reasons no doubt obscure to us but perfectly intelligible could we but perceive the workings of divine providence. Of course the many-worlds theory has since been developed in various ways – such as Kripkean modal logic – that can scarcely be construed in theological terms or as a displaced (quasi-secularized) version of Leibniz's original doctrine.[43] Nevertheless, there are some current uses of the theory, for instance as deployed by Alvin Plantinga, which do quite explicitly make the case for many-worlds as a means of reviving and refining traditional (e.g. ontological) proofs of God's existence.[44] They can do so with at least some measure of plausibility because the theory makes room for any number of alternative 'worlds', most of which lie beyond the range of human perception or intellectual grasp but all of which require some superordinate principle – on this interpretation, some sovereign intelligence or presiding deity – whereby to discriminate their various levels of necessary or contingent truth.

V

In short, the conceptual idiom of many-worlds theory lends itself very easily to just those kinds of a priori metaphysical commitment which in principle cannot be confirmed or infirmed by any kind of empirical evidence. This is clearly not the case with Kripke or the early Putnam since on their interpretation the theory has to do with issues in semantics and modal logic, especially as concerns the status of natural-kind terms and the way that they function in descriptive or causal-explanatory statements like those typically advanced in the natural sciences.[45] Thus one central claim of the Kripke–Putnam 'new' theory of reference is that it offers an alternative to earlier descriptivist theories – such as those provided by Frege and Russell – according to which our ability to pick out samples of (for instance) 'gold' or 'water' depends on our possessing the relevant criteria or applying the correct range of identifying attributes.[46] Their purpose was partly to explain how certain statements – e.g. 'gold is the metallic element with atomic number 79' or 'water is the substance with molecular composition H_2O' – were not merely tautologous (as in the standard text-book instance: 'all bachelors are unmarried men'), but could serve to convey information to someone who had previously lacked those particular items of scientific knowledge. That is to say, they *appeared* to be tautologous since in each case their surface form was that of a straightforward identity assertion between two terms with precisely equivalent extension or scope of reference. However, this appearance was deceptive, Frege argued, in so far as it concealed the underlying logico-semantic form whereby such statements could in fact possess genuine informative content.

Hence Frege's canonical distinction between 'sense' and 'reference', applied to one particular (now famous) example of the kind – 'the Morning Star is identical with the Evening Star' – but equally relevant to the above sorts of case.[47] Both of these expressions have the self-same referent (i.e. the planet Venus), which might seem to make the statement purely tautologous. However, they do not have the same sense, that is to say, the same range of descriptive attributes or identifying criteria. For the Morning Star and the Evening Star can of course be observed at different times by the same individual or by different observers, some of whom may know that both descriptions refer to the planet Venus but to others of whom that statement may come as a new piece of astronomical knowledge. So likewise with the instances of 'gold' and 'water', the one referentially equivalent to 'metallic element with atomic number 79', the other to 'substance with the molecular composition H_2O'. For here also the statements in question cannot be construed as tautologous or purely analytic, i.e. as true by definition since their subjects and predicates are extensionally equivalent or identical in scope of reference. After all, they each once served to articulate a new scientific discovery and can still communicate fresh knowledge to a lay person or neophyte scientist.

Thus at one time – before the series of advances in atomic theory from Dalton to Mendeleev's periodic table of the elements – 'gold' had been defined in some such terms as 'a yellow ductile metal soluble in *aqua regia*'. With the advent of that theory, it became possible to specify its microstructural attributes with a greater degree of precision ('atomic number = 79') and in such a way as to avoid confusing it with samples of 'fool's gold', or iron pyrites.[48] The point of the descriptivist theory as advanced by Frege and Russell is that cases like these cannot be explained – cannot be assigned a convincing logico-semantic interpretation – unless by distinguishing 'sense' from 'reference' and holding that advances in knowledge come about through the attribution of new predicates (criteria, identifying features, 'definite descriptions', etc.) to the objects or substances concerned. For it will otherwise always be possible to argue – as followers of Wittgenstein do with depressing regularity – that the truth claims of science amount to nothing more than a species of disguised tautology, since in the 'language-game' of present-day physics it is a merely definitional or analytic truth that 'water = H_2O' or 'gold is the metallic element with atomic number 79'.[49] What allows such statements to retain their informative content is precisely the distinction between sense and reference, or the fact that we can assign different microstructural predicates to 'gold', 'water', etc. and thereby achieve a greater degree of classificatory precision or depth-ontological grasp.

So the descriptivist theory goes some way towards answering Frege's initial query: what makes the difference between straightforward tautologies of the type 'A = A' and informative (non-trivial) identity-statements of the type 'A = B'. Where the latter stand apart is in virtue of referring to certain properties or features of the physical world while also conveying significant information through the sense of those terms that figure in our various theories, hypotheses, empirical observations, etc. However – as I said above – some philosophers, Kripke and Putnam among them, have challenged the standard descriptivist account since it appears to give rise to counter-intuitive or downright nonsensical conclusions. Thus: what if we had been altogether wrong about gold or water and these substances turned out (as a result of further, more advanced scientific research) *not* to possess any of the properties hitherto attributed to them?[50] Should we then have to say – absurdly – that 'gold is not gold' or that 'water is not water'? Other such Kripkean examples have to do with proper names in the usual sense of that phrase, rather than the technical extended sense (including natural-kind terms) in which it is applied by Kripke and Putnam. Thus, for instance, according to the descriptivist theory, when we use the name 'Aristotle' we are referring to just that unique historical figure of whom it is known that he lived in Athens, was a pupil of Plato, authored certain texts, acted as tutor to Alexander the Great, and so forth. Yet what if we discovered that none of these predicates applied since the person who actually bore that name – whose identity was fixed from the moment of conception when his father's sperm fertilized his mother's egg – had in fact led a wholly different career and left no such works to posterity? Or again, what if a researcher at last came up with decisive evidence that somebody else (not William Shakespeare – maybe Francis Bacon) was the author of *Hamlet*, *King Lear*, the Sonnets, and the rest of *Shakespeare's Collected Works*? On the descriptivist theory it seems that we should then have to conclude that 'Aristotle was not Aristotle' or 'Shakespeare was not Shakespeare'.

Since these are manifestly absurd conclusions – so Kripke's argument runs – then there must be something awry with that theory as construed by philosophers in the tradition descending from Frege and Russell. What is needed is a full-scale alternative account that would secure adequate stability of reference throughout the various changes of sense – or shifts in the range of descriptive criteria – which have so far affected or which might yet affect our beliefs concerning the particular items or individuals in question. Thus, according to Kripke, reference is fixed through an act of inaugural 'baptism', that is, the kind of act whereby somebody once announced that '*this* is a sample of gold', or whereby Aristotle's parents declared, 'we name this child Aristotle'. From then on the name passes down through a causal 'chain' of transmission which ensures that it always refers back to *that* sort of substance (gold) or *just that* particular person (Aristotle) across and despite any subsequent changes in our state of knowledge regarding them.

Kripke derives some far-reaching philosophical consequences from this argument. Among them is his claim that there exist certain a posteriori necessary truths, truths (that is to say) which might not have held in some alternative 'possible world' – a world where Aristotle was never conceived or where gold had no place in the periodic table of elements – yet which none the less obtain as a matter of necessity in the world we actually inhabit. Also, he thinks that it helps to resolve certain long-standing issues in epistemology and philosophy of science, especially with regard to natural-kind terms and the status of causal explanations. For these latter are counterfactual-supporting, in the sense that they apply – or possess genuine explanatory force – in just that range of possible worlds (our own included) where the laws of physics are such that a certain event could not have occurred in the absence of some other, causally connected antecedent event. In this respect they differ from a priori necessary truths – i.e. truths of logic or those established on purely definitional (analytic) grounds – whose necessity consists precisely in their holding across *all* possible worlds no matter what adjustments might be made elsewhere with respect to the structure of physical reality, the range of existent natural kinds, or the propagation of causal influences. Yet they also occupy a realm quite distinct from purely contingent matters-of-fact concerning events in our particular world which might just as well have taken some different course without contravening any law of nature or entailing any change to the established order of a posteriori necessary truth.

Thus the many-worlds idiom in its Kripkean form is a useful clarificatory device for examining the implicit logical structures and the range of ontological commitments involved in counterfactual-conditional reasoning or causal explanations more generally.[51] That is to say, it offers an alternative means for distinguishing the various realms of metaphysical, epistemic, causal, and logical possibility (or necessity) that can otherwise very easily get mixed up and generate the sorts of categorical confusion which – as I have argued – are rife in many quarters of quantum-theoretical debate. For there are other, more speculative uses of the many-worlds hypothesis that abjure such a realist constraint upon their range of counterfactual conjectures and which thereby produce all manner of extravagant or profligate ontologies. Among them are Deutsch's multiverse theory and also David Lewis's kindred doctrine concerning the plurality of worlds, a doctrine that harks back beyond Leibniz to the Neoplatonists in its unflinching commitment to the literal existence of all those alternative 'realities' entailed by forms of modal or counterfactual reasoning.[52] Then again, there is Plantinga's re-statement of the ontological proof for God's existence couched in terms of a many-worlds thesis that purports to vindicate that proof through the argument that some such being must exist as a matter of absolute necessity given his perfections as traditionally defined and the need to posit a world wherein those perfections would be realized.[53]

So it is not hard to see why theologically inclined commentators on quantum theory might favour the many-worlds interpretation as a means of defusing the old 'science versus religion' controversy, or claiming to reconcile religious faith with the most advanced theories of present-day particle physics. At their simplest, such claims amount to a version of the 'God-of-the-gaps' argument, i.e. the notion that quantum 'reality' is just as mysterious as the subject-matter of theology, so that the two must be somehow deeply (mysteriously) interconnected.[54] There is a similar and equally dubious line of thought – more often associated with the 'many-minds' interpretation – which claims that such puzzling quantum phenomena as the 'observer-induced' collapse of the wavepacket must hold an answer to the problem of consciousness since both are so resistant to received ('classical') modes of scientific understanding.[55] However, it is not so much these caricatural versions of the case that present a problem for anyone concerned with clarifying the issue as regards the relation between quantum physics, modal logic, and the many-worlds theory. Rather, it is the question how that theory might square with a realist ontology or worldview according to which counterfactual possibilities would remain within the realm of this-world causal-explanatory conjecture and not entail the *actual existence* of all those manifold parallel universes.

VI

It is this alternative that Deutsch rejects out of hand as a mere 'philosophical' device for avoiding the full implications of his own highly bizarre and yet – as he claims – strictly 'inescapable' multiverse ontology. Still, it might just be that the despised philosophers have a few good reasons – arguments from modal logic among them – for upholding the principle that entities ('worlds' especially) should not be multiplied beyond the strict requirements of explaining what is there to be explained. Deutsch himself at one point invokes Occam's Razor (p. 78) to the effect that we ought not to complicate explanations unnecessarily since, if we do, 'the unnecessary complications themselves remain unexplained'. However, he then goes on to observe that 'whether an explanation is or is not "contrived" or "unnecessarily complicated" depends on all the other ideas and explanations that make up one's world-view', which in Deutsch's case clearly leaves room for a loose application of the principle. He also remarks at one point that '[m]ost logically possible universes are not present at all – for example, there are no universes in which the charge on an electron is different from that in our universe [i.e. a negative charge], or in which the laws of quantum physics do not hold' (p.

276). So to this extent at least Deutsch accepts that there do exist certain real-world physically specifiable constraints on the degree to which other 'universes' can diverge from our own without overstepping the boundary between what makes sense according to the universal (transworld-applicable) laws of physics and what may be conceived only as a matter of abstract, hypothetical, or indeed science-fiction possibility.

However – as I have said – these constraints seem to go by the board when Deutsch introduces his multiverse ontology and his related idea that 'virtual reality' is the closest we can get to 'reality' itself since, in multiverse terms, our own is just one (highly selective) construct out of the manifold worlds that undoubtedly exist beyond our epistemic ken. As he puts it:

> [t]he connection between the physical world and the worlds that are renderable in virtual reality is far closer than it looks. We think of some virtual-reality renderings as depicting fact, and others as depicting fiction, but the fiction is always an interpretation in the mind of the beholder. There is no such thing as a virtual-reality environment that the user would be compelled to interpret as physically impossible.
>
> (Deutsch, p. 119)

These sentences display a good deal of argumentative as well as terminological slippage. Thus in one sense the very definition of 'virtual reality' is such as to require that the user (or beholder) should interpret it as representing a physically possible world, even if only by suspending disbelief for the sake of some imaginative reward. Likewise, there is a sense in which all 'virtual-reality renderings' depend upon certain real-world technological resources – computers, simulators, surround-sound systems, multiscreen televisual techniques, etc. – in order to generate their effects. In which case, 'every virtual-reality generator, running any programme in its repertoire, is rendering *some* physically possible environment' (p. 119). To that extent, granted, there is 'no such thing as a virtual-reality environment that the user would be compelled to interpret as physically impossible' (ibid). But this does not mean – as Deutsch's argument apparently requires it to mean – that the *worlds represented* by way of those effects must also be construed as genuinely real or as contravening none of the laws of physics. After all, one can readily envisage such a world (novelistic, cinematic, multimedia-induced, or maybe created by implanting probes in a willing subject's brain) where the charge on electrons was positive or wherein there occurred all manner of physically impossible events.[56]

What Deutsch seeks to do in the above passage is lower our resistance to the multiverse theory by pointing out – in familiar post-Kantian vein – that *all* reality-renderings are 'virtual' on one definition of the term. For of course there is a sense in which we cannot have direct or unmediated access to the so-called 'external world', that is to say, cannot acquire any knowledge of it save through our various (inherently selective) means of perceptual and cognitive grasp. Thus Deutsch draws the lesson that our factual as well as our fictive 'renderings' are always 'in the mind of the beholder' and are hence likewise subject to the double condition that (1) they tell us *directly* far less than there is to know about reality, while (2) they enable us – at least once apprised of the multiverse theory – to pass far beyond those limits on the range of commonsense-intuitive or naive realist thought. This is why, as Deutsch says, 'the existence of virtual reality does not indicate that the human capacity to understand the world is inherently limited, but, on the contrary, that it is inherently unlimited' (p. 103). And again, '[i]t is no anomaly brought about by the accidental properties of human sense-organs, but is a fundamental property of the multiverse at large' (ibid). For on this account there is no longer any clear distinction between the various ontological domains – real-world, counterfactual, hypothetical, virtual, and fictive – which are all necessarily 'in the mind of the beholder' but are all subject to the ultimate constraint of a multiverse reality that by very definition lies beyond our utmost powers of cognitive grasp. To challenge the theory on just these grounds is merely to show that we are still in the grip of a naive pre-quantum or single-universe metaphysic that ignores the extent to which *all* knowledge, in the physical sciences and elsewhere, exceeds the furthest limits of 'direct' (i.e. this-world) epistemic access. In short, '[w]e understand the fabric of reality only by understanding theories that explain it'. Therefore, 'since [those theories] explain more than we are immediately aware of, we can understand more than we are immediately aware that we understand' (p. 103). At which point Deutsch can present his theory as another in the great line of descent from Galileo to Einstein: a massively counter-intuitive hypothesis that is sure to provoke equally massive commonsense-intuitive resistance. Nevertheless, he asserts, its truth is 'inescapable' given the fact that the basic laws of quantum physics – like the negative charge on the electron – are among those that necessarily hold across all possible worlds.

Now the main problem with this – as I have said – is that it erects a vast structure of metaphysical argument ('metaphysical' in the strictest sense of that term) on one highly contentious interpretation of QM theory. Moreover, it is one that pre-emptively rejects any rival account, such as Bohm's, that would entail no such drastic revision to our basic concepts of physical reality, causal explanation, and what counts as an adequate or plausible scientific hypothesis. At this stage it is worth returning to the Kripke–Putnam variety of possible-worlds argument since it offers a striking contrast with Deutsch's approach to these issues.[57] Thus, in Deutsch's case, the approach is to take some item of puzzling physical evidence (such as quantum multipath deflection or photon interference effects), devise a 'virtual reality' scenario around that item of evidence, and then proceed to interpret the results in the strongest possible ontological-realist terms. This is why his theory bears a marked resemblance to that of philosophers, like David Lewis, who have arrived at it on purely 'philosophical' grounds, whatever Deutsch's professed reluctance to be counted among their company. For Kripke and Putnam, conversely, the main purpose of using possible-worlds logic is to specify just those kinds of entity and just those kinds of causal explanation concerning them that properly belong within the scope of this-world (scientifically warranted) knowledge. Where the possible-worlds approach comes in is through certain modal-logical considerations which in turn have to do with the nature and scope of causal-explanatory statements and the role played in them by counterfactual-conditional forms of reasoning.[58] Here also such reasonings may take the form of far-fetched conjectures or 'science-fiction' thought experiments that go well beyond the normal bounds of scientific method and practice.[59] Nevertheless, their chief purpose is – in a literal sense – to bring these conjectures back 'down to earth' by deriving certain consequential truths as regards the nature of this-world physical reality and the logic of causal explanations.

Deutsch remarks in passing (p. 101) that '[i]f Bishop Berkeley or the Inquisition had known of virtual reality, they would probably have seized upon it as the perfect illustration of the deceitfulness of the senses, backing up their arguments against scientific reasoning'. Other writers on quantum theory – Karl Popper among them – have perceived a clearly marked line of descent from the kinds of instrumentalist opt-out clause forced upon Copernicus and Galileo by the Church authorities, and the orthodox (Copenhagen) QM doctrine which likewise enjoins a policy of wise disengagement from vexing issues of reality and truth.[60] As I have said, Deutsch has nothing but contempt for this 'pragmatic-instrumentalist' outlook, or what he calls 'the practice of using scientific theories without knowing or caring what they mean' (p 329). However, it is far from clear that Deutsch's extreme ontological-realist commitment to the truth of his many-worlds hypothesis can possibly be reconciled with a scientifc realism which respects the operative scope and constraints of this-world scientific enquiry.

Endnotes

1 David Deutsch, *The Fabric of Reality* (Harmondsworth: Penguin, 1997); also Paul Davies, *Other Worlds* (London: Dent, 1980); Bryce S. DeWitt and Neill Graham (eds.), *The Many-Worlds Interpretation of Quantum Mechanics* (Princeton, NJ: Princeton University Press, 1973); J.A. Wheeler and W.H. Zurek, *Quantum Theory and Measurement* (Princeton, NJ: Princeton University Press, 1983).

2 See Erwin Schrödinger, *Letters on Wave Mechanics* (New York: Philosophical Library, 1967); also John Gribbin, *In Search of Schrödinger's Cat: quantum physics and reality* (New York: Bantam Books, 1984).

3 Niels Bohr, *Atomic Theory and the Description of Nature* (Cambridge: Cambridge University Press, 1934) and *Atomic Physics and Human Knowledge* (New York: Wiley, 1958); Werner Heisenberg, *Physics and Philosophy* (New York: Harper & Row, 1958) and *Across the Frontiers* (Harper & Row, 1974); also John Honner, *The Description of Nature: Niels Bohr and the philosophy of quantum physics* (Oxford: Clarendon Press, 1987); Dugald Murdoch, *Niels Bohr's Philosophy of Physics* (Cambridge University Press, 1987); Patrick A. Heelan, *Quantum Mechanics and Objectivity: a study of the physical philosophy of Werner Heisenberg* (The Hague: Nijhoff, 1965).

4 In this connection see especially Albert Einstein, B. Podolsky and N. Rosen, 'Can Quantum-Mechanical Description of Reality be Considered Complete?', *Physical Review*, series 2, Vol. 47 (1935), pp. 777–80; Niels Bohr, article in response under the same title, *Physical Review*, Vol. 48 (1935), pp. 696–702; also Bohr, 'Conversation with Einstein on Epistemological Problems in Atomic Physics', in P.A. Schilpp (ed.), *Albert Einstein: philosopher–scientist* (La Salle, IL: Open Court, 1969), pp. 199–241; J.S. Bell, *Speakable and Unspeakable in Quantum Mechanics: collected papers on quantum philosophy* (Cambridge University Press, 1987); Arthur Fine, *The Shaky Game: Einstein, realism, and quantum theory* (Chicago: University of Chicago Press, 1986); Tim Maudlin, *Quantum Non-Locality and Relativity: metaphysical intimations of modern science* (Oxford: Blackwell, 1993); Michael Redhead, *Incompleteness, Nonlocality and Realism: a prolegomenon to the philosophy of quantum mechanics* (Oxford: Clarendon Press, 1987).

5 David Bohm, *Causality and Chance in Modern Physics* (London: Routledge & Kegan Paul, 1957); David Bohm and B.J. Hiley, *The Undivided Universe: an ontological interpretation of quantum theory* (London: Routledge, 1993); also David Z. Albert, 'Bohm's Alternative to Quantum Mechanics', *Scientific American*, No. 270 (May 1994), pp. 58–63; F.J. Belinfante, *A Survey of Hidden Variable Theories* (Oxford: Pergamon Press, 1973); S.V. Bhave, 'Separable Hidden Variables Theory to Explain the Einstein–Podolsky–Rosen Paradox', *British Journal for the Philosophy of Science*, Vol. 37 (1986), pp. 467–75.

6 For a detailed account of Bohm's hidden-variable theory and its fortunes at the hands of the orthodox QM 'establishment', see James T. Cushing, *Quantum Mechanics: historical contingency and the Copenhagen hegemony* (Chicago: University of Chicago Press, 1994).

7 See entries under Note 1, above.

8 See for instance Bas C. van Fraassen, *Quantum Mechanics: an empiricist view* (Oxford: Clarendon Press, 1992).

9 Deutsch, *The Fabric of Reality* (op. cit.), p. 53; further references given by 'Deutsch' and page number in the text.

10 See Jon Elster, *Logic and Society: contradictions and possible worlds* (Chichester: Wiley, 1978); Jeremy Hawthorn, *Plausible Worlds: possibility and understanding in history and the social sciences* (Cambridge: Cambridge University Press, 1991).

11 See for instance David Lewis, *Counterfactuals* (Oxford: Blackwell, 1973); Jerome S. Bruner, *Actual Minds, Possible Worlds* (Cambridge, MA: Harvard University Press, 1986); M. Loux (ed.), *The Possible and the Actual* (Ithaca, NY: Cornell University Press, 1979); Alvin Plantinga, *The Nature of Necessity* (Oxford: Oxford University Press, 1974); Robert C. Stalnaker, *Inquiry* (Cambridge, MA: M.I.T. Press, 1987).

12 See for instance Thomas A. Pavel, *Fictional Worlds* (Cambridge, MA: Harvard University Press, 1987); Ruth Ronen, *Possible Worlds in Literary Theory* (Cambridge: Cambridge University Press, 1994).

13 See especially Linda Hutcheon, *A Poetics of Postmodernism: history, theory, fiction* (London: Routledge, 1988) and Brian McHale, *Postmodernist Fiction* (London: Methuen, 1987).

14 See especially Simon Schama, *Dead Certainties (Unwarranted Speculations)* (New York: Alfred A. Knopf, 1991); also Alex Callinicos, *Theories and Narratives: reflections on the philosophy of history* (Cambridge: Polity Press, 1995); Richard Campbell, *Truth and Historicity* (Oxford: Oxford University Press, 1992); Maurice Mandelbaum, *The Anatomy of Historical Knowledge* (Baltimore: Johns Hopkins University Press, 1977).

15 See Note 11, above; also G. Hughes and M. Cresswell, *An Introduction to Modal Logic* (London: Methuen, 1968); Saul Kripke, *Naming and Necessity* (Oxford: Blackwell, 1980); L. Linsky (ed.), *Reference and Modality* (Oxford: Oxford University Press, 1971); David Wiggins, *Sameness and Substance* (Blackwell, 1980).

16 See Peter Hylton, *Russell, Idealism and the Emergence of Analytic Philosophy* (Oxford: Clarendon Press, 1990).

17 David Lewis, *On the Plurality of Worlds* (Oxford: Blackwell, 1986).

18 See J.S. Bell, *Speakable and Unspeakable in Quantum Mechanics* (op. cit.); Peter Forrest, *Quantum Metaphysics* (Oxford: Blackwell, 1988); Henry Krips, *The Metaphysics of Quantum Theory* (Oxford: Clarendon Press, 1987); Tim Maudlin, *Quantum Non-Locality and Relativity* (op. cit.); Alastair I.M. Rae, *Quantum Physics: illusion or reality?* (Cambridge: Cambridge University Press, 1986); Michael Redhead, *Incompleteness, Nonlocality and Realism* (op. cit.).

19 See for instance the essays collected in A.J. Ayer (ed.), *Logical Positivism* (New York: Free Press, 1959).

20 For the best-known and most intransigent expression of this view, see Rudolf Carnap, 'The Elimination of Metaphysics through Logical Analysis of Language', in Ayer (ed.), *Logical Positivism* (op. cit.), pp. 60–81.

21 See Note 5, above.

22 David Hume, *An Enquiry Concerning Human Understanding*, ed. L.A. Selby-Bigge, rev. P.H. Nidditch, 3rd edn (Oxford: Clarendon Press, 1975).

23 P.F. Strawson, *Individuals: an essay in descriptive metaphysics* (London: Methuen, 1959) and *The Bounds of Sense: an essay on Kant's Critique of Pure Reason* (Methuen, 1966).

24 See Notes 5 and 6, above.

25 For further discussion see F.J. Belinfante, *Measurements and Time Reversal in Objective Quantum Theory* (Oxford: Pergamon Press, 1975); also H.D. Zeh, *The Physical Basis of the Direction of Time* (Berlin: Springer, 1989).

26 See for instance David Deutsch and Michael Lockwood, 'The Quantum Physics of Time Travel', *Scientific American*, March 1994.

27 Cushing, *Quantum Mechanics: historical contingency and the Copenhagen hegemony* (op. cit.).

28 See also Paul Forman, 'Weimar Culture, Causality, and Quantum Theory, 1918–1927: adaptation by German Physicists and Mathematicians to a hostile intellectual environment', in *Historical Studies in the Physical Sciences*, Vol. 3 (1971), pp. 1–115 and 'The Reception of an Acausal Quantum Mechanics in Germany and Britain', in S.H. Mauskopf (ed.), *The Reception of Unconventional Science, AAAS Selected Symposium*, No. 25 (1979), pp. 11–50.

29 Max Horkheimer and T.W. Adorno, *Dialectic of Enlightenment*, trans. John Cumming (New York: Seabury Press, 1972); Edmund Husserl, *The Crisis of European Sciences and Transcendental Phenomenology*, trans. D. Carr (Evanston, IL: Northwestern University Press, 1970).

30 See also Jürgen Habermas, *Knowledge and Human Interests*, trans. J. Shapiro (London: Heinemann, 1971).

31 Immanuel Kant, *Critique of Pure Reason*, trans. N. Kemp Smith (London: Macmillan, 1974) and *Critique of Practical Reason*, trans. L.W. Beck (Indianapolis: Bobbs-Merrill, 1965).

32 Ludwig Wittgenstein, *Philosophical Investigations*, trans. G.E.M. Anscombe (Oxford: Blackwell, 1953); also Thomas Morawetz, *Wittgenstein and Knowledge* (Amherst, MA: University of Massachusetts Press, 1978) and Derek L. Phillips, *Wittgenstein and Scientific Knowledge: a sociological perspective* (London: Macmillan, 1977).

33 G.E.M Anscombe, 'Causality and Determination', in Ernest Sosa (ed.), *Causation and Conditionals* (Oxford: Oxford University Press, 1975), pp. 63–81.

34 See entries under Note 3, above.

35 See Note 2, above.

36 Paul Davies, *About Time: Einstein's unfinished revolution* (London: Viking, 1995).

37 David Lewis, *On the Plurality of Worlds* (op. cit.).

38 G.W. Leibniz, *Discourse on Metaphysics*, trans. P. Lucas and L. Grint (Manchester: Manchester University Press, 1953); *Monadology*, trans. E. Latta (Oxford: Oxford University Press, 1972); *New Essays Concerning Human Understanding*, trans. A.G. Langley (La Salle, IL: Open Court, 1926); *Theodicy*, trans. E.M. Huggard (London: Routledge & Kegan Paul, 1952).

39 Cited by Peter Gibbins, *Particles and Paradoxes: the limits of quantum logic* (Cambridge: Cambridge University Press, 1987), p. 9.

40 H.G. Alexander (ed.), *The Leibniz–Clarke Correspondence, together with extracts from Newton's Principia and Opticks* (Manchester: Manchester University Press, 1976).

41 For discussion of the Leibniz–Newton controversy, see Max Jammer, *Concepts of Space: the history of theories of space in physics* (Cambridge, MA: Harvard University Press, 1954); also Patrick Heelan, *Space-Perception and the Philosophy of Science* (Berkeley & Los Angeles: University of California Press, 1983); Christopher Ray, *Time, Space and Philosophy* (London: Routledge, 1991); Wesley C. Salmon, *Space, Time, and Motion: a philosophical introduction*, 2nd edn (Minneapolis: University of Minnesota Press, 1980); Richard Swinburne, *Space and Time* (London: Macmillan, 1981); Swinburne (ed.), *Space, Time and Causality* (Dordrecht: D. Reidel, 1983).

42 Albert Einstein, *Relativity: the special and the general theory* (London: Methuen, 1954); also R.B. Angel, *Relativity: the theory and its philosophy* (Oxford: Pergamon, 1980); J.R. Lucas and P.E. Hodgson, *Spacetime and Electro-Magnetism* (Oxford: Clarendon Press, 1990).

43 Kripke, *Naming and Necessity* (op. cit.); also entries under Notes 11 and 15 (above) and Stephen Schwartz (ed.), *Naming, Necessity, and Natural Kinds* (Ithaca, NY: Cornell University Press, 1977).

44 See for instance Alvin Plantinga, *The Nature of Necessity* (Oxford: Oxford University Press, 1970); *Does God Have a Nature?* (Milwaukee: Marquette University Press, 1986).

45 See Notes 11, 15 and 43, above.

46 See especially Gottlob Frege, 'On Sense and Reference', in Max Black and P.T. Geach (eds), *Selections from the Philosophical Writings of Gottlob Frege* (Oxford: Blackwell, 1952), pp. 56–78; Bertrand Russell, 'On Denoting', *Mind*, Vol. 14 (1905), pp. 479–93 and 'Knowledge by Acquaintance and Knowledge by Description', in *Mysticism and Logic* (London: Allen & Unwin, 1963), pp. 152–67; also Wolfgang Carl, *Frege's Theory of Sense and Reference: its origins and scope* (Cambridge: Cambridge University Press, 1994).

47 Frege, 'On Sense and Reference' (op. cit.).

48 For discussion of these cases from a different (anti-descriptivist) viewpoint, see Kripke, *Naming and Necessity* (op. cit.)

49 See for instance Phillips, *Wittgenstein and Scientific Knowledge* (op. cit.); also David Bloor, *Wittgenstein: a social theory of knowledge* (New York: Columbia University Press, 1983).

50 See Kripke, *Naming and Necessity* (op. cit.).

51 See also Rom Harré and E.H. Madden, *Causal Powers* (Oxford: Blackwell, 1975); Wesley C. Salmon, *Scientific Explanation and the Causal Structure of the World* (Princeton, NJ: Princeton University Press, 1984); Brian Skyrms, *Causal Necessity* (New Haven: Yale University Press, 1980); E. Sosa (ed.), *Causation and Conditionals* (op. cit.); M. Tooley, *Causation: a realist approach* (Blackwell, 1988).

52 Lewis, *On the Plurality of Worlds* (op. cit.).

53 Plantinga, *The Nature of Necessity* (op. cit.).

54 Thus, for instance, Cushing cites the following passage from a letter by Wolfgang Pauli:

> Causality, i.e., the possibility of describing natural phenomena in a rational way, is an expression of God's intelligence. The simple idea of deterministic causality must, however, be abandoned and replaced by the idea of statistical causality. For some physicists ... this has seemed a very strong argument for the existence of God and an indication of His presence in nature.

See Cushing, *Quantum Mechanics: historical contingency and the Copenhagen hegemony* (op. cit.), p. 151.

55 For some clear-headed discussion of this and other (often far from clear-headed) ideas, see Euan Squires, *The Mystery of the Quantum World*, 2nd edn (Bristol and Philadelphia: Institute of Physics Publishing, 1994).

56 For a careful and philosophically-informed treatment of these issues, see Ruth Ronen, *Possible Worlds in Literary Theory* (op. cit.).

57 See Notes 11, 15 and 43, above.

58 See Kripke, *Naming and Necessity* (op. cit.); also Hilary Putnam, 'Is Semantics Possible?' and 'Meaning and Reference', in Schwartz (ed.), *Naming, Necessity, and Natural Kinds* (op. cit.), pp. 102–18 and 119–32.

59 On this topic see especially James Robert Brown, *The Laboratory of the Mind: thought experiments in the natural sciences* (London: Routledge, 1991) and Roy Sorensen, *Thought Experiments* (New York: Oxford University Press, 1992).

60 Karl Popper, *Quantum Theory and the Schism in Physics* (London: Hutchinson, 1982).

6 Putnam's progress

Quantum theory and the flight from realism

I

Up to now, my argument has been concerned mainly with issues in the interpretation of quantum mechanics and with the response to those issues by practising physicists (among them Bohr, Einstein, Schrödinger, Bell, Bohm, and most recently Deutsch) who have taken a range of philosophical positions in this regard. From now on there will be a fairly marked shift of emphasis towards what philosophers have had to say and regarding the extent to which quantum-theoretical debates have influenced the course of recent developments in epistemology, metaphysics, and philosophy of logic. Most significant here is the way that some thinkers – including Hilary Putnam – have retreated from a strong causal-realist position of the kind that I described at some length in Chapter 5 when contrasting it with Deutsch's ontologically extravagant many-worlds hypothesis. Putnam offers a particularly striking example since in his case the various stations of the cross (as one is tempted to call them) have been signalled in a series of books and articles that record his growing disenchantment with causal realism and also – if less explicitly – its source in certain quantum-related issues and concerns.[1]

Briefly summarized, Putnam's journey has taken him from a realist outlook premised on the existence of an objective (mind-independent) world comprising various objects, attributes, microstructural features, causal dispositions, etc., to an outlook of so-called 'internal realism' according to which those various items can only be construed as relative to some favoured descriptive framework or conceptual scheme. This latter approach is likewise characteristic of orthodox (Copenhagen) quantum theory, along with the related empiricist or positivist doctrine which holds that it cannot make sense – at least as regards events at the subatomic level – to posit the existence of an 'objective' reality apart from the act of observation or measurement.[2] There has been much debate, among physicists and philosophers alike, concerning these quantum-theoretical claims and the prospects for an alternative (realist) construal that would not require such a drastic break with the methods and assumptions of 'classical' physics.[3] However, my main purpose here is to examine their bearing on Putnam's history of well-documented visions and revisions, starting out with his early (pre-1970) defence of a causal-realist position which deploys various kinds of counterfactual reasoning – or ingenious thought experiments – to secure that position against sceptical rejoinders.[4]

Thus – to take perhaps the best-known example – we may imagine a planet called Twin Earth, which contains large quantities of a substance which they (the denizens of Twin Earth) refer to as 'water' and which moreover happens to possess all the same phenomenal attributes as terrestrial water.[5] That is to say, it is normally encountered in liquid form, has the same freezing-point and boiling-point, is colourless and odourless unless due to impurities, falls as rain under certain atmospheric conditions, fills up the oceans and rivers, is used for washing, bathing, drinking, dissolving various other substances, and so forth. The only difference between it and terrestrial water is that Twin-Earth 'water' has the chemical formula XYZ whereas its This-Earth counterpart of course has the chemical formula H_2O. Now – Putnam conjectures – if a spaceship from Earth were to visit planet Twin Earth, then the first reports back would most likely contain some such message as 'there is a whole lot of water here!'. However, this would soon be subject to correction when the mistake showed up through chemical analysis. So the next message would say something like, 'On Twin Earth the word "water" means XYZ'. And of course the same process might be expected to occur in reverse. Thus:

> symmetrically, if a spaceship from Twin Earth ever visits Earth, then the supposition at first will be that the word 'water' has the same meaning on Twin Earth and on Earth. This supposition will be corrected when it is discovered that 'water' on Earth is H_2O, and the Twin Earthian spaceship will report: 'On Earth the word "water" means H_2O'.[6]

Putnam's point – in brief – is that these really were mistakes (and likewise genuine subsequent corrections) since what 'water' means to the inhabitants of each planet is fixed by its chemical constitution and not by phenomenal or surface descriptive criteria such as those listed above. Thus Twin-Earth water is not 'water' as terrestrials use that term, and the visitors from Twin Earth are simply wrong when they describe as 'water' what they think (by all appearances) to be a sample of XYZ, but what is actually a sample of H_2O.

Or again, we could perform another thought experiment by taking the story back a couple of centuries to a time when nobody on Earth knew that water molecules contained atoms of hydrogen and oxygen, and nobody on Twin Earth had any idea that 'water' had the chemical formula XYZ. Now if this were the case, then it might perhaps be thought – at least on the descriptivist theory – that any two speakers living on the different planets who referred to 'water' at the time in question would have entertained identical beliefs concerning it, that is with regard to its observable properties. Yet surely this is wrong, Putnam argues, since we can now assert – with the benefit of scientific hindsight – that in fact they were referring to different substances (H_2O and XYZ) whatever their beliefs and despite the exactly coextensive range of criteria by which they would have picked out samples of the kind. Thus, for any such pair of individuals, it is a necessary truth on Putnam's alternative (causal) theory of reference that 'they understood the term "water" differently in 1750 although they were in the same psychological state, and although, given the state of science at the time, it would have taken their scientific communities about fifty years to discover that they understood the term "water" differently'.[7]

Clearly, one could dream up as many versions of this Twin-Earth thought experiment as there exist substances or natural kinds for which one can invent plausible counterparts with the same surface characteristics but different molecular constitutions. Putnam's other main example is that of aluminium and molybdenum, metals which – for the sake of argument – he takes to be indistinguishable except by an expert metallurgist. Suppose further that the extent of these mineral deposits varies inversely on the two planets, aluminium being as rare on Twin Earth as molybdenum on Earth, and molybdenum as common on Earth as aluminium on Twin Earth; also that the words are switched on Twin Earth with 'aluminium' referring to molybdenum (i.e. the terrestrial metal with just that atomic structure), and of course vice versa. In which case the spaceship visitors from Earth, when informed that Twin-Earth pots and pans were made out of 'aluminium', would most likely accept this information on trust since at first they could have no good reason to doubt it. However, as Putnam duly notes, there is one vital difference between this and the example of 'water' when pushed back to a time before the advent of atomic-molecular theory. For '[w]hereas in 1750 no one on either Earth or Twin Earth could have distinguished water from "water", the confusion of aluminium with "aluminium" involves only a part of the linguistic communities concerned'.[8] In other words, there are at least a few people around with the required expertise – the up-to-date knowledge of metals and their distinctive (i.e. microstructural) features – to sort out this confusion and specify how the terms in question should in future be used. Thus, '[a]n Earthian metallurgist could tell very easily that "aluminium" was molybdenum, and a Twin Earthian metallurgist could tell very easily that aluminium was "molybdenum"'.[9] Here again – as in the case of 'water' – there may well be others (perhaps the vast majority) on both planets who possess no such expert knowledge and who would therefore have just the same beliefs, mental contents, psychological states, etc., when confronted with counterpart samples. All the same, Putnam argues, aluminium and molybdenum have the same extensions (that is, an identical scope of reference) on Earth and Twin Earth, those extensions being fixed by the microstructure of the two different metals quite aside from the local range of descriptive criteria, or from whatever presumably goes on in the minds of the speakers concerned. In short, 'the psychological state of the speaker does not determine the extension (or the "meaning", speaking preanalytically) of the word'. Or again, as Putnam yet more snappily puts it, 'cut the pie any way you like, "meanings" just ain't in the head'.[10]

This argument is aimed against the standard descriptivist theory which holds (after Frege) that 'sense determines reference', or that the precondition for successfully referring to items such as 'water' or 'aluminium' is that we pick them out in virtue of possessing the appropriate range of definitions or descriptive criteria.[11] It is also effective against Wittgensteinian arguments to the effect that scientific truth claims like 'water = H_2O' are really just tautologies since they are true by definition – and hence possess no informative content – in the 'language-game' of present-day physical science.[12] Putnam's point, and his purpose in devising the above thought experiments, is that the descriptivist theory demonstrably fails on two main counts. Thus (1) it offends our intuitive sense as regards what is right and wrong in the matter of identifying natural kinds like water and aluminium, as distinct from other such real-world kinds or from their various imaginary (possible-world) counterparts such as XYZ or Twin-Earth 'aluminium'. Moreover, (2) it offers no credible explanation of how science has advanced from the primitive stage of picking them out by surface features or phenomenal appearance to the stage of more precisely distinguishing between them by methods of chemical decomposition or atomic-molecular analysis. Of course it is important for Putnam's theory that such knowledge should not be confined within the limits of a tiny expert scientific community but should rather be shared – in principle at least – among all those speakers with a practical interest in making the relevant distinctions. However, all that is needed to secure this extended franchise is what he calls the 'linguistic division of labour', that is to say, the fact that there will always be specialists around who do have the requisite knowledge (e.g. to distinguish gold from iron pyrites or aluminium from molybdenum), and whose special expertise feeds back into communal usage through various channels. What is certainly not required is that everyone – physicists, chemists, bankers, economists, jewellers, purchasers of wedding-rings, etc. – should know that gold is the metallic element with atomic number 79 and furthermore be able to perform the right sorts of test as and when needed.

Thus 'everyone to whom gold is important for any reason has to acquire the word "gold"; but he does not have to acquire the method of recognizing whether something is or is not gold …. [h]e can rely on a special subclass of speakers'. And again:

> [t]oday it is obviously necessary for every speaker to be able to recognize water (reliably under normal conditions), and probably most adult speakers even know the necessary and sufficient condition 'water is H_2O', but only a few adult speakers could distinguish water from liquids that superficially resembled water. In case of doubt, other speakers would rely on the judgement of these 'expert' speakers …. [I]n this way the most recherché fact about water may become a part of the social meaning of the word even though unknown to almost all speakers who acquire the word.[13]

What is crucial here – as likewise in his Twin-Earth thought experiments – is Putnam's point that 'meanings just ain't in the head', or again (in more technical terms) that the extension or scope of referring expressions is fixed not by what individual speakers happen to mean by them, but rather by their actually picking out genuine, real-world samples of the kind intended. Moreover, those samples are defined as such not only by appeal to shared customs of usage – as on Wittgenstein's language game theory – but according to what is actually the case with regard to their atomic configuration, molecular structure, chemical make-up, etc. Of course, the Wittgensteinian can always retort that there exist as many language games as cultural 'forms of life', some of them encompassing the widest range of human concerns and interests, whereas others – like those of metallurgy or chemistry – are intelligible only to speakers who possess the right kinds of special expertise. However, this still makes it a question of the way in which speakers (specialists included) customarily talk about these things, rather than the fact – as in Putnam's theory – that such talk is warranted only on condition that it picks out genuine, objectively existent features of physical reality. Thus, for Putnam, what counts as a valid statement concerning such microstructural features (e.g. 'water = H_2O' or 'gold is the metallic element with atomic number 79') is decided not so much by the communality or the language game in question but by the existence of experts, fit though few, who can vindicate the truth of that statement through advanced investigative techniques and can thereby distinguish H_2O from XYZ, or gold from iron pyrites, or aluminium from molybdenum.

II

In short, Putnam's theory preserves the single most important tenet of scientific realism, namely its principle that what renders our beliefs true or false is the way things stand in reality, as opposed to the Wittgenstein-derived idea that 'reality' can be construed only in accordance with the various beliefs enshrined in our various language games or communal 'forms of life'. Some commentators – Gregory McCulloch among them – have argued for a different interpretation of Wittgenstein, one that would avoid this conflict and bring him out much more in line with Putnam's position.[14] After all, there is a strong prima facie resemblance between Putnam's case that meanings 'just ain't in the head' and Wittgenstein's well-known arguments against the idea of a 'private language' or the notion that speakers have some kind of uniquely privileged epistemic access to their own meanings, thoughts, and beliefs.[15] McCulloch's point – briefly summarized – is that the best way to interpret these arguments is to push them as far as possible in the direction of a so-called 'in-the-world Wittgensteinianism' which accords with Putnam's emphasis on the mind-independent (or non-belief-relative) status of referring expressions. This argument holds that our language games – or some of them, especially those of the physical sciences – hook up with the world through their proven capacity to track certain features of it which we discover through various procedures of jointly empirical and theoretical enquiry, as for instance through technologically enhanced observation of subatomic or molecular structures informed by a knowledge of post-Daltonian physics and the periodic table of elements. On this view a great deal of our knowledge is 'external' in the sense that, in order to *count* as knowledge, it must involve reference to objects and events which exist outside the private realm of individual beliefs, mind states, psychological contents, assenting dispositions, or whatever. That is to say, those beliefs have determinate content – and can thus be assigned definite truth values – only in so far as they are taken to 'track' the particular objects and events in question, or only on condition that their content co-varies with the properties of a real-world (e.g. subatomic or molecular) physical domain.[16]

Such was the point of Putnam's various Twin-Earth thought experiments, designed to show that the truth or falsehood of beliefs concerning 'water', 'gold', 'aluminium' (etc.) is ultimately fixed by the microstructure of those substances themselves and not by the range of more-or-less adequate descriptive or identifying criteria that we or others may happen to apply. However, this is crucially not to say that reality is 'mind-independent' in the sense of existing outside and beyond all reach of human knowledge or epistemic access. McCulloch makes the point with exemplary care when he defines 'externalism' as the view 'that some things which are external to the human individual are not exterior to the mind'.[17] What he means is that the tracking relation (or co-variance condition) is sufficient to keep the mind reliably in touch with so-called 'external' reality, even though that reality is perfectly objective and in no way dependent on beliefs, psychological states, mental representations, etc. This point is crucial because sceptics of various colour – orthodox QM theorists among them – have most often started out from the argument that if reality is indeed (as the typecast 'realist' would have it) completely external to our furthest powers of perceptual or conceptual grasp, then *ex hypothesi* we can have no knowledge concerning it, and realism is thereby shown up as an incoherent or self-refuting doctrine.[18]

Thus, according to Michael Williams in a fine recent book on this topic, scepticism typically gets a hold through the thought: 'if the world is an objective world, statements about how things appear must be logically unconnected with statements about how they are; this lack of connection is what familiar thought-experiments dramatically illustrate'.[19] And again: '[t]o realize our vulnerability to scepticism we need only recognize the simple logical point that our experience could be just what it is and *all our beliefs about the world could be false*'.[20] However – and this is McCulloch's point in the above-quoted sentence – although realism does involve the claim that 'some things ... are external to the human individual', we need not (and should not) conclude from this that they are also 'exterior to the mind'.[21] For it is precisely the virtue of externalist approaches based on the causal theory of reference that they avoid the old dualism between world and mind that has been such a bugbear of epistemology from Descartes to Kant and beyond. What enables them to do so is the modal-realist argument – as in Putnam's variations on the Twin-Earth theme – that understanding necessarily 'tracks real essence', or that knowledge is inherently tied up with the quest for a better, deeper, more adequate grasp of physical objects and their various defining (e.g. microstructural) properties.

So the upshot of Putnam's externalist argument that 'meanings just ain't in the head' is that we need to adopt a wide, rather than a narrow conception of psychological states, one that makes room for a great many items which fix the content of veridical belief but which find no place in more traditional (dualist) approaches to the problem of knowledge. This also means that the realist is or should be in no way committed to the truth of our present-best scientific theories or – in Nicholas Rescher's phrase – the 'ontological finality of science as we have it'.[22] For it is just the point of such reality-tracking conceptions that they leave the way open to further refinements or advances in knowledge. In McCulloch's words, 'our prescientific understanding of substance-words is supposed to be already sensitive to future or possible scientific discovery'.[23] Thus – to take another example from Putnam – lemons were once identified 'prescientifically' through some such description as, 'fruits with a yellow peel whose juice has a sour taste when diluted in water'. Nowadays they are identified for scientific purposes by reference to their chromosomal structure, and the taste is explained by their having a certain acid content, this latter more precisely specifiable in subatomic terms through the knowledge that acids are proton donors.[24] Such knowledge has to do with the essential properties of lemons and acids, as distinct from those phenomenal features (yellowness, sourness, etc.) which can sometimes mislead – if for instance we are confronted with a green or a sugar-impregnated lemon – but which none the less enable us to pick them out in a fairly reliable fashion. Still, it would be premature and no part of the well-advised realist's case to argue that these are the *ultimate* properties which mark an end-point to the quest for a deeper, more adequate account of their molecular or subatomic structure. For if the history of science has one great lesson to impart it is the fact that a large amount of our current best knowledge will be subject to eventual revision or refinement in consequence of further such advances in the means of observation and causal-explanatory grasp.[25]

This is not the place for a detailed account of the stages in Putnam's subsequent retreat – or (as he tells it) his subsequent progress – from the realist position outlined above to a standpoint of so-called 'internal realism' that pretty much rejects every main tenet of that earlier approach.[26] To late Putnam it is simply self-evident that the sceptic will always win out as soon as the realist makes any claim concerning the existence of an objective or mind-independent 'external world'. For at this point the way is wide open for the sceptic to ask how the realist could possibly be entitled to assert such a claim if indeed it is the case – as her argument requires – that its truth value is determined by something that lies altogether outside and beyond the scope of human knowledge.[27] It is then but a short distance to the sceptic's standard knock-down rejoinder, namely (in the words of Barry Stroud) that 'all possible experience is equally compatible with the existence and the non-existence of the world', or again – as Thomas Nagel similarly argues – that the quest for objective reality and truth must always stand 'under the shadow of scepticism'.[28] In fact there is widespread agreement among recent writers on this topic that realism and scepticism are like two partners in a long-drawn marital strife, the one hopelessly attached to the other through a habit of perverse mutual dependence that keeps them both arguing without any prospect of eventual release or some kind of amicable settlement.[29] This quarrel goes back at least to Descartes and his various experiments in hyperbolic doubt, the purpose of which was of course to secure some ultimate, indubitable anchor-point for knowledge and truth, but whose outcome – as witness the record of debate since then – was to offer the sceptic a point of leverage for all manner of yet more ingenious counter-hypotheses. Among the latter is Putnam's celebrated thought experiment where he asks how I can possibly *know* that I am not a brain in a vat whose entire experience – whose thoughts, perceptions, memories, desires, and every last item of sensory input – is in fact computer simulated under the control of an evil scientist with powers equivalent to those of Descartes' demiurgic *malin génie*.[30]

Realists and anti-realists will of course divide on the issue of how we should best interpret this latest re-edition of Cartesian themes. Putnam himself was at just this time on the cusp of his turn from 'metaphysical' to 'internal' (i.e. framework-relative) realism so there is some room for doubt in the matter. Indeed, as he remarks, 'the question of "Brains in a Vat" would not be of interest, except as a kind of logical paradox, if it were not for the sharp way in which it brings out the difference between these philosophical perspectives'.[31] To this extent it is clear that Putnam now rejects any form of 'metaphysical realism', a position which he thinks must always be vulnerable to the sceptic's standard response. In short, he is already well along the road to his later (post-1980) espousal of a framework-relativist 'internal realism' which effectively concedes the impossibility of envisaging a world outside or beyond our various descriptions, conceptual schemes, investigative interests, etc.[32] On the other hand, it is just as clear that Putnam's chief purpose with this thought experiment is to refute the sceptic by showing that we *cannot* consistently believe – or intelligibly entertain the conjecture – that we might after all be so mistaken as regards the external world and our own situation in it. Thus, '[t]he existence of a "physically possible world" in which we are brains in a vat (and always were and will be) does not mean that we might really, actually, possibly *be* brains in a vat. What rules this possibility out is not physics but *philosophy*' (Putnam's italics).[33] For we could not even raise such sceptical questions unless from the real-world situated standpoint of creatures with various (mostly reliable) sources of knowledge – sensory inputs, perceptual powers, epistemic modalities, and so forth – which alone provide a contrastive basis for just such excursions into hyperbolic doubt.

In which case the upshot of Putnam's fable – as with his earlier Twin-Earth hypotheses – is to strengthen the argument for an externalist or 'wide' psychological content view, one that would render such doubts superfluous, along with the entire Descartes-derived problematic of knowledge and representation. On this view the only reason why realism had so long seemed to labour under the shadow of scepticism was that realist philosophers had typically opted for a narrow (i.e. internalist) construal where beliefs were thought of as somehow existing in a private realm whose connections to the world – or whose justificatory criteria – were at best highly tenuous and at worst totally inexplicable. Once abandon that idea, Putnam suggests, and the shadow will be lifted straight away since it then becomes clear that what renders our beliefs true or false is not their possessing or failing to possess some mysterious 'correspondence' relation with objects and events outside ourselves. Rather, it is the fact of their tracking or failing to track some relevant portion of physical reality which itself determines their truth content in virtue of (say) its genetic constitution, chemical make-up, molecular structure, subatomic configuration, or whatever. Yet there is clearly a problem – if not for Putnam then at least for his more devoted exegetes – in explaining how this externalist approach could possibly be reconciled with the doctrine of 'internal realism' which first makes its appearance in connection with the brain-in-a-vat thought experiment.[34] For it is hard to see how this could work as an antidote to sceptical doubt if the argument is weakened to the point of conceding that reality is 'internal' *not* to our minds (or 'psychological states') but rather to our various descriptive schemes, conceptual frameworks, etc.

III

I shall return to this question later on in the context of debates within quantum theory and what I take to have been their motivating influence at various stages of Putnam's retreat to a (so-called) internal-realist position. Meanwhile – by way of contrast – I shall say some more about the kinds of countervailing externalist approach that he once defended with great vigour and which others have since then continued to develop in broadly similar directions.[35] These offer an alternative not only to mainstream (post-Fregean) descriptivist theories of sense and reference, but also to the long tradition of epistemological thought – going back to Plato's *Theaetetus* – which explicates the concept of knowledge in terms of 'justified true belief'.[36] Thus, in order for something to count as a genuine item of knowledge, it should satisfy the following three criteria: (1) it must be true, (2) we must believe it to be true, and (3) our belief should be adequately grounded, i.e. based on adequate evidence, arrived at through a valid process of reasoning, or known as a matter of self-evident, a priori truth.[37] On the face of it, this seems to capture pretty well what we mean by genuine (veridical) 'knowledge' as opposed to mere opinion, belief, private conviction, and the like. Still, there are certain problems with it, as shown by fictive counter-examples where the above three conditions are fully satisfied but where they do not yield knowledge in anything like our usual (intuitive as well as philosophical) sense of that term. Thus we might believe X, and X might be true, and we might moreover have rational grounds for our belief, and yet it just happens that they are not the *right* grounds in this particular case.

Edmund Gettier offered several such instances in a well-known essay, and they have since become a challenge for philosophers to produce ever more ingenious examples of the kind.[38] Often these involve some fairly large stretches of counterfactual imagining, designed to show how it could conceivably come about – within the bounds of real-world epistemic possibility – that somebody was placed in this highly unusual predicament. However, their remoteness from everyday experience is no reason to count them irrelevant or merely of specialized 'philosophic' interest. McCulloch makes the point with admirable force when he remarks that 'people who become irritated at such [thought-experimental] procedures simply show that they have no proper grasp of what enquiry is all about'.[39] That is, they fail to grasp how the logic of enquiry in various fields – from the physical sciences to history, sociology, and the humanistic disciplines – depends on the kinds of explanatory argument that would simply not exist without this appeal to subjunctive, hypothetical, or counterfactual-conditional modes of reasoning.

Of course it is important (indeed a main point of the exercise) to distinguish clearly between these various 'possible worlds' or orders of counterfactual imagining. To confuse them is to run the risk of erecting a whole elaborate metaphysical system – like the 'many-worlds' interpretation of quantum mechanics – on the basis of some highly disputable data and a fixed conception of what reality must be like in order to meet those same metaphysical requirements.[40] So counterfactual hypotheses have to be constrained by a due recognition of the difference between causal explanations, scientific thought experiments, modal-logical arguments, metaphysical speculations, science-fiction scenarios, and so forth. However, there is a perfectly legitimate place on this scale for the sorts of jointly logical and epistemological enquiry represented by Putnam's thought experiments and by Gettier's ingenious counter-examples to the standard account of knowledge as justified true belief. What these very effectively bring out is the failure of a certain traditional way of thinking about issues in epistemology and philosophy of science to offer any possible escape route from the vicious circle that holds realism and scepticism in a kind of endlessly self-supporting yet mutually destructive embrace.

An alternative approach – much favoured of late – is to make justification dependent on the existence of some causal-explanatory link between beliefs and objects of belief.[41] This approach is often called 'externalist' since, as we have seen, it breaks with the traditional ('internalist') idea that knowledge is a certain distinctive state of mind, one that can be accessed only by the person who knows at first hand what it is to be in that particular state. On the contrary, it is argued that what properly counts as knowledge is just what that person is entitled to assert on the basis of their having been exposed to the right sorts of causal stimuli or their having drawn appropriate conclusions from the right kinds of knowledge-conducive learning experience. So an outside observer would be just as well – perhaps even better – placed to judge whether that person's beliefs qualified as genuine knowledge or whether they failed to come up to the required justificatory standard. Externalism is therefore very much a part of the widespread present-day dissatisfaction with Cartesian ideas about mind, knowledge, and privileged (first-person) epistemic access. Indeed, it challenges just about every aspect of Descartes' epistemology, including his religiously inspired belief that non-human animals could not be held to 'know' anything on account of their merely mechanical nature as soulless creatures devoid of intelligence and possessing nothing more than a fixed repertoire of reflex responses to physical stimuli. For if one accepts the externalist argument which severs the link between knowledge and (epistemological) justification, then there is simply no reason – prejudice apart – for denying that animals may possess a whole range of truth-tracking capacities which match (and in some respects clearly surpass) the best ability of human trackers.

From one point of view this may appear to be a vote of no confidence in the superior powers of the human intellect to know and understand what is going on around it in the so-called 'external world'. But from another – less tied to traditional ideas of what constitutes genuine knowledge – it is a welcome release from the self-imposed travails of Cartesian sceptical doubt. For it helps to explain, in Michael Williams's words, why 'foundationalism is a presupposition of scepticism, not a by-product'.[42] In other words, what first got us into that fix was the idea of knowledge as a special (problematical) relation between mind and world that could only be explained or made good by establishing some a priori grounds in the very nature of human understanding. What can therefore best get us out of it – so this argument runs – is an externalist approach that effectively reverses that order of dependence and locates the criteria for knowledge and truth in the a posteriori discoverable nature of those various real-world objects and events which themselves determine the truth value of our various assertions, beliefs, conjectures, or scientific theories concerning them.

So the point of Putnam's brain-in-a-vat hypothesis – contrary to widespread report – is to draw the sting of epistemological scepticism by showing it to rest on a false (narrow) conception of mental content and an equally false (Cartesian or dualist) understanding of the relation between mind and world or knowledge and objects of knowledge. It thus belongs to a class of thought experiments which are not just disguised forms of syllogistic reasoning (i.e. strictly non-informative logical deductions from given premises) but which aim to tell us something of genuine, non-trivial import concerning the world and our actual or potential knowledge of it. Other instances would include Galileo's classic thought experiments with gravity-induced fall and the motion of bodies on an inclined plane, and also – more directly relevant in the present context – Einstein's series of exchanges with Bohr on the topics of quantum measurement and nonlocality.[43] What distinguishes these modes of investigative thought is their claim to discover some objective property of the physical world whose existence follows necessarily from the adoption of certain well-established scientific principles which are taken to apply in other fields of advanced theoretical research where as yet there is no direct observational evidence. Of course it is a matter of dispute among historians of science just which was the order of priority, in Galileo's case, between trials of this sort conducted in the 'laboratory of the mind' and his other, 'real-world' physical experiments like those reputedly carried out at the Leaning Tower of Pisa.[44] The case is somewhat clearer – in this respect at least – as concerns quantum nonlocality since here the results in question were arrived at first by thought-experimental derivation from known QM principles and then borne out – under controlled laboratory conditions – with the advent of more sophisticated measuring equipment.[45] Nevertheless, the same argument holds in each case: that it is possible to derive certain truths concerning physical properties or laws of nature through the conduct of scientific thought experiments whose conclusions are in no sense analytically or conceptually 'contained in' their premises, and which thus go beyond any merely tautologous process of syllogistic reasoning.

Such procedures are sometimes dismissed by empirically minded critics as exceeding the bounds of proper (observationally constrained) theory construction and therefore as having no legitimate place either in the conduct of scientific enquiry or in the kinds of speculative reconstruction offered by philosophers/historians of science. However, this ignores the extent to which *all* explanatory arguments, whether in the physical or the social sciences, depend upon modes of subjunctive-conditional reasoning which alone make it possible to distinguish mere instances of observed regularity (or Humean 'constant conjunction') from instances of genuinely causal, i.e. counterfactual-supporting explanation.[46] Hence – to repeat – McCulloch's sharp comment that 'people who become irritated at such procedures simply show that they have no proper grasp of what enquiry is all about'.[47] Least of all can the scientific realist afford to reject this powerful source of supporting arguments for the existence of a (largely) mind-independent object domain whose physical features, causal dispositions, molecular or subatomic structures, etc., are precisely what determine the truth content of our various descriptive and explanatory theories. For the early (causal-realist) Putnam this was a case amply borne out not only by the sheer self-evidence of scientific progress to date but also by a range of condign thought experiments – like the Twin-Earth conjectures summarized above – which supported the realist claim by grounding it in premises that were strictly indispensable to any form of scientific-investigative thought. And yet, as we have seen, Putnam then retreated to a stance of so-called 'internal realism' which relativized truth to its operative role within various possible descriptive 'frameworks' or conceptual schemes. In so doing he renounced his entire earlier line of argument from the existence of real-world (causal-explanatory) constraints on the range of what could properly and justifiably count as an adequate scientific theory, ontology, or mode of reasoning on the evidence.

This is why Putnam takes issue with Ian Hacking as regards the latter's supposedly naive belief that interpretative problems in quantum mechanics can be addressed in terms that make sense in the language of 'classical' (pre-quantum) space–time physics.[48] I discussed their difference of views at some length in Chapter 2, but it may help to focus the issues at this point if I offer a brief summary. Hacking had described the impression made upon him by a scientist colleague who recounted an experiment concerning the change of electrical charge on a supercooled niobium ball. 'Now how does one alter the charge on the niobium ball? "Well, at that stage", said my friend, "we spray it with positrons to increase or decrease the charge". From that day forth I've been a scientific realist. *So far as I'm concerned, if you can spray them they are real.*'[49] Thus, according to Putnam, Hacking subscribes to the view that there exist certain 'distinct things' called positrons whose effect in altering the charge on a niobium ball is sufficient to warrant belief in their existence as subatomic particles of matter. Yet surely it is the case, according to quantum field theory, that such 'particles' possess neither definite position, momentum nor distinct (numerical) identity, except in so far as those values are obtained by arranging the experiment in some particular way and thus producing just one among the range of possible conjoint outcomes.[50] In a different set-up – as Putnam remarks – one could arrange things in such a way that 'one "sprayed" the niobium ball, not with three positrons, and not with four positrons, but with a *superposition of three and four positrons!*' (Putnam's italics).[51] And this problem for Hacking arises not only in the context of quantum field theory but also on any understanding of the wave–particle dualism since 'elementary quantum mechanics already tells us that we cannot think of positrons as having trajectories or as being, in general, *reidentifiable*'.[52]

I have focused on Putnam's argument *contra* Hacking because it offers such a clear and typical example of the way that quantum mechanics is routinely adduced as a knock-down case against realist approaches in epistemology and philosophy of science. Also, it gains a certain added poignancy from the fact that Putnam was himself once attached to just the kind of viewpoint that he here rejects out of hand, whereas Hacking is not so much a hard-line realist as a qualified instrumentalist, one who believes that 'if you can spray them they are real', but who otherwise resists any temptation to swell the ontology of particle physics with all manner of hitherto unobserved or as-yet hypothetical entities. What is symptomatic about Putnam's response is the way that it shows him swinging right across from his earlier, highly developed and sophisticated realist position to a quantum-derived sceptical stance which endorses the orthodox (Copenhagen) theory as a matter of 'elementary' knowledge. Yet surely it is a more elementary precept – or one with a far stronger claim to that title – which holds that we should grant credence to scientific theories in proportion to the amount of evidence we possess in their favour or the degree to which they have been borne out through repeated observation, experimental testing, theoretical refinement and critique, comparative assessment against rival theories, and so forth. Conversely – on the same principle – we should withhold or at any rate grant only limited credence where the candidate theory is one (such as orthodox quantum mechanics) that entails a great number of unresolved problems and paradoxes, or which comes into conflict with other, more basic and well-established principles of scientific knowledge.[53] If this approach is applied to Putnam's argument in the passage on Hacking cited above, then it must surely appear that he (Putnam) has jumped to some highly dubious conclusions on the basis of minimal evidence.

The greatest irony in all this is that Putnam's earlier realist approach to issues in epistemology and philosophy of science offered the strongest possible grounds for adopting just such a qualified attitude to the claims of orthodox quantum theory. That is, it explained how scientific knowledge in various fields could be thought of as advancing through stages of increased precision and depth-explanatory grasp in the usage of terms like 'atom', 'molecule', 'gold', 'water', 'acid', and 'lemon'.[54] Such terms become attached to the objects or substances in question through an act of designative naming and thereafter retain a certain fixity of reference *throughout and despite* any later advances in knowledge concerning their various chemical properties, causal attributes, subatomic configurations, chromosome structures, and so forth. Thus for Locke it was perfectly rational, given the state of scientific knowledge in his time, to adopt a sceptical attitude as regards the possibility of advancing from 'nominal' to 'real' definitions of substances (like gold or water) which as yet could only be identified according to their surface characteristics or directly observable features.[55] What was needed in order for this to come about was just the kind of scientific progress achieved by Dalton's atomic theory of the elements and the consequent transformation of chemistry into a physics-based discipline with far greater powers of depth-ontological and causal-explanatory grasp.

Even so, there remained other branches of science – e.g. molecular biology – which had to wait much longer to attain that status despite such progress having already been made in fields that were in some sense more 'basic' or 'fundamental'. Thus, as Putnam remarks:

> even if I have a description in, say, the language of particle physics of what are in fact the chromosomal properties of a fruit, I may not be able to tell that it is a lemon because I have not developed the theory according to which (1) those physical–chemical characteristics are the chromosomal structure-features (I may not even have the notion 'chromosome'); and (2) I may not have discovered that chromosomal structure is the *essential* property of lemons.[56]

His point is that the process of advancement from 'nominal' to 'real' definitions (or essences) is one that occurs at different rates in different fields of knowledge and which may leave some in a state of comparative underdevelopment. Thus at times the most rational course is to adopt an agnostic stance – or a measure of epistemological reserve – with regard to the status of certain items which figure in the current best theories of science yet whose reality might yet be challenged. Such was the case with atoms during the two millennia from Democritus to Dalton, when the atomist hypothesis was a matter of largely metaphysical conjecture. This issue remained in some degree open, even during the half-century or so after Dalton when its truth was increasingly borne out by a range of theoretical considerations and various kinds of indirect evidence.[57] Only with Perrin's famous experiments and through the work of others (Einstein among them) on Brownian motion did the atomic-molecular hypothesis achieve the kind of jointly observational and theoretical warrant that placed it beyond reasonable doubt on any but the most sceptical of views.[58] Thus it was still possible, quite late in this history, for an eminent physicist such as Mach to refuse on principle to credit the reality of atoms, since belief in their existence – although supported by every indication so far – still exceeded the limits of plain observation or direct experimental proof.[59] On the other hand, Mach's instrumentalism was already a minority view (at least as regards the existence of atoms) and one that thereafter retained its appeal more among sceptically inclined philosophers of science – including the logical positivists – than among practising scientists.[60]

IV

Putnam's early causal-realist theory of reference seems to me quite simply the best we have when it comes to explaining not only how terms such as 'molecule', 'atom' and 'electron' acquire scientific currency but also how they come to be used with an increasing degree of conceptual precision and depth-ontological grasp. This is the chief advantage of that theory compared with others in the mainstream (descriptivist) mode descending from Frege and Russell. On their account 'sense determines reference' in so far as we cannot refer to anything – whether macroscopic objects or subatomic particles – unless we are in possession of the relevant descriptive criteria.[61] What enables us to pick out samples of various kinds (gold, lemons, acids, atoms, or molecules) is precisely our knowledge of the standard definitions which apply in each case – by communal assent – and which thus ensure that we are properly referring to the same sorts of thing. However, this theory runs into trouble for the reasons that I have mentioned above, namely (1) that it fails to explain the continuity of reference to natural-kind terms across sometimes radical shifts in the range of scientifically acceptable criteria, and (2) that it opens the way to other, more extreme (e.g. Wittgensteinian) arguments for supposing our use of referential terms to be wholly dependent on their operative role within certain language games, conventions, cultural 'life-forms', or whatever.[62] Thus descriptivism may start out – as it did with Frege and Russell – as a theory that aims to specify the conditions and hence to secure the possibility of accurate usage by revealing the logico-semantic structure that underlies various (proper and deviant, e.g. fictive or referentially empty) forms of natural-language expression. But in so doing, it yields ground to the Wittgensteinian idea that reference can only be a matter of following the communally sanctioned linguistic 'rules' that decide what shall count as a sample instance of this or that kind.

In a typical exchange along these lines – one I witnessed recently – a realist may point to the glass of water in front of him and declare, 'this transparent liquid substance is *water* and can be identified as such (should we have need to check) by its possessing the molecular structure H_2O'. To which the Wittgensteinian will routinely object, 'that is a mere tautology!'. And if the realist then protests that the disputed statement is far from tautologous – that it represents an item of jointly empirical and theoretically informed knowledge that once required scientific investigation and which by no means follows from the mere definition of 'water' – then the Wittgensteinian will bounce straight back with the standard response. That is, he or she will make some remark to the general effect, 'no doubt "water = H_2O" (just as "gold is the metallic element with atomic number 79") but only within the language-games of modern physics and chemistry'.[63] From here it is but a short step to those varieties of cultural-relativist thinking – some of them explicitly indebted to Wittgenstein – which treat the 'language-game' of physical science as possessing absolutely no privileged status or superior epistemic warrant as compared to the language games of (say) religion, astrology, magic, or rain dance ritual.[64]

Of course it was just this feature of Wittgenstein's later philosophy that Russell so strenuously rejected, believing as he did that the world and all its physical furniture existed quite apart from our current theories concerning it, but that the sciences offered our best means of access to truths that were not just an all-too-human product of superstitious fantasy or wishful thinking.[65] Yet it is not hard to see how Russell's theory of descriptions – like Frege's theory of sense and reference – could be given the kind of late-Wittgensteinian gloss that Wittgenstein himself retroactively applied to his own early (Frege–Russell-influenced) thoughts about language, logic, and truth.[66] That is to say, there was always a possibility of pressing the argument one stage further and holding not only that 'sense determines reference' but that sense and reference must likewise be construed as intelligible only within some language-game or cultural form of life. Indeed, it is an article of faith for some Wittgensteinians – Baker and Hacker conspicuous among them – that Frege, Russell, and their analytic progeny failed to take the point of Wittgenstein's autocritique and were hence condemned to labour fruitlessly in the arid wastes of logico-linguistic analysis.[67]

Putnam's causal theory of reference avoids this particular nemesis and also (as I have said) provides the best means of explaining how issues in philosophical semantics relate to issues of ontology, epistemology, and philosophy of science. This it does by anchoring reference in the act of designation by which terms are first used ostensively, i.e. to pick out some particular object, and thereafter continue to signify that same object throughout and despite any subsequent advances in science which enable us to specify more exactly both its nature (genetic constitution, molecular structure, subatomic configuration, etc.) and also the sense of such terms when correctly applied. Thus:

> [i]f I describe something as a *lemon,* or as an *acid,* I indicate that it is likely to have certain characteristics (yellow peel, or sour taste in dilute water solution, as the case may be); but I also indicate that the presence of those characteristics, if they are present, is likely to be accounted for by some 'essential nature' which the thing shares with other members of the natural kind. What the essential nature is is not a matter of language analysis but of scientific theory construction; today we would say it was chromosome structure, in the case of lemons, and being a proton-donor, in the case of acids.[68]

The crucial point here – as against descriptivist theories – is Putnam's claim that getting things right in this regard is 'not a matter of linguistic analysis' but rather of 'scientific theory construction'. For what ultimately fixes the reference of terms such as 'lemon' and 'acid' is their *real* as opposed to their *nominal* essence, or the underlying structures in virtue of which they are correctly picked out as genuine samples of the kind, as distinct from the past or present range of (perhaps inadequate) descriptive criteria by which we and others have so far happened to define them. This is why – to repeat the passage from McCulloch – 'our prescientific understanding of substance-words is supposed to be already sensitive to future or possible scientific discovery'.[69] It is also what Putnam intends to convey by his homely dictum that 'meanings just ain't in the head'.[70]

This argument gains a good deal of credibility when applied to issues in the history and philosophy of science. Thus, for instance, it explains why scientists were justified in adopting an instrumentalist outlook with regard to the atomist hypothesis just so long as that hypothesis was based very largely on conjecture or metaphysical speculation, and as yet lacked the strong theoretical warrant – and at last the observational evidence – which accrued to it from the mid-nineteenth century on. More precisely, they were justified in taking this position from their own epistemological standpoint (given the state of knowledge at the time), although it can also be said that their use of the term 'atom' was 'already sensitive to future or possible scientific discovery', and should therefore be seen as strongly borne out by the subsequent history of atomic and subatomic physics. The same goes for later advances in knowledge – with regard to (say) electrons, protons, positrons, or neutrinos – where the existence of some yet more elusive particle was at first a matter of informed theoretical conjecture (in the case of neutrinos arrived at through detection of unexplained energy loss) and was then subject to a range of more decisive observational-predictive tests. What is far from clear is why quantum mechanics should be thought of as marking the end of the road so far as such progress is concerned.

J.S. Bell makes this point with his usual good sense in relation to the de Broglie/Bohm pilot-wave theory and the strong reluctance, on the part of so many physicists, to accept it as a possible (indeed the most likely) solution to the quantum paradoxes. Thus:

> [w]hile the founding fathers agonized over the question: 'particle' *or* 'wave'? de Broglie in 1925 proposed the obvious answer: 'particle' *and* 'wave'. Is it not clear from the smallness of the scintillation on the screen that we have to do with a particle? And is it not clear, from the diffraction and interference patterns, that the motion of the particle is directed by a wave? De Broglie showed in detail how the motion of a particle, passing through just one of two holes in a screen, could be influenced by waves propagating through both holes. And so influenced that the particle does not go where the waves cancel out, but is attracted to where they cooperate. This idea seems to me so natural and simple, to resolve the wave–particle dilemma in such a clear and ordinary way, that it is a great mystery to me that it was so generally ignored.[71]

What seems to have created this widespread resistance to de Broglie's theory – as likewise to Bohm's elaboration of it – is the deep-laid orthodox QM belief that no such solution could possibly be forthcoming given the inherent strangeness of quantum phenomena and their non-translatability into the language or conceptual framework of classical physics. There is a parallel here with Locke's scepticism as regards the possibility of our ever being able to make the decisive advance from 'nominal' to 'real' definitions or essences. In a well-known passage of the *Essay*, Locke justifies this sceptical attitude by pointing to the current state of the physical sciences, chemistry in particular, and remarking that despite all the recent advances scientists still had to work with nominal definitions and appeared no closer to explaining the ultimate nature of things.[72] From one point of view – that later adopted by Hume and his sceptical progeny – this was quite simply the condition of all human knowledge and could never be resolved or overcome by any advance towards a better understanding of those real definitions or essences. From another – that espoused by the early Putnam and by most scientific realists – it is absurd to suppose that such philosophic problems should remain unaffected by subsequent progress in our knowledge of the various essences (e.g. molecular and subatomic structures) that have shown Locke's attitude to reflect nothing more than the limits of physical science in his time.[73] On this account understanding 'tracks real essence', even if it points beyond the current best scientific theories or the furthest capacities of presently attainable experiment and observation.

Thus it is fair to say that Mach erred on the cautious side – or adopted an overscrupulous attitude – in refusing to credit the existence of atoms at a time when the evidence tended so strongly to support a realist view. Where instrumentalism had once been a sensible policy (given the lack of such evidence), it later became something more like a sceptical *parti pris* maintained as a matter of doctrinal commitment and increasingly at odds with the working assumptions of most physicists. Einstein himself started out by espousing a Machian instrumentalist approach according to which the chief demands of a scientific theory – such as special relativity – were that it should be observationally and predictively adequate without any further (ontological) commitment to the reality 'behind' phenomenal appearances.[74] However, he abandoned this doctrine in his later writings, chiefly on account of his deep dissatisfaction with the orthodox quantum theory and his belief that instrumentalism had often been used – by physicists like Bohr and Heisenberg – as a means of protecting that theory from any challenge on alternative (rational and realist) grounds.[75] Karl Popper repeated this charge in his book *Quantum Theory and the Schism in Physics*, where he traced the argument back to its sources in the way that religious apologists such as Osiander and Bellarmine had sought to defuse any threat posed by the heliocentric hypothesis by providing an instrumentalist gloss for the works of Copernicus and Galileo.[76] For no dispute need arise between Christian doctrine and the new science if the latter just renounced its ontological ambitions – its claim to describe the actual structure of the solar system – and presented itself rather as a purely instrumental means of 'saving the phenomena'.

Of course theological motives play no role in most modern versions of the instrumentalist thesis, apart from those few philosophers – such as Pierre Duhem – who have espoused it quite explicitly in the interests of avoiding any clash between science and religious belief.[77] On the other hand, as Popper is quick to point out, such arguments do leave room for just the kind of blandly accommodating strategy in face of acute doctrinal or interpretative problems that typified their early deployment. Thus it may seem ironic that a thinker such as Quine – committed as he is to a 'naturalized' epistemology or an outlook of hard-line physicalism – should find himself standardly linked with Duhem via the so-called 'Duhem–Quine thesis' with respect to ontological relativity.[78] But the coupling will perhaps appear less strange if one reflects on the extent to which Quine's doctrine – if taken *au pied de la lettre* – lends itself to uses that can easily admit all manner of 'strong' relativist proposals for making adjustments at this or that point in the overall 'web of belief', and thereby avoiding any threatened clash between rival paradigms, conceptual schemes, interpretative frameworks, etc.[79] In short, instrumentalism in its various modern forms still carries something of its old capacity to head off awkward questions by adopting the line that science should properly rest content with 'saving the phenomena' and should not get drawn into issues of a deeper (ontological or causal-explanatory) import.

It is just this strategy that is ruled out by the externalist theory of reference, that is to say, the argument that understanding 'tracks real essence' and that 'some things which are external to the human individual are not exterior to the mind'.[80] For it can then be maintained – as against current instrumentalist doctrines or Quinean appeals to ontological relativity – that what ultimately decides the truth value of our various theories, hypotheses, observation statements, and so forth, is the extent to which their terms make reference to real-world (physically existent) objects and their essential properties which in turn serve to fix the appropriate criteria for using those terms. Thus it may be rational at certain stages in the development of this or that science – say chemistry in Locke's time – for scientists to adopt an instrumentalist position, or warn against the fallacy of misplaced concreteness involved in attempts to pass from 'nominal' to 'real' definitions or essences. But this caution is no longer justified when the joint results of empirical observation and more advanced theoretical understanding have brought it about that science is able to explain a whole range of phenomena that would lack an adequate causal-explanatory account were it not for the existence of just those underlying structures, properties, causal dispositions, etc. This applies even more to present-day philosophers of science – such as Bas van Fraassen – who erect Locke's scruple into a full-scale programme of 'constructive empiricism', i.e. the doctrine that science should *always on principle* rest content with saving phenomenal appearances, eschewing talk of causes or 'laws of nature', and above all imposing a strict veto on claims concerning the existence or reality of 'unobservable' items.[81] For such arguments conspicuously fail to explain (1) how genuine advances come about in our knowledge of just such entities, and (2) how the reference of terms like 'molecule', 'atom' or 'electron' can be construed as genuinely truth-tracking, or 'already sensitive to future or possible scientific discovery'.[82] In other words they represent an attitude of a priori scepticism that controverts all the evidence of scientific progress to date as well as pre-emptively concluding against the realist principle of inference to the best (most causally adequate) explanation.

V

Quantum mechanics is a much-discussed test case here since it has seemed to involve such a drastic challenge to precisely those principles of scientific realism and causal-explanatory warrant. Thus Putnam, for one, can now be found citing the 'evidence' of strange quantum phenomena (such as nonlocality and wave–particle superposition) as an argument against the kind of realist ontology that formed the basis of his own early work in philosophical semantics and philosophy of science. So it is that he takes Hacking to task – in a passage I have cited already – for assuming (naively) that one can talk about 'particles' as if they were numerically distinct or possessed determinate position and momentum, rather than 'existing' only as wave-like probability functions or as 'spread over' some space-like distribution of possible coordinate values.[83] This argument is typical of the way that certain (albeit deeply puzzling) aspects of quantum physics have been made to do service in the cause of various instrumentalist, anti-realist, or van-Fraassen-type 'constructive empiricist' approaches. However, there is something decidedly suspect about taking a branch of science so rife with unresolved problems and paradoxes and using it to cast doubt on principles of scientific method that have elsewhere proven their validity beyond reasonable question. What confuses matters still more is the constant slippage from one to another interpretation of QM theory and its supposed consequences for our understanding of science and the physical world. This ambivalence is especially marked in Bohr's writings on the topic where it is often impossible to decide just where he wants to draw the line between ontological and epistemological issues. Thus his philosophizing tends to slide imperceptibly from questions of the type: 'what is it *in the nature* of quantum "reality" that makes it appear so strange from our classically-informed scientific standpoint?', to questions of the type, 'What is it about our classically-informed concepts and categories that prevents us from understanding the nature of quantum reality'?[84] And this latter line of argument sometimes leans over into the quasi-mystical idea that quantum 'reality' is indeed so unthinkably strange – so remote from our utmost powers of conceptual grasp – that it points beyond the limits of human reason to a realm of deep paradoxical truths which inherently elude the logic of classical (bivalent) truth and falsehood.

Hence the debate among philosophers of science – Putnam included – with regard to the possibility of devising an alternative or 'deviant' quantum logic that would accommodate the wave–particle dualism and other such anomalous findings.[85] Hence also Bohr's well-known proposal that we should give up the language of univocal state descriptions and adopt a generalized 'principle of complementarity' that would allow us to avoid forcing the issue between rival (incommensurable) theories or descriptive frameworks.[86] Perhaps this 'solution' may be thought preferable to the kind of irrationalist afflatus that sometimes overtakes zealous exponents and which – as some commentators have argued – can plausibly be traced back to certain aspects of the cultural-intellectual climate in Europe during the period after 1918.[87] However, there are other, more decisive reasons for rejecting the claim that these interpretative problems with quantum theory must be taken to necessitate a drastic change in our conception of physical reality, our canons of logical reasoning, or our very idea of scientific knowledge as a rational and truth-seeking enterprise. One is the existence of a rival interpretation (Bohm's 'hidden-variables' theory) which resolves most of the conceptual dilemmas bequeathed by the orthodox account while also explaining why the EPR–Bell paradox of remote simultaneous 'entanglement' between widely separated particles entails no necessary conflict with the principles of special relativity or indeed those of causal realism on a suitably modified (i.e. nonlocal) construal.[88] Another is the fact that quantum mechanics has made possible a range of uniquely convincing explanations for phenomena that would otherwise be quite beyond the reach of scientific understanding. These include signal advances in the theory of spectral analysis, chemical bonding, condensed-state or quantum coherence phenomena (such as superconductivity and superfluidity), and other fields of research.[89] For those advances cannot be explained except on the assumption that quantum mechanics *does* in fact capture some salient aspects of subatomic reality and, moreover, that the various interpretative problems are artefacts of our currently limited understanding rather than ultimate mysteries residing in the quantum-physical nature of things. Otherwise we should have to admit what Einstein found so unthinkable, namely the prospect of a radical break with the entire existing structure of scientific knowledge, from the ground rules of logic or evidential reasoning to the basic idea that there exists a (largely) mind-independent physical object domain which decides the truth value of our various statements, theories, or conjectures concerning it.

Thus the most rational option is to think of present-day quantum theory by analogy with the atomic-molecular hypothesis at a time – around the mid-nineteenth century – when as yet that hypothesis was open to challenge on respectable scientific grounds despite its good standing as a matter of indirect warrant or inference to the best explanation. Clearly there is a sense in which quantum mechanics has to be accepted as the sole current theory capable of 'explaining' a whole wide range of well-attested physical phenomena, from wave–particle dualism to the series of later results thrown up by EPR–Bell type delayed-choice experiments to test the existence of remote simultaneous correlation effects.[90] In light of all this – not to mention the spectacular success of various present-day advanced technologies developed on quantum principles – it could scarcely be rational to reject QM altogether as a false theory or one that lacks adequate empirical warrant. However, there remains a considerable gap between acceptance of the theory in its basic (uninterpreted or ontologically neutral) form and the kinds of strictly metaphysical construal that have been placed upon it by thinkers of various persuasion. These latter range all the way from Bohr's complementarity doctrine to various forms of anti-realism or framework relativism, among them Putnam's argument – *contra* Hacking – that quantum mechanics has undermined the case for any realist ontology premised on the notion of an objective, observer-independent physical domain.

What is especially striking about Putnam's claim is the fact that it marks such a drastic change of mind when compared with his earlier causal-realist position on issues in ontology, epistemology, and philosophical semantics. Of course, as Putnam sees it, this is not so much a philosophical retreat as a signal advance in his thinking brought about by reflection on the problems with any realist approach to these issues, whether at the subatomic (quantum) level or indeed as applied to objects and events in the macrophysical realm. For it is now his firm conviction that there is simply no escape from the logic of anti-realism – i.e. the manifest impossibility of our ever having access to a world 'outside' or 'beyond' our epistemic ken – except by endorsing the internalist case that any such 'reality' must always be construed as relative to this or that framework, paradigm, conceptual scheme, or whatever.[91] Still one may reply that Putnam senior has somehow missed or forgotten the point of his own strongest arguments against this form of (so-called) 'internal realism'. That point – briefly put – is the realist case that our claims are more or less warranted depending on the extent to which our terms, descriptions, conceptual schemes (and so forth) actually manage to specify or 'track' those features of objective reality that must finally decide what shall count as a true or scientifically adequate theory.

Richard Boyd puts the realist case to telling effect when he argues (in company with others like Hartry Field) that the use of certain referential terms is subject to a process of ongoing conceptual refinement that allows us – in retrospect – to distinguish various stages in the emergence of any such candidate theory.[92] Thus there may be cases of 'partial reference' where the term in question is vaguely defined or where it later shows up – as for instance by comparing Newton's with Einstein's definitions of 'mass' – as a usage that is valid only within some restricted referential domain but which none the less possesses a truth value when those operative limits are kept clearly in mind. In other words, we need not be driven to a Kuhnian relativist conclusion by reflection on the fact that a term like 'mass' underwent such a sizeable shift of meaning in the passage from Newtonian to Einsteinian physics. Rather, we should see that the different senses of the terms – 'rest mass', 'absolute mass', and 'relative mass' – can be separated out with some degree of precision in their various contexts of usage and can thus provide a means of comparing and contrasting the two theories in point of their conceptual adequacy. Thus, according to Boyd, we should view this as part of 'the ongoing process of continuous accommodation of language to the world in the light of new discoveries about [the world's] causal powers'.[93] And again, we should adopt as a general maxim: '[a]lways inquire, in the light of the best available knowledge, in what ways your current beliefs about the world might plausibly be incomplete, inadequate, or false, and design observations or experiments with the aim of detecting and remedying such possible defects'.[94] This combines the two basic realist principles: (1) that reality is 'verification-transcendent' in the sense that it – rather than our current best theories or means of observation – sets the standards for what should ultimately count as a true or valid belief; and (2) that the content (or truth value) of our various descriptive-explanatory schemes is relative *not* to the belief system currently or locally in place but rather to the best available theory once all the evidence is in.

So it is – in the words of S.P. Mohanty, drawing on Field and Boyd – that there occur 'crucial instances of "partial" denotation, where terms in a particular (scientific) theory refer imprecisely to the world, but the imprecision is removed through advances in the science rather than through refinement of our purely linguistic habits and practices'.[95] This was also

Putnam's cardinal point in linking the progress of scientific knowledge to our better understanding of the real (depth-ontological) nature of things as distinct from the various nominal definitions that left no room for evaluative comparison between theories, and which thus opened the way to Kuhnian or other such cultural-relativist doctrines. It is ironic, therefore, that he should now feel compelled to renounce that position for reasons that appear to derive at least partly from his thinking about issues in the quantum-theoretical domain. For if there is one principle that emerges clearly from Putnam's earlier work, it is that the truth content of scientific theories is determined by the way things stand in reality rather than those theories themselves deciding what shall count as 'reality' or 'truth' relative to some framework, paradigm, conceptual scheme, or whatever. To take quantum theory – on the orthodox (Copenhagen) account – as entailing a shift from the former to the latter point of view can scarcely be warranted given the signal lack of any current interpretation that comes even close to resolving these deep-laid epistemological problems.

Endnotes

1   This shift comes most clearly into view if one compares the realist position adopted in Putnam's *Philosophical Papers*, Vols. 1 and 2 (Cambridge: Cambridge University Press, 1975) with the 'transitional' phase represented by Vol. 3, *Realism and Reason* (Cambridge, 1983), and then with such recent works as *The Many Faces of Realism* (La Salle, IL: Open Court, 1987), *Realism With a Human Face* (Cambridge, MA: Harvard University Press, 1990) and *Renewing Philosophy* (Harvard University Press, 1992).

2   See for instance Niels Bohr, *Atomic Theory and the Description of Nature* (Cambridge: Cambridge University Press, 1934); *Atomic Physics and Human Knowledge* (New York: Wiley, 1958); Werner Heisenberg, *Physics and Philosophy* (New York: Harper & Row, 1958); Patrick A. Heelan, *Quantum Mechanics and Objectivity: a study of the physical philosophy of Werner Heisenberg* (The Hague: Martinus Nijhoff, 1965); John Honner, *The Description of Nature: Niels Bohr's philosophy of quantum physics* (Oxford: Clarendon Press, 1987).

3   See especially David Z. Albert, *Quantum Mechanics and Experience* (Cambridge, MA: Harvard University Press, 1993); J.S. Bell, *Speakable and Unspeakable in Quantum Mechanics: collected papers on quantum philosophy* (Cambridge: Cambridge University Press, 1987); Arthur Fine, *The Shaky Game: Einstein, realism, and quantum theory* (Chicago: University of Chicago Press, 1986); David Bohm, *Causality and Chance in Modern Physics* (London: Routledge & Kegan Paul, 1957); Bohm and B.J. Hiley, *The Undivided Universe: an ontological interpretation of quantum theory* (London: Routledge, 1993); James T. Cushing, *Quantum Mechanics: historical contingency and the Copenhagen interpretation* (Chicago: University of Chicago Press, 1994); Evadro Agazzi (ed.), *Realism and Quantum Physics* (Amsterdam: Rodopi, 1997); Tim Maudlin, *Quantum Non-Locality and Relativity: metaphysical intimations of modern science* (Oxford: Blackwell, 1993); Karl Popper, *Quantum Theory and the Schism in Physics* (London: Hutchinson, 1982); Michael Redhead, *Incompleteness, Nonlocality and Realism: a prolegomenon to the philosophy of quantum mechanics* (Oxford: Clarendon, 1987).

4   Putnam, 'Is Semantics Possible?' and 'Meaning and Reference', in Stephen Schwartz (ed.), *Naming, Necessity, and Natural Kinds* (Ithaca, NY: Cornell University Press, 1977), pp. 102–18 and 119–32.

5 Putnam, 'Meaning and Reference' (op. cit.).
6 Ibid, p. 121.
7 Ibid, p. 122.
8 Ibid, p. 123.
9 Ibid, p. 123.
10 Ibid, p. 124.
11 See Gottlob Frege, 'On Sense and Reference', in Max Black and P.T. Geach (eds.), *Selections from the Philosophical Writings of Gottlob Frege* (Oxford: Blackwell, 1952), pp. 56–78; also Bertrand Russell, 'On Denoting', in *Mind*, Vol. 14 (1905), pp. 479–93.
12 Ludwig Wittgenstein, *Philosophical Investigations*, trans. G.E.M. Anscombe (Oxford: Blackwell, 1953).
13 Putnam, 'Meaning and Reference' (op. cit.), p. 125.
14 Gregory McCulloch, *The Mind and Its World* (London: Routledge, 1995).
15 Wittgenstein, *Philosophical Investigations* (op. cit.), Sections 242–315.
16 See also Alvin Goldman, *Epistemology and Cognition* (Cambridge, MA: Harvard University Press, 1986) and *Empirical Knowledge* (Berkeley & Los Angeles: University of California Press, 1988); Hilary Kornblith, *Inductive Inference and its Natural Ground* (Cambridge, MA: M.I.T. Press, 1993) and Kornblith (ed.), *Naturalizing Epistemology* (M.I.T. Press, 1985).
17 McCulloch, *The Mind and Its World* (op. cit.), p. 189.
18 Among recent discussions of this topic, see especially A.C. Grayling, *The Refutation of Scepticism* (London: Duckworth, 1985); Christopher Hookway, *Scepticism* (London: Routledge, 1992); Alan Musgrave, *Common Sense, Science and Scepticism: a historical introduction to the theory of knowledge* (Cambridge: Cambridge University Press, 1993); Barry Stroud, *The Significance of Philosophical Scepticism* (Oxford: Oxford University Press, 1984).
19 Michael Williams, *Unnatural Doubts: epistemological realism and the basis of scepticism* (Princeton, NJ: Princeton University Press, 1996), p. 56.
20 Ibid, p. 74.
21 McCulloch, *The Mind and Its World* (op. cit.), p. 189.
22 Nicholas Rescher, *Scientific Realism: a critical reappraisal* (Dordrecht: D. Reidel, 1987), p. 61.
23 McCulloch, op. cit., p. 163.
24 Putnam, 'Is Semantics Possible?' (op. cit.).
25 For further discussion, see J. Aronson, R. Harré and E. Way, *Realism Rescued: how scientific progress is possible* (London: Duckworth, 1994); David Hamlyn, *Experience and the Growth of Understanding* (London: Routledge & Kegan Paul, 1978); Imre Lakatos and Alan Musgrave (eds.), *Criticism and the Growth of Knowledge* (Cambridge: Cambridge University Press, 1970); Karl R. Popper, *Objective Knowledge* (Oxford: Clarendon Press, 1972); Nicholas Rescher, *Scientific Progress* (Oxford: Blackwell, 1979); Wesley C. Salmon, *Scientific Explanation and the Causal Structure of the World* (Princeton, NJ: Princeton University Press, 1984); Peter J. Smith, *Realism and the Progress of Science* (Cambridge University Press, 1981).
26 See Note 1, above.
27 For the most influential statement of this anti-realist case, see Michael Dummett, *Truth and Other Enigmas* (London: Duckworth, 1978); also Crispin Wright, *Realism, Meaning and Truth* (Oxford: Blackwell, 1987) and – in a somewhat more qualified vein – *Truth and Objectivity* (Cambridge, MA: Harvard University Press, 1992); Michael Luntley, *Language, Logic and Experience: the case for anti-realism* (Duckworth, 1988); N. Tennant, *Anti-Realism and Logic* (Oxford: Clarendon, 1987); Kenneth P. Winkler, 'Scepticism and Anti-Realism', *Mind*, Vol. 94 (1985), pp. 36–52; Timothy Williamson, 'Knowability and Constructivism: the logic of anti-realism', *Philosophical Quarterly*, Vol. 38 (1988), pp. 422–32. For opposing arguments see

especially D.M. Armstrong, *Universals and Scientific Realism*, 2 vols. (Cambridge: Cambridge University Press, 1978); Michael Devitt, *Realism and Truth*, 2nd edn (Oxford: Blackwell, 1987); Frank B. Farrell, *Subjectivity, Realism and Postmodernism: the recovery of the world in recent philosophy* (Cambridge University Press, 1996); Christopher Norris, *New Idols of the Cave: on the limits of anti-realism* (Manchester: Manchester University Press, 1997); Smith, *Realism and the Progress of Science* (op. cit.); Gerald Vision, *Modern Anti-Realism and Manufactured Truth* (London: Routledge, 1988).

28 Cited in Williams, *Unnatural Doubts* (op. cit.), pp. 43 and 226.

29 See entries under Notes 18 and 19, above.

30 Hilary Putnam, *Reason, Truth and History* (Cambridge: Cambridge University Press, 1981).

31 Ibid, p. 49.

32 See entries under Note 1, above.

33 Putnam, *Reason, History and Truth* (op. cit.), p. 15.

34 See for instance Peter Clark and Bob Hale (eds.), *Reading Putnam* (Oxford: Blackwell, 1993).

35 See entries under Note 16, above.

36 Plato, *Theaetetus*, trans. Robin A.H. Waterfield (Harmondsworth: Penguin, 1987).

37 See especially R.M. Chisholm, *Theory of Knowledge*, 2nd edn (Englewood Cliffs, NJ: Prentice-Hall, 1977); Keith Lehrer, *Theory of Knowledge* (London: Routledge, 1990); Paul K. Moser, *Knowledge and Evidence* (Cambridge: Cambridge University Press, 1989); Moser and A. van der Nat (eds.), *Human Knowledge: classical and contemporary approaches* (New York: Oxford University Press, 1987).

38 Edmund L. Gettier, 'Is Justified True Belief Knowledge?', in *Analysis*, Vol. 23 (1963), pp. 121–3; also Robert Shope, *The Analysis of Knowledge* (Princeton, NJ: Princeton University Press, 1983).

39 McCulloch, *The Mind and Its World* (op. cit.), p. 174.

40 See especially David Deutsch, *The Fabric of Reality* (Harmondsworth: Penguin, 1997); also Bryce S. DeWitt and Neill Graham (eds.), *The Many-Worlds Interpretation of Quantum Mechanics* (Princeton, NJ: Princeton University Press, 1973); J.A. Wheeler and W.H. Zurek (eds.), *Quantum Theory and Measurement* (Princeton University Press, 1983).

41 See Notes 14 and 16, above.

42 Williams, *Unnatural Doubts* (op. cit.), p. 356.

43 See James Robert Brown, *The Laboratory of the Mind: thought experiments in the natural sciences* (London: Routledge, 1991); Roy Sorensen, *Thought Experiments* (Oxford: Oxford University Press, 1992); Albert Einstein, B. Podolsky and N. Rosen, 'Can Quantum-Mechanical Description of Reality be Considered Complete?', *Physical Review*, series 2, Vol. 47 (1935), pp. 777–80; Niels Bohr, article in response under the same title, *Physical Review*, Vol. 48 (1935), pp. 696–702; Bohr, 'Conversation with Einstein on Epistemological Problems in Atomic Physics', in P.A. Schilpp (ed.), *Albert Einstein: philosopher–scientist* (La Salle, IL: Open Court, 1969), pp. 199–24.

44 See especially Alexandre Koyré, *Galilean Studies*, trans. J. Mepham (Brighton: Harvester, 1978).

45 See Note 43, above; also J.S. Bell, *Speakable and Unspeakable in Quantum Mechanics: collected papers on quantum philosophy* (Cambridge: Cambridge University Press, 1987); James T. Cushing and Ernan McMullin (eds.), *Philosophical Consequences of Quantum Mechanics: reflections on Bell's Theorem* (Notre Dame, IN: University of Notre Dame Press, 1989); A. Aspect, P. Grangier and C. Roger, 'Experimental Realization of the E–P–R Paradox', *Physical Review*, Vol. 48 (1982), pp. 91–4.

46 See for instance – from a range of philosophic and disciplinary perspectives – Jerome S. Bruner, *Actual Minds, Possible Worlds* (Cambridge, MA: Harvard

University Press, 1986); Jon Elster, *Logic and Society*: *contradictions and possible worlds* (Chichester: Wiley, 1978); Rom Harré and E.H. Madden, *Causal Powers* (Oxford: Blackwell, 1975); Geoffrey Hawthorn, *Plausible Worlds*: *possibility and understanding in history and the social sciences* (Cambridge: Cambridge University Press, 1991); David Lewis, *Counterfactuals* (Oxford: Blackwell, 1973); M. Loux (ed.), *The Possible and the Actual* (Ithaca, NY: Cornell University Press, 1979); Alvin Plantinga, *The Nature of Necessity* (Oxford: Oxford University Press, 1974); Ernest Sosa (ed.), *Causation and Conditionals* (Oxford University Press, 1975); Robert C. Stalnaker, *Inquiry* (Cambridge, MA: M.I.T. Press, 1987).

47 McCulloch, *The Mind and Its World* (op. cit.), p. 174.

48 Putnam, *Pragmatism*: *an open question* (Oxford: Blackwell, 1995); Ian Hacking, *Representing and Intervening*: *introductory topics in the philosophy of natural science* (Cambridge: Cambridge University Press, 1983).

49 Hacking, *Representing and Intervening* (op. cit.), p. 23.

50 See Notes 2 and 3, above; also Harvey R. Brown and Rom Harré (eds.), *Philosophical Foundations of Quantum Field Theory* (Oxford: Clarendon Press, 1988); Paul Teller, *An Interpretive Introduction to Quantum Field Theory* (Princeton, NJ: Princeton University Press, 1994).

51 Putnam, *Pragmatism*: *an open question* (op. cit.), p. 59.

52 Ibid, p. 59.

53 See especially Michael Devitt, *Realism and Truth* (op. cit.); also entries under Notes 16 and 25, above.

54 See Note 4, above.

55 John Locke, *An Essay Concerning Human Understanding*, 2 vols. (New York: Dover, 1959); also M.R. Ayers, 'Locke versus Aristotle on Natural Kinds', *Journal of Philosophy*, Vol. 77 (1981), pp. 247–72.

56 Putnam, 'Is Semantics Possible?' (op. cit.), p. 105.

57 See especially Martin Gardner, 'Realism and Instrumentalism in Nineteenth-Century Atomism', *Philosophy of Science*, Vol. 46 (1979), pp. 1–34.

58 J. Perrin, *Atoms*, trans. D.L. Hammick (New York: Van Nostrand, 1923); also Mary Jo Nye, *Molecular Reality* (London: MacDonald, 1972); Salmon, *Scientific Explanation and the Causal Structure of the World* (op. cit.); Robin Waterfield, *Before Eureka*: *the presocratics and their science* (Bristol: The Bristol Press, 1989).

59 See Ernst Mach, *The Science of Mechanics*: *a critical and historical account of its development*, trans. T.J. McCormack (La Salle, IL: Open Court, 1960); also – for defences and criticisms of the instrumentalist approach – Pierre Duhem, *To Save the Phenomena*: *an essay on the idea of physical theory from Plato to Galileo*, trans. E. Dolan and C. Maschler (Chicago: University of Chicago Press, 1969); C.J. Misak, *Verificationism*: *its history and prospects* (London: Routledge, 1995); Bas C. van Fraassen, *The Scientific Image* (Oxford: Clarendon Press, 1980) and *Quantum Mechanics*: *an empiricist view* (Clarendon Press, 1992).

60 See especially Misak, *Verificationism* (op. cit.); also Oswald Hanfling (ed.), *Essential Readings in Logical Positivism* (Oxford: Blackwell, 1981); Nicholas Rescher (ed.), *The Heritage of Logical Positivism* (Lanham: University Press of America, 1985).

61 See Note 11, above.

62 See Wittgenstein, *Philosophical Investigations* (op. cit.) and *On Certainty*, ed. G.E.M. Anscombe and G.H. von Wright (Oxford: Blackwell, 1969); also David Bloor, *Wittgenstein*: *a social theory of knowledge* (New York: Columbia University Press, 1983); Derek L. Phillips, *Wittgenstein and Scientific Knowledge*: *a sociological perspective* (London: Macmillan, 1977).

63 The speaker on this occasion was Sean Sayers, the event a University of Wales staff/student colloquium at Gregynog, and the line of questioning mainly pursued by members of the Swansea Philosophy Department. Should this footnote by any chance come to their notice then let me recommend a careful reading of

Sayers, *Realism and Reason: dialectic and the theory of knowledge* (Oxford: Blackwell, 1985).

64 See Note 62, above; also Wittgenstein, *Remarks on Frazer's Golden Bough*, ed. Rush Rhees, trans. A.C. Miles (Nottingham: Brynmill Press, 1979) and *Culture and Value*, ed. R. Rhees (Oxford: Blackwell, 1980); Peter Winch, *The Idea of a Social Science and its Relation to Philosophy* (London: Routledge & Kegan Paul, 1958); Norman Malcolm, *Wittgenstein: a religious point of view?* (London: Macmillan, 1993).

65 See for instance Bertrand Russell, *The Scientific Outlook* (London: Allen & Unwin, 1931); *Human Knowledge: its scope and limits* (New York: Simon & Schuster, 1948); *On the Philosophy of Science* (Indianapolis: Bobbs-Merrill, 1965).

66 See entries under Notes 12, 62 and 64, above; also Wittgenstein, *Tractatus Logico-Philosophicus*, trans. D.F. Pears and B.F. McGuiness (London: Routledge & Kegan Paul, 1961); Norman Malcolm, *Nothing is Hidden: Wittgenstein's criticism of his early thought* (Oxford: Blackwell, 1986).

67 See especially G.P. Baker and P.M.S. Hacker, *Language, Sense and Nonsense: a critical investigation into contemporary theories of language* (Oxford: Blackwell, 1984).

68 Putnam, 'Is Semantics Possible?' (op. cit.), p. 104.

69 McCulloch, *The Mind and its World* (op. cit.), p. 163.

70 Putnam, 'Meaning and Reference' (op. cit.), p. 124.

71 Bell, *Speakable and Unspeakable in Quantum Mechanics* (op. cit.), p. 191.

72 Locke, *An Essay Concerning Human Understanding* (op. cit.), Book III, Chapter vi, Section 8.

73 See entries under Notes 16, 25 and 27, above.

74 See especially Einstein, 'Autobiographical Notes', in P.A. Schilpp (ed.), *Albert Einstein: philosopher–scientist* (op. cit.), pp. 3–105; also entries under Notes 43 and 45, above.

75 Einstein, 'Autobiographical Notes' (op. cit.).

76 Karl Popper, *Quantum Theory and the Schism in Physics* (London: Hutchinson, 1982); see also Note 59, above.

77 See Duhem, *To Save the Phenomena* (op. cit.); also *The Aims and Structure of Physical Theory*, trans. Philip Wiener (Princeton, NJ: Princeton University Press, 1954); Stanley L. Jaki, *Uneasy Genius: the life and work of Pierre Duhem* (The Hague: Martinus Nijhoff, 1984); R.N.D. Martin, *Pierre Duhem: philosophy and history in the work of a believing physicist* (La Salle, IL: Open Court, 1991).

78 See especially W.V. Quine, 'Two Dogmas of Empiricism', in *From a Logical Point of View*, 2nd edn (Cambridge, MA: Harvard University Press, 1961), pp. 20–46; also Sandra G. Harding (ed.), *Can Theories Be Refuted? essays on the Duhem–Quine thesis* (Dordrecht: D. Reidel, 1976).

79 See Christopher Norris, 'Ontological Relativity and Meaning-Variance: a critical-constructive review', in *Against Relativism: philosophy of science, deconstruction and critical theory* (Oxford: Blackwell, 1997), pp. 66–100.

80 McCulloch, *The Mind and its World* (op. cit.), p. 189.

81 See Note 59, above.

82 McCulloch, op. cit., p. 163.

83 See Note 48, above.

84 See Note 2, above; also *The Philosophical Writings of Niels Bohr*, 3 vols. (Woodbridge, CT: Ox Bow Press, 1987).

85 See especially Putnam, 'How to Think Quantum-Logically', *Synthèse*, Vol. 29 (1974), pp. 55–61 and his papers collected in *Mathematics, Matter and Method* (Cambridge: Cambridge University Press, 1979); also Enrico Beltrametti and Bas C. van Fraassen (eds.), *Current Issues in Quantum Logic* (New York: Plenum, 1981); G. Birkhoff and J. von Neumann, 'The Logic of Quantum Mechanics', *Annals of Mathematics*, Vol. 37 (1936), pp. 823–43; Rachel Wallace Garden, *Modern Logic and Quantum Mechanics* (Bristol: A. Hilger, 1983); Martin Gardner, 'Is

Quantum Logic Really Logic?', *Philosophy of Science*, Vol. 38 (1971), pp. 508–29; Susan Haack, *Deviant Logic: some philosophical issues* (Cambridge University Press, 1974); Peter Gibbins, *Particles and Paradoxes: the limits of quantum logic* (Cambridge University Press, 1987); Peter Mittelstaedt, *Quantum Logic* (Princeton, NJ: Princeton University Press, 1994).

86 See Note 2, above; also Henry J. Folse, *The Philosophy of Niels Bohr: the framework of complementarity* (Amsterdam: North-Holland, 1985).

87 See especially Paul Forman, 'Weimar Culture, Causality, and Quantum Theory, 1918–1927: adaptation by German Physicists and Mathematicians to a hostile intellectual environment', in *Historical Studies in the Physical Sciences*, Vol. 3 (1971), pp. 1–115 and 'The Reception of an Acausal Quantum Mechanics in Germany and Britain', in S.H. Mauskopf (ed.), *The Reception of Unconventional Science, AAAS Selected Symposium*, No. 25 (1979), pp. 11–50; also James T. Cushing, *Quantum Mechanics: historical contingency and the Copenhagen hegemony* (op. cit.).

88 See Notes 2, 3 and 45, above.

89 See especially Euan Squires, *The Mystery of the Quantum World*, 2nd edn (Bristol and Philadelphia: Institute of Physics Publishing, 1994).

90 See Note 3, above.

91 See Note 1, above.

92 See especially Richard Boyd, 'Observation, Explanatory Power, Simplicity: toward a non-Humean account', in Peter Achinstein and Owen Hannaway (eds.), *Observation, Experiment, and Hypothesis in Modern Physical Science* (Cambridge, MA: M.I.T. Press, 1985) and 'Metaphor and Theory-Change: what is "metaphor" a metaphor for?', in A. Ortony (ed.), *Metaphor and Thought*, 2nd edn (Cambridge: Cambridge University Press, 1993), pp. 481–532; also Hartry Field, 'Theory Change and the Indeterminacy of Reference', *Journal of Philosophy*, Vol. 70 (1973), pp. 462–81.

93 Boyd, 'Metaphor and Theory-Change' (op. cit.), p. 523.

94 Ibid, p. 523.

95 Satya P. Mohanty, *Literary Theory and the Claims of History* (Ithaca, NY: Cornell University Press, 1997), p. 71.

7 Can logic be quantum-relativized?

Putnam, Dummett and the 'great quantum muddle'

I

It is often assumed – without much in the way of supporting argument – that quantum mechanics has thrown a huge problem into any philosophical defence of scientific realism that would claim to have kept abreast of developments in the physical sciences. If this premise is not made explicit, then it enters, none the less, as a kind of tacit presumption to the effect that values such as truth, objectivity, and rational warrant can be thought of only as relative (or 'internal') to some given theoretical framework or conceptual scheme. For we now have it on the word of many QM theorists that quantum phenomena are *irreducibly* uncertain or probabilistic; that there is no real-world (objective) physical domain apart from various localized acts of observation–measurement; and, moreover, that these problems cannot be treated as epistemological in nature (i.e. put down to the limits of our current best knowledge), but must rather be viewed as in some sense intrinsic to the very nature of quantum-physical 'reality'.[1] In which case – so the argument runs – we should accept something like the 'many-worlds' interpretation of quantum physics, if not in the full-fledged realist sense (i.e. that there *actually exist* as many parallel 'worlds' or 'universes' as alternative outcomes of every wavepacket collapse[2]), then at least in the sense that our world-versions will be just as many and various. According to some quantum theorists, David Deutsch among them, this latter is merely a kind of shifty instrumentalist tactic adopted so as to avoid facing up to the strictly inescapable (albeit massively counter-intuitive) truth of many-worlds QM.[3] For others of a more agnostic disposition – those in the orthodox line of descent from Bohr and Heisenberg – the quantum paradoxes remain deeply puzzling but are best managed by accepting just such an instrumentalist outlook and declining to speculate on the reality 'behind' quantum appearances.[4]

Even so, as Einstein was quick to object, this outlook involves a radical break with some chief principles of thought and method that had characterized the conduct of physical enquiry from Galileo down. In particular, it has to renounce the idea – strongly reaffirmed by Einstein in his series of debates with Bohr – that a 'complete' physical theory should do more than merely 'save the phenomena', or prove itself adequate in predictive–

observational terms.[5] That is, it should meet the twin requirements of (1) describing an objective, observer-independent physical reality, and (2) describing it in such a way as to pass beyond phenomenal appearances to a rational or causal-explanatory grasp of its underlying principles. Einstein had himself once espoused an instrumentalist position – much influenced by Ernst Mach – but later renounced it very largely as a result of his deep and abiding dissatisfaction with orthodox QM theory.[6] Nevertheless, it was the latter (Copenhagen-instrumentalist) approach that came to dominate thinking on this topic, not only among quantum physicists but also among those – philosophers included – who took QM to have raised insuperable problems for any realist ontology or epistemology. Whence the currently widespread idea that modern physics has broken altogether with the 'naive' objectivist belief in a world that somehow exists quite apart from the various theories, conceptual schemes, or interpretative frameworks that we bring to bear in our efforts to describe or explain it.

This latter line of argument is espoused by some Kuhnians – if not (or not without occasional qualms) by Kuhn himself[7] – and is pushed to an extreme by 'strong' descriptivists such as Richard Rorty and Nelson Goodman.[8] On their view, quite simply, there exist as many 'worlds' as there exist different languages or descriptive schemes by which to impose some selective ontology (e.g. those of Azande witchcraft, of 'commonsense' realism, Newtonian astronomy, or present-day particle physics) on the inchoate flux of experience. Others again – Quine and late Putnam among them[9] – purport to avoid such relativist extremes while none the less tending in a similar direction by adopting what Davidson famously criticized as the 'very idea of a conceptual scheme', or what Popper has more pithily described as the 'myth of the framework'.[10] Add to these the 'constructive-empiricist' approach of a philosopher of science like Bas van Fraassen – an anti-realist approach in all but name – and one will gain some idea of the range of arguments that are currently deployed to that effect.[11] Moreover, as I have said, it is often the case that quantum mechanics is invoked in support of various anti-realist positions. What suits it for this role is the way that it inverts the realist order of priority between ontology and epistemology by transferring the burden of unresolved doubts, paradoxes, uncertainties, etc., from our restricted *knowledge* of whatever transpires in the quantum-physical domain to the *very nature* of quantum 'reality' construed – in vaguely Kantian terms – as a noumenal thing-in-itself.[12] Whence the idea (congenial to some) that physics has itself at last come around to the kind of anti-realist or constructivist outlook that is something like the default philosophy of many thinkers in the present-day social and human sciences.

Among hermeneutic theorists especially there is a belief that quantum mechanics spells the end of that old 'metaphysical' conception of scientific method that strove to draw a firm categorical line between subject and object, mind and world, or epistemology and ontology.[13] We may recall, in this context, that Heidegger dedicated one of his works to Heisenberg; also that he wrote admiringly of Heisenberg and Bohr on account of their willingness to 'hold out in the questionable', that is, their acceptance of uncertainty and their courage in venturing beyond the inherited dualisms of Western techno-metaphysics.[14] So far as I know, there is nothing on record to tell us whether or not they appreciated the compliment. Still it is ironic that quantum mechanics on the orthodox (Bohr/Heisenberg) interpretation was itself closely tied to the logical-positivist programme which hermeneutic theorists – following Heidegger – have standardly viewed as the last chapter in that dead-end history of thought.[15] That is to say, it was precisely the positivist refusal to assign a realist ('metaphysical') content to empirical data or observation statements that was taken by physicists such as Bohr and Heisenberg as lending support to their own instrumentalist views.[16] As so often, there is a curious convergence of opposite extremes whereby a strongly reactive movement of thought – Heideggerian depth-hermeneutics – turns out to have sources and premises in common with that which it so strenuously disavows.[17] In this case the shared premise is a doctrinaire rejection of scientific realism that can lead either to the positivist veto on 'metaphysical' (e.g. causal-explanatory) talk or, conversely, to a Heidegger-inspired hermeneutic approach which thinks to go beyond the merely ontic domain of physical realia to a realm of authentic (depth-ontological) Being and truth. Moreover, there is the well-documented fact of Bohr's early attraction to the writings of Kierkegaard – via his teacher Harald Høffding – and his exposure to various irrationalist currents of thought that responded to the post-1918 crisis of confidence in science and technological progress.[18] At any rate, orthodox QM theory can be seen to have taken rise in a context of strong intellectual and cultural pressures which predisposed thinkers like Bohr and Heisenberg against any possible challenge mounted by alternative (realist) interpretations.[19]

From the realist viewpoint, on the other hand, there is something irrational in the very idea that a branch of science so fraught with unresolved puzzles and paradoxes should be thought to require a wholesale change in our basic conceptions of truth, knowledge, and physical reality. After all, it could well be argued that ever since its inception quantum theory has been in a state of more-or-less permanent crisis, a protracted version of what Kuhn describes as 'pre-revolutionary' science but without – as yet – any sign of a breakthrough to the new (post-revolutionary) paradigm. In this sense it resembles the state of knowledge with regard to chemistry when Locke concluded that 'real essences' were unknowable in their very nature and that science must therefore rest content with 'nominal' definitions or essences.[20] To the realist, Locke's attitude will appear fully justified in epistemological terms, i.e. as the most rational attitude to adopt at a time – the mid-to-late seventeenth century – when chemistry was as yet a fledgling science with highly restricted theoretical resources and powers of detailed observation. However, she will also wish to claim (speaking ontologically with benefit of scientific hindsight) that Locke's scepticism can *now* be seen as unwarranted in view of various later developments, among them the signal progress achieved in explaining chemical properties with reference to advances in molecular and atomic theory.

This is just what it means for terms such as 'molecule', 'atom', or 'electron' to pass from a stage of relatively ill-defined usage where their referential status is insecure and their warrant instrumental at best, to a stage of mature scientific understanding where they are taken to possess both realist credentials and strong descriptive-explanatory power.[21] It is also what is meant by the externalist claim that such terms are 'sensitive to future development' in so far as their true reference was *fixed all along* by certain objectively existent features of the world (natural-kind attributes, molecular structures, subatomic configurations, and so forth) which awaited subsequent discovery.[22] That is to say – *contra* Locke – we are not forever confined to knowledge of mere phenomenal appearances or nominal definitions, since there is a 'real essence' which those terms can properly be said to have tracked through the progress from nominal to real definition.[23] Moreover, we have every reason to believe that this process will continue beyond our present (no doubt very partial) state of scientific understanding. For anti-realists of various colour, this thought typically gives rise to a kind of sceptical meta-induction, namely the idea that since most scientific 'knowledge' up to now has turned out false or any rate subject to later refinement or correction, therefore it is overwhelmingly likely that the same must apply to our present-best theories, truth-claims, reality ascriptions, and so forth. But here again the realist will take just the opposite view, pointing out (in Nicholas Rescher's well-chosen phrase) that realism is perfectly compatible with – and indeed presupposes – the 'ontological non-finality of science as we have it'.[24] For it is precisely her point that we could have no grasp of such basic notions as scientific truth, knowledge, or progress if we lacked this sense of the standing possibility that a great deal of what we currently believe might yet be shown false (or inadequate) as a result of future developments.[25]

Hence – to repeat – the widely accredited shift from instrumentalism to realism as the dominant scientific outlook in respect of molecules, atoms, electrons, and more recently a whole new range of subatomic or subnuclear particles (such as quarks) whose existence is now strongly borne out by the best evidence to hand.[26] In other words, there comes a point where it is no longer rational – where it involves an excess of doctrinal scepticism or fixed anti-realist conviction – to persist with such scruples despite and against the advancement of scientific knowledge. But it is often claimed that quantum mechanics marks a radical break with this entire way of thinking about what constitutes 'knowledge' or 'progress' in the physical sciences. For here we have a theory of singular predictive and instrumental power, but one that has so far proved recalcitrant to any kind of realist or causal-explanatory treatment. In this unusual circumstance perhaps the most rational option is indeed to adopt a qualified instrumentalist outlook, one that withholds ontological commitment to the various descriptive terms ('wave', 'particle', 'superposition', etc.) which allow us to talk about quantum phenomena in a fairly intelligible way but which – like Locke's nominal definitions – cannot be thought to capture 'real essence' or to limn the ultimate features of subatomic reality. On one construal this is just the point of Bohr's own instrumentalist approach: to prevent us from confusing ontology with epistemology, or the question 'What exactly is the nature of quantum reality?' with the question 'How must we think about it in order to make adequate sense of the observed phenomena, statistical data, predictive hypotheses, and so forth?'.[27] However, Bohr's philosophy – and that of his followers – more often works out as a kind of hybrid instrumentalist-realist doctrine whereby these problems of interpretation are effectively raised into a full-scale quantum ontology which treats them as pertaining to the ultimate nature of things and hence as blocking any possible advance towards a properly realist alternative theory of the type proposed by dissident thinkers such as Bohm.[28] It thus goes against one of the basic precepts of scientific reasoning: that we should always, as a matter of rational procedure, attach less weight to theories or hypotheses whose terms are comparatively ill-defined than to those whose terms have achieved a high measure of referential or descriptive-explanatory warrant.

On this view there is no reason to accept the strangely inverted 'realist' logic whereby the puzzles of quantum nonlocality, superposition, and so forth are used as a basis on which to erect whole structures of arcane metaphysical speculation – such as Deutsch's many-worlds theory – which far exceed the available evidence.[29] Nor can erstwhile defenders of realism (Hilary Putnam prominent among them) be justified in backing off from that position partly in response to the sorts of problem thrown up by orthodox quantum mechanics.[30] For those problems are epistemological in nature, having to do with limits on our present-best powers of observation, measurement, or conceptual grasp. Thus they cannot be held up as counter-evidence to the ontological-realist case, i.e. that there exist certain properties of the quantum domain which objectively determine the truth value of our various statements concerning them *whatever our current state of knowledge in that regard*.[31] Then again, such statements may be 'undecidable' on the best evidence to hand. But this is no reason – as Popper argues – to accept Bohr's typically vague yet extravagant claim that such limits on our current knowledge 'entail the necessity of a final renunciation of the classical ideal ... and a radical revision of our attitude towards the problem of physical reality'.[32] Nor does it possess the kind of knock-down anti-realist force that Putnam clearly intends when he maintains – against Ian Hacking – that 'positrons do not in general have a definite number', that one can spray the niobium ball 'not with three positrons, and not with four positrons, but with a superposition of three and four positrons', and moreover 'that we cannot think of positrons as having trajectories or as being, in general, reidentifiable'.[33] For Putnam is here staking his case on one, highly debatable interpretation of some less-than-conclusive quantum 'evidence' which may perhaps complicate matters with regard to Hacking's choice of example but which cannot in any way undermine the case for a scientific realism premised on the existence of objective (verification-transcendent) truths. After all, we have much less reason to accept the completeness or finality of orthodox quantum theory – beset as it is with so many unresolved conceptual dilemmas – than to accept the basic realist proposition that there are still many truths about the physical world that exceed our present-day powers of scientific understanding. Indeed, it is quite possible that these will turn out to include certain truths about quantum physics which provide a solution to its long-standing problems, perhaps by offering decisive support for Bohm's hidden-variables theory.

This argument follows directly from Putnam's earlier externalist and causal-realist approach to issues in philosophy of science.[34] That he has now seen fit to abandon that approach is, I think, a strong indication of the extent to which anti-realism in various forms – and often under different names – has lately gained ground in many quarters of Anglo-American philosophic debate. Very often this persuasion has gone along with the appeal to quantum mechanics (on the orthodox construal) as a test case for realist claims and one which presumptively shows them up as unable to cope with the latest evidence provided by physical science. It seems to me that this case is far from clear and that it rests on just the kind of dubious warrant – the rush to far-reaching ontological claims on shaky epistemological grounds – that has typified much of the current debate about quantum theory and its supposed implications for our knowledge of the physical world. Such is the position of those, like Putnam, for whom the interpretative problems with orthodox QM have either prompted or at any rate strongly reinforced their rejection of a realist outlook. For these thinkers the conclusion is inescapable: if quantum mechanics is right (that is to say, empirically adequate and borne out by its high measure of predictive success), then there is simply no upholding the basic tenets of that classical realist worldview – endorsed by scientists from Galileo to Einstein – which assumed the existence of a physical object domain with determinate (observer-independent) values of location, momentum, numerical identity, and so forth. Hence Putnam's famous retreat from the externalist causal realism of his early writings to a position of (so-called) 'internal realism' according to which there exist as many possible valid descriptions or theories as there exist frameworks or object languages whereby to accommodate the range of putative realia.[35] This appears to be Quine's position also in those passages of his work – notably 'Two Dogmas of Empiricism' – where quantum mechanics is adduced in support of the argument from variant conceptual schemes to a doctrine of full-scale ontological relativity.[36]

These responses are basically sceptical in character and involve the idea that quantum physics has created insuperable problems for any realist theory (such as Putnam once espoused) where the truth value of scientific statements is a function of their actually *getting things right* with respect to various determinate aspects of the macro- and microphysical world. On the other hand, there are theorists – Deutsch among them – who accept QM on account of its massive empirical–predictive success but who refuse to adopt an instrumentalist line or to rule out the kinds of ontological issue (i.e. with regard to the ultimate reality 'behind' quantum appearances) which find no place in the orthodox theory.[37] Thus Deutsch is implacably realist when it comes to the existence – the physical reality – of those multiple branching 'universes' or counterpart worlds which supposedly spring into being whenever there occurs a collapse of the wavepacket or an instance of the shadow effect brought about by particle interference phenomena. What both parties – sceptics and realists – implicitly take for granted is the completeness of classical QM theory as a matter of basic physics, that is to say, the idea that any valid interpretation must either forego ontological commitments (since the theory is incompatible with realism in whatever guise), or envisage an alternative 'reality' which squares with the theory at no matter what cost in terms of far-fetched metaphysical conjecture. This 'completeness' argument of course goes back to Bohr's debates with Einstein where the issue was precisely whether orthodox QM might at length give way to some deeper, more adequate interpretation that would satisfy Einstein's realist requirements and avoid the various conceptual problems encountered with the standard model.[38] Thereafter it became an article of faith among upholders of the orthodox view – Heisenberg especially – that quantum mechanics was indeed 'complete' in the sense that no alternative account (such as Bohm's hidden-variables theory) could possibly support a realist interpretation while also matching the well-proven QM formalisms and predictions.

That this resistance went far beyond anything required on purely scientific or observational grounds is a fact borne out by Cushing's amply documented study and also by Popper's polemical but none the less detailed and philosophically astute treatment of the topic.[39] Thus Popper cites Heisenberg as more or less dismissing Einstein's objections since he (Einstein) 'did not want to acknowledge that quantum mechanics represented a finally valid, and even less a complete, description of these phenomena'.[40] Cushing likewise presents many statements which show how unthinkable it was – from an orthodox standpoint – that some alternative theory might yet supersede the established view or resolve its problems by producing a realist interpretation consistent with the QM observational–predictive results. This seems to me one of the great unsupported dogmas in recent philosophy of science and one that has exerted an influence far exceeding its rational or evidential warrant. Perhaps it is the case, as Popper writes, that on the standard version 'hidden variables could not exist in quantum mechanics or, according to a slightly different account, ... [that] the existence of hidden variables contradicted quantum mechanics'.[41] However, there is good reason to think that orthodox QM is an immature theory which lacks, as yet, an interpretation that would save it from conflict with the basic principles of scientific method and reasoning. No doubt any Bohm-type realist theory will need to take account of Bell's theoretical proof – since borne out by experiment – that local realism has to be abandoned on the evidence of remote particle 'entanglement' and other such quantum phenomena.[42] But it is also clear from Bohm's statements that this cannot be taken as a knock-down argument against the realist position just so long as the 'no first signal' rule obtains, that is to say, so long as there is no possibility of exploiting those phenomena in order to convey information at superluminal velocity between remotely situated observers.[43]

II

In short, realism remains very much a live option despite the many announcements to contrary effect that have issued from quantum physicists and theorists since Einstein's debates with Bohr. It is an option represented on the one hand by a theory such as Bohm's which challenges the supposed completeness or adequacy of orthodox QM, and on the other by a more fundamental argument concerning the existence of an objective (theory-independent) physical domain whose structure and attributes must be thought of as determining the truth or falsehood of our various scientific theories, including the various rival interpretations of quantum mechanics. On the first point Popper has a typically trenchant objection to the orthodox (Copenhagen) account: that its premises are self-contradictory and hence logically self-refuting. Thus he cites Heisenberg as stating in the plainest terms that 'the discontinuous "reduction of the wave packets" *which cannot be derived from Schrödinger's equation*, is ... a consequence of the transition from the possible to the actual' (Popper's italics).[44] But in that case what are we to make of Heisenberg's equally confident (not to say dogmatic) claim for the 'completeness' of orthodox QM theory? After all, as Popper remarks, 'it is certainly not possible to insist on the one hand that the formalism is complete and to insist on the other that its application to "the actual" actually demands a step *which cannot be derived from it*'.[45]

This is just one of the many contradictions, anomalies, non sequiturs, and instances of false or circular reasoning that Popper denominates the 'great quantum muddle', and which he finds constantly prefigured in the statements of Bohr and Heisenberg. Most often – he argues – they result from the confusion between epistemology and ontology, or the limits of our knowledge with regard to certain perplexing quantum phenomena and the notion that those limits are somehow intrinsic to the very nature of so-called quantum 'reality'. In Bohr's case – as Cushing shows in more detail – there was a gradual slide from the one to the other position, that is to say, from an argument that the QM paradoxes reflected our presently limited state of knowledge to an argument that they could not *conceivably* be resolved by any future scientific advance since they captured something irreducibly strange about the ultimate truth of quantum mechanics. Popper makes a similar point when he remarks on the dogmatic scepticism of orthodox QM theory and the way that it raises interpretative problems – problems peculiar to quantum mechanics on just this orthodox construal – into a full-scale doctrine concerning the limits of scientifically attainable knowledge. Thus:

> [i]f we wish to look more closely at this last and, I believe, most important issue – the problem of *understanding* our theories – then we may discern that the orthodox party represents, in its attitude towards quantum theory, a philosophical theory of the nature of science which implies the futility of the dissenters' attempt to understand. It is the view that there is nothing there to be understood: that we can do no more than *master the mathematical formalism*, and learn *how to apply it*.[46]

Popper calls this the 'end-of-the-road' hypothesis, the idea that orthodox QM must be treated as 'the last, the final, the never-to-be-transcended revolution in physics'. From which it follows – again on the orthodox account – that the uncertainty relations (as theorized by Heisenberg) must henceforth be taken as pertaining to quantum 'reality' and not to the limits of physically possible measurement or the restrictions imposed by our 'classical' framework of descriptive-explanatory concepts.[47] In Bohr's writing it is often hard to be sure just which interpretation is intended, although Cushing discerns a fairly marked shift from the 'weak' (epistemological) thesis to his 'strong' (ontological) version of the claim that quantum reality *cannot be other* than the way it is represented according to the orthodox theory. In which case – given the completeness assumption – it follows necessarily that the paradoxes attendant on orthodox QM cannot be resolved or 'transcended' by a deeper, more adequate interpretation involving the appeal to hidden variables or to anything that finds no place in the standard model.

Hence Heisenberg's complaint – cited by Popper – that Einstein, in mistakenly holding out for the possibility of just such a theory, 'did not want to acknowledge that quantum mechanics represented a finally valid, and even less a complete, description of these phenomena'.[48] I think that Popper is right to treat this as a curious (not to say perverse) doctrinal persuasion and one that contravenes some of the most basic principles of scientific reason. What it amounts to, in effect, is a kind of ultra-realism with regard to the orthodox account which takes that theory to dictate the terms for any possible description of quantum 'reality', even though the description thus arrived at must somehow be thought of as blocking the way to a realist construal in anything like the usual ('classical') sense. Another, marginally less paradoxical way of making this point would be to say that orthodox QM theory is an instrumentalist doctrine but one which treats instrumentalism not as a *faute de mieux* philosophy adopted for lack of definite evidence concerning the reality of quantum phenomena but rather as an accurate representation of uncertainties that characterize the very nature of (so-called) quantum reality.

Bohr himself appears to have started out from a moderate or non-dogmatic instrumentalist position according to which those uncertainties might yet be resolved through the advent of a better, more complete interpretation that showed them to reflect the conceptual limits of that earlier (orthodox) theory. Thus in an article of 1949 he can still be found asking 'whether the renunciation of a causal mode of description of quantum processes ... should be regarded as a temporary departure from ideals to be ultimately revived or whether we are faced with an irrevocable step towards obtaining the proper harmony between analysis and synthesis of physical phenomena'.[49] His phrasing here – as so often – is extremely elusive and obscure. Nevertheless, the first disjunct appears perfectly compatible with Einstein's or Bohm's hopes of achieving an alternative (realist and causal-explanatory) theory, whereas the second seems to leave it an open question whether the proposed 'analysis and synthesis' might yet find room for an interpretation along broadly realist lines. At any rate, Bohr had not yet espoused that doctrinaire 'end-of-the-road' philosophy which treated the orthodox QM account as a priori incapable of substantive criticism or revision. At this stage the renunciation of classical realism '*appeared* to be the only way open to proceed with the immediate task of co-ordinating the multifarious evidence concerning quantum pheneomena'.[50] However, as Cushing remarks, '[Bohr] later changed the modality of expression and stated that the 'renunciation of the visualization of atomic phenomena is *imposed upon us*'. And again, in similar vein, 'we are not dealing with an arbitrary renunciation of a more detailed analysis of atomic phenomena, but with a recognition that such an analysis is *in principle* excluded'.[51]

Thus Bohr can be seen to have swung right across from an outlook of moderate instrumentalism regarding the limits of presently attainable knowledge to a dogmatic instrumentalism which entailed nothing less than the *absolute impossibility* that quantum mechanics could ever be reconciled with any form of realist ontology. And in the process – as Popper more than once remarks – he laid himself open to the obvious charge of reasoning illogically from the limits of current scientific knowledge to the way things must ultimately stand with respect to quantum-physical reality, no matter how counter-intuitive (or downright unthinkable) the resultant 'physical' worldview. Popper sees this as a chief contributory factor in the 'great quantum muddle', namely the fallacy of misplaced concreteness whereby the presence of irreducible uncertainties (or statistical probabilities) in our predictive–observational knowledge is taken as intrinsic to the object domain over which those uncertainties range. In demographic terms, it is the kind of fallacy involved 'in taking a distribution function, i.e. a statistical measure function characterizing some *sample space* (or perhaps some "population" of events), and treating it as *a physical property of the elements of the population*'. Demographers who commit this fallacy also fall prey to some basic category mistakes, relying as they do 'on the very unsafe assumption that "my" probability of living in the South of England is, like "my" age, one of "my" properties – perhaps one of my physical properties'.[52] In much the same way – Popper argues – physicists who attribute intrinsic 'uncertainty' to phenomena such as the wave–particle dualism are simply mistaking a contingent limit on our knowledge of the sample 'population' for an absolute limit on knowledge in general or (more radically) an intrinsic feature of quantum-physical reality.

It was on this account chiefly that Einstein took issue with Bohr and refused to accept the in-principle 'completeness' of orthodox QM theory. Thus, '[t]he statistical character of the present theory would then have to be a necessary consequence of the incompleteness of the description of the systems in quantum mechanics, and there would no longer exist any ground for the supposition that a future ... physics must be based upon statistics'.[53] However, it is the orthodox interpretation that has exerted a powerful influence on recent anti-realist trends in philosophy of science. Such arguments typically start out from the assumption – stated or assumed – that the Bohr/Heisenberg orthodox theory has carried the day against Einstein, Bohm, and other proponents of an alternative (realist) view. Thus the measurement problem is treated as proof positive that – in Cushing's words – 'a physical system cannot (*in principle*) possess even a definite (but merely unknown to us) value of the physically observable attribute of position'.[54] Yet of course this conclusion already goes far beyond anything warranted by the quantum evidence, since the latter is perfectly consistent with the view – the far less problematical or counter-intuitive view – that these uncertainties result from determinate limits on our powers of precise or continuous measurement, rather than attaching to the very nature of a quantum 'reality' beyond our best powers of conceptual-explanatory thought. 'Taking such a theory seriously', Cushing remarks, 'as an actual representation of the physical world ... requires that we accept a rather bizarre ontology, one that may not even be conceptually coherent.'[55] Yet it *has* been taken seriously by a good many physicists and philosophers who accept it not only as a 'complete' description of microphysical reality but also as carrying large implications for what we are entitled – or not entitled – to assume in connection with objects and events in the macrophysical domain.

Thus, for instance, it is hard to imagine an extreme anti-realist and ontological relativist such as Nelson Goodman either venturing his theses or gaining any measure of credence were it not for the widespread (if vague) idea that they find support in the most advanced quarters of theoretical physics. There is a nicely indicative series of exchanges in the volume *Starmaking*, where Goodman defends his views against various critics, among them Putnam and Israel Scheffler.[56] At one point the challenge arises: surely he (Goodman) cannot be claiming that we human observers 'make' stars in the same way that we 'make' constellations by projecting or imposing a humanly recognisable shape onto the otherwise unmappable heavens?[57] Yes, Goodman responds: we do make stars just as we make the constellations since until we christened them with individual names – indeed until we called them 'stars' – there were no such bodies in the firmament.[58] Come now, says Scheffler: the stars existed long before there were human beings around to observe and to name them. However, Goodman is unruffled by this since on his (purportedly Kantian) account space and time are also human constructs and it is thus strictly meaningless to refer to celestial bodies as somehow 'existing' before the time or beyond the scope of human observation. So it cuts no ice when Scheffler commonsensically protests that the mere fact of their not being *called* 'stars' is quite beside the point, ontologically speaking. For on Goodman's view it is nothing less than self-evident that '[w]e make a star as we make a constellation, by putting its parts together and marking off its boundaries'.[59] And again, lest his point not be taken: 'that we can make the stars dance, as Galileo and Bruno made the earth move and the sun stop, not by physical force but by verbal invention, is plain enough'.[60]

This seems to me about as far from the truth as any philosopher could get by pushing the linguistic turn to a point where everything becomes a construct of our various languages, discourses, descriptive schemes, conceptual paradigms, etc. All the same, it is a doctrine that other philosophers – Rorty conspicuous among them – come very close to endorsing, despite their anxious (and somewhat disingenuous) treatment of Goodman as the Man Who Said the Tactless Thing.[61] Thus Rorty is no less committed to the view that, since 'reality' is always under some description or other, therefore reality is made and not discovered, or 'discovered' only in the notional sense that we constantly devise new descriptions, theories, or language games whereby to express our particular interests at this or that stage in the ongoing 'cultural conversation'.[62] It is also far from clear why the later Putnam should take issue with Goodman's claim, arguing as he now does – clean against his own previous convictions – that the only sort of realism worth defending is a realism 'internal' (for which read 'relative') to some culture-specific framework of beliefs or given conceptual scheme.[63] In Putnam's case – as I have argued above – there is evidence that this shift of views came about at least partly as a result of his prolonged engagement with issues in the philosophy of quantum mechanics.[64] With Goodman, it enters more as a kind of enabling background awareness, a sense of how far one can tolerably push the relativist line of argument given the degree of uncertainty attaching to claims about 'objective' truth or the existence of a real-world object domain apart from our various descriptive interests and purposes.

However, there is also a more specific connection between Goodman's ideas about 'starmaking' and the quantum-based astrophysical conjectures advanced by John Wheeler. Thus, according to Wheeler, it follows from the results of post-EPR and Bell-type delayed-choice experiments[65] that science has to reckon not only with the phenomenon of remote superluminal (i.e. faster-than-light) particle 'communication' but also with that of retroactive causality or of 'past' events that are somehow brought about by our present choice of measurement parameter.[66] Moreover, such effects can take place across massive distances of space and time, as for instance – Wheeler's example – through our making some particular radiotelescope observation whose results must be taken as effectively *deciding* what occurred billions of light-years ago. Thus, there was once thought to exist a pair of quasars (A and B) which had hitherto always been observed at a six-second angle of separation, and which astronomers had therefore assumed to occupy different celestial locations. But it was subsequently shown – through red-shift spectral analysis – that both sets of incoming data derived from a single source, although one had been deflected *en route* by a powerful 'gravitational lens' (a nearby massive galaxy) which delayed its arrival and hence produced the illusory separation effect. On Wheeler's account, this entire astrophysical system – quasar, galaxy, and radiotelescope + observer – should be thought of by analogy with what transpires on a smaller (laboratory) scale when scientists conduct delayed-choice experiments to establish the occurrence of nonlocal interaction or remote particle entanglement.[67] In this case the galaxy acts like a spin-polarizer inserted into the path of a photon at some point after its emission from source. It thus enables us to determine or predict (1) the measurement value actually obtained for that particle; (2) the value that can be known to obtain for any other particle (at whatever space-like distance) with which it had previously interacted; and (3) the antecedent history of the entire system right back to its supposed point of origin some billions of light-years ago.[68] This analogy can then be extended – again with reference to Bell's paradoxical results – by asking what would happen if the telescope were modified (or its orientation changed) by some adjustment comparable to the insertion of a semitransparent mirror into the photon path. Just as in the Bell–Aspect experiments, so here: such a change would necessarily have the effect of splitting the two photon streams so as to 'force the next photon that arrives from the quasar to have followed both the possible trajectories or to have followed one of them, respectively'.[69]

Thus to Wheeler it seems an inescapable conclusion that 'we are able to influence the past even on time scales comparable to the age of the universe'.[70] For there is (so he argues) no other way to explain how we can now, through some particular choice of measurement, decide just which of the outcomes will be realized *along with its entire co-implicated history of past astrophysical events.* Thus the way that the radiotelescope is set up will itself retroactively decide whether or not the observed results are such as can only be explained by the intervening presence of a galaxy which acts as a strong gravitational lens and which thus produces the illusory 'evidence' of two distinct quasars. Furthermore, any answer to the question whether that galaxy 'exists' and must therefore be thought to have exerted such an influence is itself dependent on our possessing both the instrumental means (the telescope) and the kinds of theoretical knowledge required to arrive at the correct – one-quasar-only – solution. So it is we terrestrial observers in the here-and-now who effectively decide that the galaxy was (or *shall have been*) 'inserted' into the astrophysical system, just as it is the galaxy – like the spin-polarizer in EPR/Bell-type experiments – which effectively decides (or *shall have decided*) the truth of the matter as between the double and single-quasar hypotheses.

In short, there is no avoiding the implication – at least from Wheeler's quantum-astrophysical viewpoint – that past effects can have present causes or that 'light from the quasar [which] passed the gravitational lens billions of years ago' can now be observed under various possible (experimentally realized) conditions that somehow determine what *must have been the case* with respect to both the quasar and the galaxy. It is therefore mistaken, Wheeler concludes,

> to think of the past as 'already existing' in all detail .... What we have the right to say of past space–time, and past events, is decided by choices – of what measurements to carry out – made in the near past and now. The phenomena called into being by these decisions reach backward in time in their consequences ..., back even to the earliest days of the universe. Registering equipment operating in the here and now has an undeniable part in bringing about that which appears to have happened. Useful as it is under everyday circumstances to say that the world exists 'out there' independent of us, that view can no longer be upheld. There is a strange sense in which this is a 'participatory universe'.[71]

There are many conceptual confusions in this passage, not least the apparently unnoticed slide from Wheeler's strong (ontological) claim that past events are 'called into being' or 'brought about' by subsequent acts of observation–measurement to his far less extreme (epistemological) claim that what 'appears to have happened' will always be dependent on the kinds of observation we are able to make or the kinds of 'registering equipment' which we happen to have at our disposal. Or again: such factors must of course play a role in 'what we have the right to say' concerning events in the remote astrophysical past, although this does not mean – as Wheeler would have it – that our statements and theories 'reach backward in time' to decide what shall actually have occurred from 'the earliest days of the universe'.

So, depending on which phrases are taken to carry most emphasis, the passage can be read *either* as expressing a valid (though trivial) truth about the scope and limits of human knowledge, *or* as propounding a far more dramatic (though distinctly underargued) thesis with regard to retroactive causation, the observer-dependent nature of reality, and so forth. I have cited it here as a further striking example of the way that certain quantum-theoretical conjectures have carried across into other areas of recent philosophical debate. In Goodman's case one can find echoes of just about every school of QM interpretation, from the orthodox (Bohr–Heisenberg) instrumentalist line to the many-worlds theory propounded by Deutsch and also – as in Wheeler – the notion of time as relative to various 'world-versions', some of which follow the linear sequence past–present–future, whereas others may be thought of as configuring time in altogether different ways. To object that he runs these ideas all at once and without any sense that they might involve conflicting (or at any rate rival) interpretative claims is probably to place too much weight on Goodman's arguments here. Still – as I have said –

they reflect a more widespread and often equally dubious appeal to quantum physics as having somehow undermined every version of the case for scientific realism.

The orthodox model is mainly a source of handy local metaphors, as when Goodman makes his ontological-relativist point by remarking how the physicist easily 'flits back and forth between a world of waves and a world of particles as suits his purpose'.[72] With regard to the many-worlds QM hypothesis Goodman is in some respects more cautious than a theorist like Deutsch, insisting that his 'worlds' are *not* to be construed in an overly literal (realist) sense. For then of course the question would arise, 'Where are these many actual worlds? How are they related to one another? Are there many earths all going along different routes at the same time and risking collision?' To this simple-minded question Goodman can simply reply, 'Of course not; in any world there is only one Earth; and the several worlds are not distributed in any space–time'.[73] On the strength of such passages, one could take his whole argument to involve nothing more than a moderate descriptivist claim, i.e. that we 'make' those several worlds by bringing reality under various schemes, versions, or descriptions, but only in so far as that reality exists independently of us and our beliefs concerning it. This tends to be Goodman's fallback position in the face of realist counter-arguments, as for instance when he says that 'the multiple worlds I countenance are just the actual worlds made by and answering to true or right versions'.[74] In which case – one might suppose – their truth and rightness is version-relative only in the harmless sense that atomic physicists, molecular biologists, civil engineers, and others will have reason to pick out different aspects or features of the physical world, each with an equally legitimate claim to get things right in some particular regard.[75] But elsewhere it is just impossible to interpret Goodman as holding this moderate view. Thus, '[w]hatever can be said truly of a world is dependent on the saying – not that whatever we say is true but that whatever we say truly (or present rightly) is nevertheless informed by and relative to the language or other symbol system that we use'.[76]

It is here that Goodman's argument comes closest to the kind of extreme (Wheeler-type) quantum physical worldview that would treat both present and past 'reality' as somehow quite literally *brought into being* through our various descriptions, observations, or choices of measurement parameter. Thus when Scheffler incredulously asks how we can have 'made' the stars or anything else that we surely know – by every rational standard of evidence – to be so much older than ourselves, Goodman once again comes back with the quasi-Kantian response: '[p]lainly, by making a space and time that contains those stars'.[77] Still, he concedes, there is the problem of grasping 'how a star that existed before all versions could be made by a version'. But we can get our minds around this particular obstacle simply by taking his constructivist point and seeing that our worries are themselves just a product of the world version (or the version-relative space-time framework) that orders events in a certain temporal sequence. Thus:

according to any of our trusted familiar world-versions, a star came much earlier than any version. Such a version, call it *W*, politely puts its own origin much later than the origin of the star – that is, much earlier *in this version's own time-ordering*. Yet according to a quite different version, call it *V*, (perhaps a version at a different level or metaversion) the star and everything else come into being only via a version. As we have seen, there is no ready-made world waiting to be labeled. There is no absolute time. In the time of *W*, the star comes first; in the time of *V*, the version comes first. Which is right? The answer is '*both*'.[78]

Now one might, just possibly, construe this passage in keeping with Goodman's other (more moderate) line of talk that would place the different 'versions' or time-series on the side of our various descriptive purposes, conceptual schemes, scientific interests, etc. It would then entail no such drastic revision to our basic ideas of time, space, and reality as Wheeler suggests through his quantum-astrophysical conjecture, and as Goodman likewise seems to require when he refuses to yield ontological ground to critics like Scheffler. However, this reading goes clean against his more typical claim – here and elsewhere – that it *cannot make sense* to posit an order of reality (of objects, events and their spatio-temporal relations) that would somehow stand outside or beyond those various descriptive frameworks. For there are passages in Goodman which leave no doubt as to his *really believing* that we 'make' stars (as distinct from just naming them or assigning them to constellations), and moreover that it is we human observers who effectively decide the truth or absurdity of claims like this through our adopting some particular world version, along with its associated system of temporal predicates. Maybe it is the case – and here he seems to waver for a moment – that the world of modern science is one that was 'made with great difficulty and is, like the several worlds of phenomena that also contain stars, a more or less right or real world'.[79] However, this concession counts for little when compared with Goodman's insistence on the point that every world thus made (along with all its objects and spatio-temporal attributes) is just one among the multitude of worlds that we might otherwise project.

III

I should say in all frankness – though it is probably apparent by now – that I agree with Scheffler in finding Goodman's position to be either 'absurd' (if one takes it in the 'plain and literal sense'), or hedged about with so many saving clauses and qualifications as to make no dent on the realist argument for the existence of an objective, mind-independent, or non-version-relative world.[80] All the same, his position is not so far from that of other philosophers who have been anxious to explain why they do not go along with Goodman's extreme relativist views while in fact their own arguments scarcely differ except by adopting a somewhat more cautious or circumspect manner of approach.[81] Thus Michael Dummett, for one, likewise pursues his anti-realist case to the point of claiming that our present state of knowledge can in some sense 'bring about the past' since there exist no objective ('verification-transcendent') historical truths that do not depend ultimately on what we know – or at least have the means to find out – concerning past events.[82]

In Dummett's case, as in Wheeler's, this argument derives its seeming plausibility from his constant habit of equivocation as between the weak (trivial-though-justified) and the strong (consequential though strictly unintelligible) versions of the claim. Thus there seems little reason to object if it amounts to no more than the basic epistemological point, i.e. that *what counts* as 'knowledge' or 'truth' in some given context of enquiry – whether history or the physical sciences – will always depend on our observation data, sources of evidence, verification procedures, powers of conceptual-explanatory grasp, and so forth. But the case is quite different when Dummett goes on to deny the existence of 'verification-transcendent' truths, that is to say, objective truth/falsehood values that would apply even to statements in what he calls the 'disputed class', those for which as yet we possess no conclusive evidence either way.[83] Such statements can range all the way from more-or-less informed historical conjectures to speculative theories in physical science (e.g. those of quantum mechanics) or mathematical theorems that remain as yet beyond reach of adequate proof. For Dummett, it makes no sense to claim – as the realist must – that these statements *do* have an objective truth-value and one that is unaffected by our present state of ignorance or partial knowledge concerning them. After all, how could we possibly be in a position to assert the existence of truths that *ex hypothesi* exceed our utmost powers of present verification?

For the realist, conversely, it makes no sense to suppose that past events depend on present knowledge for their actually having occurred, or that hypotheses in physics are neither true nor false unless we can decide the issue, or that Fermat's last theorem – to take a topical instance – was likewise devoid of any truth-value until someone came along with a proof. Dummett is committed to the view that the meaning of a statement is given by its method of verification, so that anyone working on Fermat's last theorem before the proof turned up must have been working in ignorance of what the theorem meant or what its proof might entail.[84] The same would apply to any scientist or historian who sought to establish the truth (or falsehood) of a speculative hypothesis while as yet – in the nature of the case – not knowing what would count as a clinching piece of evidence. In fact Dummett tends to vacillate between a strong version of the thesis that runs straight into these sorts of problem and a weaker version which just about avoids them by relaxing the conditions so as to admit statements for which there exists at least an in-principle possibility of proving them true or false. Still the main objection to this whole line of argument – in Dummett and other anti-realists – is that it relativizes truth to some given state of knowledge (whether now or at the end of enquiry), and thus pre-emptively collapses the distinction between epistemological and ontological issues. That is to say, it rules out the very idea of truths that are 'verification-transcendent', i.e. that hold good objectively whatever the restrictions on our present or indeed our best-possible future knowledge concerning them.[85]

To the realist, on the other hand, it appears nothing short of self-evident that there have been, are, and will continue to be a great many truths that we do not (perhaps cannot) know but that would be known – if one cares to think of it in this way – by an omniscient intelligence sharing none of our creaturely limits. Those limits may be remediable in the sense that they result from restricted access to the relevant information sources or from error-prone reasoning, perceptual or cognitive bias, anthropomorphic 'commonsense' illusion, and so forth. Such factors we can always hope to overcome through some further advance in scientific knowledge or better understanding of their sources and effects. Then again – for all that we can tell – human beings may be subject to various epistemic or conceptual shortcomings that set an ultimate limit on the scope of attainable scientific knowledge. This is what the realist has in mind when she argues that the truth-value of statements cannot be a matter of epistemic warrant or of whether we possess adequate criteria (evidence, grounds, documentary warrant, logical proof procedures, etc.) for establishing their truth or falsehood.[86] Still less can it be thought – as appears to follow from Dummett's anti-realist stance – that since the meaning of a statement is given by its verification conditions, therefore any statement which lacks determinate conditions of just that kind must be construed as strictly meaningless. For the upshot of this argument is to resurrect all the problems of old-style logical positivism, among them the sheer impossibility of explaining how knowledge could ever make progress (whether in mathematics, physical science, or historical and other branches of enquiry) through the testing of hypotheses whose truth or falsehood cannot be known in advance.[87]

As I have said, Dummett occasionally softens this requirement at least to the extent of allowing statements to be meaningful so long as we can grasp what it would entail (in counterfactual-conditional terms) for us to have the kind of knowledge or proof procedure that removed it from the disputed class. Even so, this concession will not satisfy the realist since it still lies open to the charge of inverting the order of dependence between truth (that which pertains to objective or veridical states of affairs) and knowledge (that which pertains to the scope and limits of human understanding). Besides, the whole point and presumptive force of Dummett's anti-realist case rests on his adopting the stronger version of the claim, that which holds – to cite a fairly typical passage – that '[w]henever a statement is true, it must be possible, in principle, for us to know that it is true, that is, for creatures with our particular restricted observational and intellectual facilities and spatio-temporal viewpoint'.[88] In which case – the realist is surely entitled to ask – what is it that distinguishes human knowledge (subject to those particular sensory-cognitive restrictions) from the kinds of knowledge attainable by creatures equipped with a different perceptual apparatus?

William Empson once made the point with respect to an earlier verificationist programme when he reviewed A.J. Ayer's book *The Foundations of Empirical Knowledge*. Thus, '[b]ees see ultra-violet light; perhaps birds feel the points of the compass; some people say atoms have dim sensations – is it logically possible for me to be an atom?'.[89] What Empson finds strange about Ayer's phenomenalist approach is not so much its anthropomorphic reduction of 'knowledge' to the register of perceptual inputs but rather its programmatic refusal to credit the existence of an objective reality some aspects of which may lie as far beyond the powers of human cognitive grasp as beyond the understanding of bees or birds. To Ayer's way of thinking, '[t]he objection to assertions about matter is that we can't conceivably observe it'. Yet phenomenalism offers no remotely adequate account of how sensations are produced, what their 'content' might be, or how we might distinguish veridical (scientifically valid) sense-data from perceptual illusions or the sorts of experience undergone by creatures responsive to a whole different range of physical stimuli. Thus, 'how are we better off by reducing it [i.e. "matter"] to sense-data which we can't conceive ourselves as having? Here again, we know less about the sense-data than we do about the things'.[90] All of which suggests to Empson that phenomenalism is logically self-refuting and – besides that – hopelessly unable to explain how we could ever gain knowledge of a physical world that was not just a construct out of our own private sensations.

Other critics of the doctrine, Popper among them, have traced its origins all the way back to Berkeley's critique of Locke and have noted the ease with which a radical empiricism can flip over – so to speak – into an equally radical form of idealist metaphysics.[91] What the two philosophies have in common is their rejection of the basic realist tenet, i.e. that our perceptual data (along with any beliefs or statements based upon them) are rendered either true or false *according to the way things stand in reality*, and not – as the verificationist would have it – according to our best observational warrant or means of checking them out. Empson puts this case in the plainest terms when he asserts as a matter of rational self-evidence that 'the universe has been sturdily indifferent for aeons to the observers to whom [on the phenomenalist argument] its reality is reduced'.[92] Professional philosophers – those of a realist persuasion – have mostly gone a longer way around in order to demonstrate the incoherence of phenomenalism and (beyond that) the problems encountered by any version of the verificationist thesis which likewise entails the reduction of truth to a matter of epistemic warrant.[93] Still they are making much the same point as Empson: that those problems must inevitably arise if one rejects the appeal to an observer-independent (which is also to say, a verification-transcendent) domain of objective reality and truth. For there will otherwise always come a stage in the argument when phenomenalism again takes the Berkeleian turn towards a solipsist conception of sense-data as the sole means of access to a notional 'reality' that exists only in the private consciousness of this or that private observer.

Dummett's is in many ways a more circumspect and, without doubt, a more logically resourceful statement of the case than anything to be found in Ayer or the proponents of old-style logical positivism. All the same, it is a version that runs into similar problems and which raises them with all the more force by pressing the anti-realist argument to its ultimate conclusion. Thus, according to Dummett, 'there are gaps in reality ... meaningful statements, which we can understand and whose truth or falsity we can therefore conceive of establishing but for which, nevertheless, the question whether they are true or false has no answer: they concern a region of reality which is simply indeterminate'.[94] This is a 'weak' statement of the case in so far as it allows such statements to be 'meaningful' (i.e. not *entirely* devoid of intelligible content) since we can at least form some conception of what might be involved in assigning them a definite truth-value despite our presently lacking any means to decide either way on the issue. Thus again:

> Of any statement concerning the past, we can never rule it out that we might subsequently come upon something which justified asserting or denying it, and therefore we are not entitled to say of any specific such statement that it is neither true nor false: but we are not entitled either to say in advance that it has to be either one or the other, since this would be to invoke notions of truth or falsity independent of our recognition of truth or falsity, and hence incapable of having been derived from the training we received in the use of these statements.[95]

One is tempted to remark of this tortuous sentence that there is nothing one can say of it, by way of clarification, that would not be open to challenge on some other, equally plausible reading. At any rate, it is hard to see how Dummett – given his anti-realist convictions – can hold the line against a full-scale (Goodman-type) strong constructivist argument that *we make the past* and everything else, stars included, according to our preferential world version or currently favoured descriptive scheme. Thus the sentence starts out on a somewhat concessive note by allowing that in cases of undecidability with respect to past events we might in future turn up some decisive piece of evidence which effectively resolved the issue and thereby assigned a truth-value to previous statements of the disputed class concerning those same events. Yet it promptly goes on to retract that concession by decreeing that we are *not* entitled to claim – in advance of such evidence – that the statements in question must be either true or false depending on what actually did or did not occur, and hence irrespective of our current *de facto* limited state of knowledge.

Dummett's entire case for anti-realism rests on the idea that these gaps in knowledge must also be construed as 'gaps in reality'. That is, they concern past events in respect of which we can frame a disjunctive statement ('X either occurred or did not occur') that the realist will take as necessarily true – since one or other must have been the case – but which Dummett regards as neither true nor false unless we have adequate grounds for deciding between them. (If this conjures up memories of Schrödinger's cat and its 'superposed' [neither alive nor dead] predicament then the linkage is by no means fortuitous, as I shall go on to argue with respect to quantum logic and its anti-realist implications.) Where such grounds are lacking for whatever reason we shall just have to say that there is no truth of the matter and, moreover, no fact or circumstance of past 'reality' against which our statement might be assessed (and the disjunct resolved either way) had we but the requisite knowledge. Thus, '[r]ealism I characterise as the belief that statements of the disputed class possess an objective truth-value, independently of our means of knowing it: they are true or false in virtue of a reality existing independently of us'. For the anti-realist, conversely, 'statements of the disputed class are to be understood only by reference to the sort of thing which we count as evidence for a statement of that class'.[96]

No doubt there is some room for manoeuvre in Dummett's phrasing here since 'the sort of thing which counts as evidence' might well be taken as covering cases where we do not possess conclusive evidence to back up a given statement but where we do possess the kind of transferable knowledge that results from our having dealt with other (perhaps better warranted) statements of a similar type. In this mode, as William Alston remarks, Dummett's thought 'recapitulates the movement of the Vienna Circle from requiring conclusive verifiability for factual meaningfulness to requiring merely "confirmability", that is, the possibility of finding considerations that count for or against the statement in question'.[97] However, this still leaves Dummett committed to the basic anti-realist premise, namely his denial that such statements can have an 'objective truth value' or one that obtains 'in virtue of a reality existing independently of us'. Hence his claim that there are 'gaps in reality' – and not just gaps in our knowledge of reality – whenever we are unable to assign such a value. For on Dummett's account reality ascriptions can extend no further than our best current (or perhaps best attainable) state of knowledge concerning the statements in question. Where the realist goes wrong – exceeds the limits of warranted assertability – is in making this strictly insupportable appeal to a realm of objective (verification-transcendent) truth-values that would not be subject to the limiting conditions of epistemic or justificatory warrant.

Thus it is fair to say that Dummett subscribes to something very like the 'old' verificationist criterion even in those passages where he seems to adopt a less restrictive or a more accommodating line. As he puts it, 'the central notions of a theory of meaning must ... be those of verification and falsification rather than those of truth and falsity'.[98] And again, 'our notions of truth and falsity ... consist merely in the conception of a situation's occurring which would thus conclusively determine its truth value'.[99] Of course this second formulation of the case is open to the more liberal reading, one that allows us to talk intelligibly about matters of truth and falsehood concerning statements in the disputed class, just so long as we can frame some adequate idea of what it would take – hypothetically speaking – to lift them out of that class. Still it leaves a vast range of 'undecidable' sentences that do not come up to this standard and which must therefore be counted as simply not candidates for the warranted ascription of truth or falsehood values. These latter include – on Alston's count – 'open-ended universal generalizations, subjunctive conditionals and sentences containing components that are explained in terms of such conditionals, and statements about the past'.[100] They would also include as-yet unproven mathematical or logical theorems that appear to possess significant (truth-evaluable) content – since mathematicians and logicians are working to prove them with at least some hope of ultimate success – but must none the less be thought of as lacking such content by Dummett's verificationist lights.

This argument derives partly from the Fregean doctrine that 'sense determines reference' and partly, as I have said, from Dummett's Wittgensteinian belief that the ability to interpret utterances of whatever kind consists in the ability 'to do just what we actually learn to do when we learn to use them, that is, in certain circumstances to recognize them as having been verified and in others as having been falsified'.[101] From which it follows – supposedly – that in the case of sentences belonging to the disputed class we must treat them as neither true nor false (i.e. as possessing no truth-evaluable content) since we lack the relevant learning experience or grasp of what would count in the proper context as adequate assertoric warrant. Thus these 'gaps in our knowledge' are also 'gaps in reality' in so far as we can frame no intelligible statement concerning reality or truth except through having learned the appropriate criteria for recognizing and evaluating statements of the kind. In any case, where we lack those criteria – as with the various instances listed above – we shall simply have to say that there is no truth of the matter and (what is more) no conceivable 'reality' that could warrant the realist in his or her claim that such statements are determinately true or false quite apart from our current state of knowledge, understanding, or logico-linguistic grasp.

Dummett puts the case most succinctly when he decrees that '[i]f a statement is true, it must be in principle possible to know that it is true'.[102] That is to say – and here the argument clearly links up with issues in quantum theory – we can never be justified in asserting the existence of a real-world, knowledge-independent object domain in respect of which we possess (as yet) only limited or maybe conflicting kinds of evidence, but whose properties we take to exist quite apart from these 'gaps' in our current state of understanding. So far as I know the only extended discussion of quantum mechanics in Dummett's work occurs in his essay 'Is Logic Empirical?' and concerns precisely this issue of epistemic warrant as a limiting condition on our statements with regard to the truth or falsehood of certain quantum-theoretical conjectures.[103] Here Dummett takes issue with Putnam over the latter's claim that empirical discoveries – such as wave–particle dualism or quantum nonlocality – may give us adequate reason to revise certain 'laws' of classical logic (e.g. that of bivalent truth/falsehood) so as to avoid anomalies in QM theory or any clash with empirical observation data.[104] It may seem odd that he should challenge Putnam on this point since Dummett is himself more than willing – as we have seen – to suspend that law in the case of certain statements, among them (presumably) some in the area of quantum mechanics, which are strictly 'undecidable' on the best evidence to hand. However, his argument has more to do with Putnam's defence of realism in the quantum domain than with Putnam's revisionist philosophy of logic. Thus, according to Dummett, it can scarcely advance our understanding of physics or anything else if the response to some perceived anomaly is to change the very ground rules that had hitherto defined what should count as valid reasoning on the evidence, and among whose more problematical entailments was precisely the anomaly in question. What this amounts to is merely a stipulative ruse for avoiding such problems by redefining the pertinence or scope of certain logical connectives and operators.

In Quine's case also – as Dummett notes – there is the somewhat throwaway statement ('Two Dogmas of Empiricism') that certain developments in quantum physics may turn out to force revisions to the 'laws of thought' supposedly enshrined in classical logic.[105] But here again this claim gets into conflict with Quine's more considered treatment of the topic where he denies that we could ever, in principle, have rational grounds for preferring such a drastic response in the face of recalcitrant (whether physical or verbal) evidence.[106] Rather, we should suspect that there must be some problem with that evidence, some alternative (logic-preserving) construal of it, or – in the case of communicative breakdowns – some localized semantic (rather than deep-laid logical) mismatch between their understanding and our own. Dummett makes the same point as follows:

> if, when this procedure [of rational hypothesis-testing] leads a given theory into apparent antinomies, this is suddenly taken, not as a ground for revising the theory, but for adjusting the rules for deriving consequences, it is not merely a natural but a justifiable reaction to feel that we no longer know what is the content of calling a theory correct or incorrect.[107]

That is to say, if *everything* is called into question – from observation statements to logical 'laws of thought' – then nothing can any longer count as good reason for rejecting this or that candidate hypothesis. In which case there could be no prospect of further advance, whether towards a better, more adequate scientific theory (one that genuinely resolved the anomalies rather than just redefining them out of existence) or again, towards a better, more adequate grasp of the problems that had hitherto blocked communicative uptake.

IV

So Dummett has a strong case, *contra* the Quine of 'Two Dogmas', for rejecting the logical-revisionist thesis as simply incoherent and (besides) as plainly incompatible with Quine's own arguments elsewhere. However, this is not his case against Putnam since here – predictably enough – it is the realist and truth-preserving aspect of Putnam's argument that Dummett finds unacceptable. That is, he takes issue *not* with the idea that certain precepts of classical logic (such as bivalence) may be revisable in regard to certain disputed statements – which is, after all, a main feature of his own position – but rather with the thesis that they might or should be revised in order to yield truth-values consistent with a realist interpretation of the given (empirical) evidence. More precisely, Putnam's proposal differs from Quine's in so far as he leaves intact the law of excluded middle (or *tertium non datur*) but thinks that the quantum anomalies can be reconciled – rendered compatible with a realist ontology – by abandoning the strictly distributive assignment of truth/falsehood values.

Dummett sees two possible ways of interpreting Putnam's claim and argues that neither can be made to square with the argument for a realist (verification-transcendent) construal of quantum phenomena. On the one hand, 'failure of the distributive law [makes] realism untenable: a particle has some position, but, since truth does not distribute over disjunction or existential quantification, there is not necessarily any one position which it has' (ILE, p. 287). On this account Putnam is simply not entitled to maintain a realist view since – as Dummett construes it – his revisionist logic undercuts any possible appeal to the criteria of classical (determinate) truth and falsehood with regard to quantum 'reality'. However, Putnam then goes on to defend a more overtly realist position that requires just those logical resources – bivalence among them – which he had previously counted 'revisable' under pressure from conflicting empirical evidence. Thus, 'all that he is doing is to introduce new senses of "&" and "v" ["and" and "or"] *alongside* the old senses, without displacing the latter' (ILE, p. 287).

In other words, Putnam shifts back and forth between a strong revisionist claim that undercuts any grounds for realism (since it yields undecidable statements in precisely Dummett's sense) and a realist argument that effectively leaves all the logical constants in place and which thus offers no possible solution to the antinomies of quantum theory (classically construed). 'If the realism is upheld', Dummett writes, 'then the two sets of logical constants must both be admitted as intelligible, when applied to statements about quantum-mechanical systems; ... [which] does not involve the abandonment, in response to experience or otherwise, of any logical law formerly held' (ILE, p. 287). Then again, Putnam might choose to grasp the other horn of this dilemma, insist that we adopt quantum logic 'considered as supplanting classical logic', and thereby relinquish the realist position that – as Dummett understands it – entails the acceptance of bivalent truth–falsehood values. However, 'such a modification of his view would blunt the cutting edge of his thesis that logic is empirical ... in that it would no longer be possible to claim that the discovery of the invalidity of the distributive law was a discovery *about the world*' (ILE, p. 288). That is to say, it would cut away his grounds for asserting that quantum phenomena such as the wave–particle dualism or the impossibility of obtaining precise simultaneous measurements of particle position and momentum required a revision of the classical law of distributed truth- and falsehood-values. For by adducing those phenomena *as evidence* – by claiming that they count decisively against the classical distributivist view – Putnam must have recourse to just the criteria that are standardly invoked in support of that view. Thus, 'the realist terms in which he [Putnam] construes statements about quantum-mechanical systems cannot but allow as legitimate a purely classical interpretation of the logical constants applied to such statements' (ILE, p. 285). On the other hand, Putnam's revisionist case with regard to classical logic – more specifically, with regard to bivalence and distributivity – leaves him with no means of explaining why the anomalies of orthodox QM should be thought of as *logically* entailing any such revision. For if one lets go of those criteria then it is far from clear why any given item of empirical evidence (such as the wave–particle dualism or the limits of precise measurement on conjugate quantum variables) should carry any *logical* implications whatsoever, least of all far-reaching implications with regard to the limits of classical logic. On the realist construal, conversely, 'Putnam is in no way rejecting the distributive law, as holding for "and" and "or" on their standard meanings; it remains as valid for him as for anyone else' (ILE, p. 287).

To Dummett's way of thinking this dilemma is strictly inescapable if one adopts a realist interpretation of quantum mechanics or of statements belonging to the 'disputed class' in whatever field of enquiry. This follows from his understanding of realism as essentially a logico-semantic thesis that entails the existence of distributed truth-values for all candidate statements and which therefore – as he claims with regard to Putnam – cannot but assign the logical connectives their standard (classical) meaning. Thus Putnam's argument necessarily fails (or runs into manifest self-contradiction) since he purports to offer a realist account of the relevant quantum-physical data by revising or suspending a logical axiom which must hold good, so Dummett believes, for any version of the realist case. However, this is clearly *not* the sense of 'realism' that Putnam has in mind. That is to say, there is a strong suspicion that Putnam is here being hoist not so much with his own contradictory petard as with one carefully fashioned by Dummett out of a logico-semantic doctrine whose chief purpose throughout all his work is to prove realism strictly untenable. What is more, the doctrine is designed to enforce this anti-realist lesson not only as concerns statements of the disputed class – such as those of orthodox quantum mechanics – but also with respect to *any* statement whose truth-conditions (and hence whose meaning) cannot be verified as a matter of direct empirical warrant. For it is Dummett's firm belief – repeated in a wide range of contexts – that 'the central notions of a theory of meaning must be those of verification and falsification rather than those of truth and falsity'.[108] And again:

even the most thoroughgoing realist must grant that we could hardly be said to grasp what it is for a statement to be true if we had no conception whatever of how it might be known to be true; there would, in such a case, be no substance to our conception of its truth condition.[109]

Now clearly this creates some large problems for realism as applied to quantum mechanics if one accepts both Dummett's verificationist approach and also – closely allied to that – his understanding of what 'realism' amounts to in logico-semantic terms. For it can scarcely be claimed that there currently exists *any* interpretation of quantum mechanics that would meet the verificationist requirement of a straightforward appeal to the empirical evidence without thereby giving rise to a range of well-known logical anomalies. On the other hand, if realism is taken to entail a commitment to bivalent (distributed) truth-values, then Putnam's case runs up against the problem – the insuperable problem, as Dummett sees it – that those anomalies cannot be resolved without abandoning precisely that commitment and letting the realist argument go by logical default.

Thus, to repeat, 'the failure of the distributive law [makes] realism untenable: a particle has some position, but, since truth does not distribute over disjunction or existential quantification, there is not necessarily any one position which it has' (ILE, p. 287). And the same would apply to all other statements in the disputed (quantum-physical) class, such as those which result from Heisenberg's uncertainty principle, from the EPR paradox on various construals, or from Bell's theorem concerning quantum nonlocality and particle 'entanglement' over large distances of space-time separation.[110] For it is clear that such statements cannot be thought of as meeting all the criteria of (1) straightforward empirical warrant, (2) conformity with the logic of distributed (decidable) truth-values, and (3) consistency with a realist worldview in anything like the required sense. Thus it might well seem that Dummett is right when he objects that Putnam cannot have his cake and eat it, i.e. that he can *either* conserve the empirical QM data and modify his logic accordingly (in which case a Dummett-type verificationist approach is perfectly in order), *or* hang on to the classical logic and thereby adopt a 'realist' position (as Dummett defines it) which generates all manner of anomaly or paradox when confronted with those same empirical data. After all, as Dummett writes:

> What is a realistic interpretation of statements of some given class? It is, essentially, the belief that we possess a notion of truth for statements of that class under which every statement is determinately either true or not true, independently of our knowledge or our capacity for knowledge. Putnam's realist doctrine plainly fits that characterization. Now, it is by no means a requirement on realism that we deny that there is any use for, let alone that there is any intelligible interpretation of, non-classical logical constants as applied to statements of the class in question. But it

is a requirement on realism that the classical two-valued constants can meaningfully be introduced. Since every statement is determinately either true or not, it must be *possible* to introduce a negation ¬ such that ¬*A* is true just in case *A* is not true, even if this is not the negation which we ordinarily employ, or even the most useful one to employ. Likewise, it must be possible to introduce two-valued conjunction and disjunction. (ILE, p. 274)

I have cited this passage at length because it goes to the heart of Dummett's anti-realist argument and also shows how he misses the point – or what I take to be the point – of Putnam's avowedly realist position as applied to quantum mechanics.

In one respect Dummett is undoubtedly right in his insistence that any form of realism must have recourse to certain classically defined logical constants, among them negation and distributed truth- and falsehood-values. For to give these up is to give up the basic realist principle that what ultimately decides the truth-content of our statements, theories, ontological commitments and so forth is whether or not they truly represent the way things stand in reality. This is where Putnam chiefly differs from the Quine of 'Two Dogmas', i.e. in requiring that we leave the law of excluded middle intact while suggesting that we abandon the distributive law in order to resolve the quantum paradoxes. Otherwise the way would indeed be open to ontological relativity or wholesale framework-relativism of the kind that Quine there endorses, whatever his subsequent changes of view in this respect.[111] All the same, Dummett has a strong case when he claims that Putnam cannot have it both ways: on the one hand arguing for a realist construal of quantum-physical statements while on the other hand adopting a revisionist line with respect to the distributive law. After all, as Peter Gibbins pointedly remarks, '[t]he program itself is odd in that quantum logic is most naturally thought of as expressing quantum-mechanical antirealism, just as quantum mechanics itself is most naturally interpreted antirealistically'.[112] And again, with particular reference to Putnam's logical-reformist proposals, '[q]uantum logic does not *resolve*, but merely *embodies* all the strange features of quantum mechanics. The quantum logician who is happy with quantum logic can adopt a quietist pose in the face of the "paradoxes". This may not seem much, but it is the best quantum logic has to offer'.[113] That is to say, it amounts to just a handy technique for shifting the burden of explanation from physical theory – where the problem ultimately lies – to 'logic' conceived as a formal rendition of empirically valid results and hence as subject to revision under pressure from those same quantum paradoxes.

Thus, as Gibbins remarks, 'Putnam's problem is not to get the *right* answer for the two-slit pattern, which is the physicist's problem. Putnam's problem is to *stop getting the wrong answer*. Avoidance of paradox is, I am afraid, typically the philosopher's but not the physicist's strategy'.[114] However, this is not the main thrust of Dummett's case against Putnam. The issue between them is posed most sharply by Dummett's assumption that the realist argument must either be epistemic in nature (in which case it falls to the verificationist objection that we can attach no meaning or truth-value to epistemically undecidable statements), or else involve a merely notional appeal to absolute (verification-transcendent) values of truth and falsehood that can have no purchase on any matter of empirical or evidential warrant, least of all statements of the 'disputed class' such as those of quantum mechanics. What is thereby excluded from consideration is the more fundamental realist claim that the truth-value of all statements – undecidable statements among them – is ultimately fixed by the way things stand in reality, quite apart from any limits imposed by our present-best methods of verification and quite apart from any logico-semantic quandaries introduced by Dummett's preferred construal of the realist case. Indeed, the whole force of his argument against Putnam comes of its narrowing the terms of debate to just these last two possibilities: the one an epistemic conception where truth can amount to nothing more than a matter of warranted assertability, the other a deluded 'realist' conception (on Dummett's understanding of realism) where truth is preemptively defined in such a way that it can have no bearing on matters beyond the specialized sphere of philosophical semantics.

Thus, according to Dummett, 'the issue whether, for quantum mechanics, classical logic should be replaced by quantum logic is an issue belonging to the theory of meaning; an affirmative answer would neither be nor be derivable from any proposition of quantum mechanics' (ILE, p. 288). For it is only within the 'theory of meaning' – that is, post-Fregean philosophy of language and logic – that we can hope to get clear about other issues in ontology, epistemology, and philosophy of science. Of course this is a main plank in Dummett's argument for the great revolution that has lately come about through the turn towards questions of meaning and truth as a basis for all future metaphysics.[115] Thus, 'the correctness of any piece of analysis carried out in another part of philosophy cannot be determined until we know with reasonable certainty what form a correct theory of meaning for our language must take'.[116] And again, in Dummett's best-known statement of the case:

> Only with Frege was the proper object of philosophy finally established: namely, first, that the goal of philosophy is the analysis of the structure of *thought*; secondly, that the study of *thought* is to be sharply distinguished from the study of the psychological process of *thinking*; and, finally, that the only proper method for analysing thought lies in the analysis of *language*.[117]

One could scarcely fault this as a general description of the analytic programme in philosophy as it developed after Frege and Russell. However, there is still the question of how far that programme succeeded in imposing its agenda and thereby – as some would argue – diverting attention from those 'other parts' of philosophy which might else have exerted an independent claim to treatment on different terms.

My point is that Dummett's whole approach to such issues as the interpretation of quantum mechanics (and, more generally, to issues concerning epistemology and philosophy of science) is one that derives from the set of priorities laid out in the above-cited passage. So it is that he can state – as if it were established pretty much beyond reasonable doubt – that any question concerning the adoption of an alternative quantum logic is one that 'belong[s] to the theory of meaning', and whose answer (should it finally appear) can 'neither be nor be derivable from any proposition of quantum mechanics' (ILE, p. 288). So it is also that Dummett can impale Putnam on the horns of an apparent dilemma which in fact results more from his (Dummett's) construal of the options than anything that Putnam is obliged to accept as a matter of non-negotiable choice. One alternative that neither sees fit to entertain is the case for a causal-realist approach as developed in Bohm's 'hidden variables' theory, a theory which avoids all the long-standing problems that divide philosophers like Putnam and Dummett.[118] Thus they both assume that it is the orthodox (Copenhagen) account that sets the philosophical agenda, which poses the crucial issue with regard to deviant or many-valued logics, and whose challenge cannot be circumvented by any such large-scale alternative proposal. Beyond that, however, Putnam and Dummett are at odds on some basic philosophical issues, above all whether questions in epistemology or philosophy of science must always wait upon a 'theory of meaning' (or more adequate form of logico-semantic analysis) before there is any prospect of resolving them.

Putnam's opposition to this view is much clearer in his early essays on the causal theory of reference, where he comes out strongly against the Frege–Dummett approach, and in favour of an externalist approach whereby the reference of certain terms (and hence the truth-value of statements containing them) is fixed to begin with – and held firm thereafter – through their applying to objects of a given sort.[119] It is less clear in his essays on quantum philosophy, since the main issue here is how to make sense of statements that lack such secure referential criteria and which thus inevitably lead into regions where his argument is exposed to Dummett's habitual line of anti-realist attack. But this is no reason to think that Putnam is devoid of further resources once Dummett has shown – to his own satisfaction – that he (Putnam) cannot make good on his realist claims without conserving the distributive law, i.e. retaining the logical connectives 'and' and 'or' in what can only be construed as their standard (classical) sense. After all, Dummett himself disavows any claim to have resolved the issue either way. On the contrary:

> no pronouncement has been made, in the present paper, either for or against the adoption of quantum logic, nor yet for or against a realistic interpretation of quantum mechanics. All that I have been concerned to

maintain is that it is inconsistent to combine, as Putnam wishes to do, a realistic interpretation with the thesis that quantum logic should supplant classical logic. (ILE, p. 287)

To this extent I think that Dummett is justified in pointing out the problems with Putnam's case. However, as I have said, it is only on Dummett's idiosyncratic understanding of 'realism' that this must be taken to have left Putnam – or any other defender of a realist approach – bereft of all possible alternative arguments. For one such alternative is precisely to hold (*contra* Dummett's verificationist account) that the truth-value of quantum-physical statements may indeed be 'undecidable' at present owing to our limited state of knowledge, but is none the less a function of whether or not they obtain in respect of quantum reality.

Of course this goes clean against the orthodox QM theory according to which any talk about quantum 'reality' must either be abandoned or heavily qualified in deference to Heisenberg's uncertainty relations or Bohr's complementarity principle.[120] However, it is precisely the realist case that this orthodox interpretation is not so much forced upon us by the very nature of quantum phenomena – whatever that could mean – but comes about rather through a basic confusion between epistemological and ontological issues. That is to say, it takes the limits of current understanding with regard to (for instance) the measurement problem or the non-commuting character of conjugate variables and treats those limits as somehow *intrinsic* to quantum-physical 'reality'. This is what Popper has chiefly in mind when he speaks of the 'great quantum muddle' and deplores the readiness of so many thinkers (physicists and philosophers alike) to adopt a whole range of drastic revisionist 'solutions' without, as yet, any clear sense of the problems that need answering.[121] It seems to me that his strictures are well founded and that philosophers – especially those of an anti-realist or ontological-relativist bent – have often done more to deepen than to clarify the various confusions involved.

Endnotes

1 For philosophically informed discussion from a variety of realist and anti-realist standpoints, see David Z. Albert, *Quantum Mechanics and Experience* (Cambridge, MA: Harvard University Press, 1993); Bernard D'Espagnat, *Veiled Reality: an analysis of present-day quantum-mechanical concepts* (Reading, MA: Addison-Wesley, 1995); Peter Forrest, *Quantum Metaphysics* (Oxford: Blackwell, 1988); John Honner, *The Description of Nature: Niels Bohr and the philosophy of quantum physics* (Oxford: Clarendon Press, 1987); Max Jammer, *Philosophy of Quantum Mechanics* (New York: Wiley, 1974); Josef M. Jauch, *Are Quanta Real? a Galilean dialogue* (Bloomington, IN: Indiana University Press, 1973); Alastair I.M. Rae, *Quantum Physics: illusion or reality?* (Cambridge University Press, 1986); A. Sudbury, *Quantum Mechanics and the Particles of Nature* (Cambridge University Press, 1986).

2 David Deutsch, *The Fabric of Reality* (Harmondsworth: Penguin, 1997); Bryce S. DeWitt and Neill Graham (eds.), *The Many-Worlds Interpretation of Quantum Mechanics* (Princeton, NJ: Princeton University Press, 1973).

3 Deutsch, *The Fabric of Reality* (op. cit.).

4 See Niels Bohr, *Atomic Theory and the Description of Nature* (Cambridge: Cambridge University Press, 1934) and *Atomic Physics and Human Knowledge* (New York: Wiley, 1958); Werner Heisenberg, *The Physical Principles of the Quantum Theory* (New York: Dover, 1949) and *Physics and Philosophy* (New York: Harper & Row, 1958); also Henry J. Folse, *The Philosophy of Niels Bohr: the framework of complementarity* (Amsterdam: North-Holland, 1985); Patrick A. Heelan, *Quantum Mechanics and Objectivity: a study of the physical philosophy of Werner Heisenberg* (The Hague: Martinus Nijhoff, 1965); Honner, *The Description of Nature* (op. cit.); Dugald Murdoch, *Niels Bohr's Philosophy of Physics* (Cambridge University Press, 1987).

5 See especially Albert Einstein, B. Podolsky and N. Rosen, 'Can Quantum-Mechanical Description of Reality be Considered Complete?', *Physical Review*, series 2, Vol. 47 (1935), pp. 777–80; Niels Bohr, article in response under the same title, *Physical Review*, Vol. 48 (1935), pp. 696–702; also Bohr, 'Conversation with Einstein on Epistemological Problems in Atomic Physics', in P.A. Schilpp (ed.), *Albert Einstein: philosopher–scientist* (La Salle, IL: Open Court, 1969), pp. 199–241; Arthur Fine, *The Shaky Game: Einstein, realism, and quantum theory* (Chicago: University of Chicago Press, 1986); Michael Redhead, *Incompleteness, Nonlocality and Realism: a prolegomenon to the philosophy of quantum mechanics* (Oxford: Clarendon Press, 1987).

6 See especially Einstein, 'Autobiographical Notes', in P.A. Schilpp (ed.), *Albert Einstein: philosopher–scientist* (op. cit.), pp. 3–105; Ernst Mach, *The Science of Mechanics: a critical and historical account of its development*, trans. T.J. McCormack (La Salle, IL: Open Court, 1960); also – for various defences or critiques of instrumentalism in philosophy of science – Pierre Duhem, *To Save the Phenomena: an essay on the idea of physical theory from Plato to Galileo*, trans. E. Dolan and C. Maschler (Chicago: University of Chicago Press, 1969); C.J. Misak, *Verificationism: its history and prospects* (London: Routledge, 1995); Bas C. van Fraassen, *The Scientific Image* (Oxford: Clarendon Press, 1980) and *Quantum Mechanics: an empiricist view* (Clarendon Press, 1992).

7 See Thomas S. Kuhn, *The Structure of Scientific Revolutions*, 2nd edn (Chicago: University of Chicago Press, 1970); also Kuhn, *The Essential Tension: selected studies in scientific tradition and change* (University of Chicago Press, 1977).

8 See for instance Richard Rorty, *Consequences of Pragmatism* (Brighton: Harvester, 1982); *Essays on Heidegger and Others* and *Objectivity, Relativism, and Truth* (both Cambridge: Cambridge University Press, 1991); *Truth and Progress* (Cambridge University Press, 1998); Nelson Goodman, *Ways of Worldmaking* (Indianapolis: Bobbs-Merrill, 1978) and *Problems and Projects* (Bobbs-Merrill, 1972); also – for various contributions by and responses to Goodman – Peter J. McCormick (ed.), *Starmaking: realism, anti-realism, and irrealism* (Cambridge, MA: M.I.T. Press, 1996).

9 See especially W.V. Quine, 'Two Dogmas of Empiricism', in *From a Logical Point of View*, 2nd edn (Cambridge, MA: Harvard University Press, 1961), pp. 20–46; also *Word and Object* (Cambridge, MA: M.I.T. Press, 1960) and *Ontological Relativity and Other Essays* (New York: Columbia University Press, 1969); Hilary Putnam, *The Many Faces of Realism* (La Salle, IL: Open Court, 1987); *Realism With a Human Face* (Harvard University Press, 1990); *Renewing Philosophy* (Harvard University Press, 1992).

10 Donald Davidson, 'On the Very Idea of a Conceptual Scheme', in *Inquiries into Truth and Interpretation* (Oxford: Oxford University Press, 1984), pp. 183–98; Karl R. Popper, *The Myth of the Framework: in defence of science and rationality*, ed. M.A. Notturno (London: Routledge, 1994).

11 See Note 6, above.
12 See especially Honner, *The Description of Nature* (op. cit.) and other entries under Notes 1, 4 and 5 above.
13 See for instance Joseph Rouse, *Knowledge and Power: toward a political philosophy of science* (Ithaca, NY: Cornell University Press, 1987); also – on a range of related topics – Richard J. Bernstein, *Beyond Objectivism and Relativism: science, hermeneutics, and praxis* (Philadelphia: University of Pennsylvania Press, 1983).
14 Heidegger, dedication to *On Time and Being*, trans. Joan Stambaugh (New York: Harper & Row, 1972); 'Modern Science, Metaphysics, and Mathematics', in *Heidegger: Basic Writings*, ed. David F. Krell (London: Routledge, 1993), pp. 271–305, p. 272; also *The Question Concerning Technology and Other Essays*, trans. William P. Lovitt (New York: Harper & Row, 1977).
15 See Rouse, *Knowledge and Power* (op. cit.); also John Loscerbo, *Being and Technology: a study in the philosophy of Martin Heidegger* (Martinus Nijhoff: The Hague, 1981); Mark Okrent, *Heidegger's Pragmatism: understanding, being, and the critique of metaphysics* (Ithaca, NY: Cornell University Press, 1988).
16 See for instance Rudolf Carnap, 'The Elimination of Metaphysics through Logical Analysis of Language', in A.J. Ayer (ed.), *Logical Positivism* (New York: Free Press, 1959), pp. 60–81; Carl Gustav Hempel, *Fundamentals of Concept Formation in Empirical Science* (Chicago: University of Chicago Press, 1972); Hans Reichenbach, *Experience and Prediction* (University of Chicago Press, 1938) and *Philosophic Foundations of Quantum Mechanics* (Berkeley & Los Angeles: University of California Press, 1944); N. Rescher (ed.), *The Heritage of Logical Positivism* (Lanham: University Press of America, 1985).
17 For further discussion, see Christopher Norris, *New Idols of the Cave: on the limits of anti-realism* (Manchester: Manchester University Press, 1997).
18 See especially James T. Cushing, *Quantum Mechanics: historical contingency and the Copenhagen interpretation* (Chicago: University of Chicago Press, 1994); also Paul Forman, 'Weimar Culture, Causality, and Quantum Theory, 1918–1927: adaptation by German Physicists and Mathematicians to a hostile intellectual environment', *Historical Studies in the Physical Sciences*, Vol. 3 (1971), pp. 1–115 and 'The Reception of an Acausal Quantum Mechanics in Germany and Britain', in S.H. Mauskopf (ed.), *The Reception of Unconventional Science*, AAAS Selected Symposium, No. 25 (1979), pp. 11–50.
19 On the realist interpretation of quantum theory, see especially David Bohm, *Causality and Chance in Modern Physics* (London: Routledge & Kegan Paul, 1957); David Bohm and B.J. Hiley, *The Undivided Universe: an ontological interpretation of quantum theory* (London: Routledge, 1993); also Evadro Agazzi (ed.), *Realism and Quantum Physics* (Amsterdam & Atlanta: Rodopi, 1997); David Z. Albert, *Quantum Mechanics and Experience* (op. cit.) and 'Bohm's Alternative to Quantum Mechanics', *Scientific American*, No. 270 (May 1994), pp. 58–63; F.J. Belinfante, *A Survey of Hidden Variable Theories* (Oxford: Pergamon Press, 1973); Cushing, *Quantum Mechanics* (op. cit.); Peter Holland, *The Quantum Theory of Motion* (Cambridge: Cambridge University Press, 1993); Dipankar Home, *Conceptual Foundations of Quantum Mechanics: an overview from modern perspectives* (New York: Plenum, 1998).
20 Locke, *An Essay Concerning Human Understanding*, 2 vols. (New York: Dover, 1969), (op. cit.), Book III, Chapter vi, Section 8.
21 See especially J. Perrin, *Atoms*, trans. D.L. Hammick (New York: Van Nostrand, 1923); Mary Jo Nye, *Molecular Reality* (London: MacDonald, 1972); Wesley C. Salmon, *Scientific Explanation and the Causal Structure of the World* (Princeton, NJ: Princeton University Press, 1984); Robin Waterfield, *Before Eureka: the presocratics and their science* (Bristol: The Bristol Press, 1989).

22 See especially Hilary Putnam, 'Is Semantics Possible?' and 'Meaning and Reference', in Stephen Schwartz (ed.), *Naming, Necessity, and Natural Kinds* (Ithaca, NY: Cornell University Press, 1977), pp. 102–18 and 119–32; also Saul Kripke, *Naming and Necessity* (Oxford: Blackwell, 1980); L. Linsky (ed.), *Reference and Modality* (Oxford: Oxford University Press, 1971); Gregory McCulloch, *The Mind and its World* (London: Routledge, 1995); David Wiggins, *Sameness and Substance* (Blackwell, 1980).

23 See also M.R. Ayers, 'Locke versus Aristotle on Natural Kinds', *Journal of Philosophy*, Vol. 77 (1981), pp. 247–72; Alvin Goldman, *Epistemology and Cognition* (Cambridge, MA: Harvard University Press, 1986) and *Empirical Knowledge* (Berkeley & Los Angeles: University of California Press, 1988); Hilary Kornblith, *Inductive Inference and its Natural Ground* (Cambridge, MA: M.I.T. Press, 1993) and Kornblith (ed.), *Naturalizing Epistemology* (M.I.T. Press, 1985).

24 Nicholas Rescher, *Scientific Realism: a critical reappraisal* (Dordrecht: D. Reidel, 1987), p. 61.

25 See also J. Aronson, R. Harré and E.Way, *Realism Rescued; how scientific progress is possible* (London: Duckworth, 1994); Imre Lakatos and Alan Musgrave (eds.), *Criticism and the Growth of Knowledge* (Cambridge: Cambridge University Press, 1970); Peter Lipton, *Inference to the Best Explanation* (London: Routledge, 1993); Nicholas Rescher, *Scientific Progress* (Oxford: Blackwell, 1979); Peter J. Smith, *Realism and the Progress of Science* (Cambridge University Press, 1981).

26 See entries under Note 21, above; also M. Gardner, 'Realism and Instrumentalism in Nineteenth-Century Atomism', *Philosophy of Science*, Vol. 46 (1979), pp. 1–34; Euan Squires, *The Mystery of the Quantum World*, 2nd edn (Bristol & Philadelphia: Institute of Physics Publishing, 1994); A. Sudbury, *Quantum Mechanics and the Particles of Nature* (op. cit.).

27 See entries under Note 4, above.

28 See Note 19, above.

29 See Note 2, above.

30 I refer here to Putnam's dispute with Ian Hacking over the latter's supposedly naive belief that QM allows us to think of 'particles' as possessing numerical identity or definite values of position and momentum. See Putnam, *Pragmatism: an open question* (Oxford: Blackwell, 1995), p. 59; Ian Hacking, *Representing and Intervening: introductory topics in the philosophy of natural science* (Cambridge: Cambridge University Press, 1983), p. 23. For Putnam's retreat to anti-realism in all but name, see also *The Many Faces of Realism* (La Salle, IL: Open Court, 1987); *Realism With a Human Face* (Cambridge, MA: Harvard University Press, 1990); *Renewing Philosophy* (Harvard University Press, 1992).

31 As my colleague Vivian Beedle more succinctly put it, 'using a specific construal of reality (orthodox QM) to repudiate "orthodox" reality implies a belief in the capacity of science to explain the world which is at odds with that repudiation' (private communication).

32 Cited in Karl R. Popper, *Quantum Theory and the Schism in Physics* (London: Routledge, 1992), p. 40.

33 Putnam, *Pragmatism: an open question* (op. cit.), p. 59.

34 See Note 22, above.

35 See Note 30, above.

36 Quine, 'Two Dogmas of Empiricism' (op. cit.).

37 Deutsch, *The Fabric of Reality* (op. cit.).

38 See entries under Note 5, above.

39 Cushing, *Quantum Mechanics* and Popper, *Quantum Theory and the Schism in Physics* (Notes 18 and 32, above).

40 Cited in Popper, *Quantum Theory* (op. cit.), p. 8.

41 Ibid, p. 11.

42 J.S. Bell, *Speakable and Unspeakable in Quantum Mechanics: collected papers on quantum philosophy* (Cambridge: Cambridge University Press, 1987); also James T. Cushing and Ernan McMullin (eds.), *Philosophical Consequences of Quantum Theory: reflections on Bell's Theorem* (Notre Dame, IN: University of Notre Dame Press, 1989); A. Aspect, P. Grangier and C. Roger, 'Experimental Realization of the E–P–R Paradox', *Physical Review*, Vol. 48 (1982), pp. 91–4.

43 See entries under Note 19, above.

44 Popper, *Quantum Theory* (op. cit.), p. 123.

45 Ibid, p. 123

46 Ibid, p. 101

47 See Werner Heisenberg, *The Physical Principles of the Quantum Theory* (New York: Dover, 1949) and *Physics and Philosophy* (New York: Harper & Row, 1958); also Patrick A. Heelan, *Quantum Mechanics and Objectivity: a study of the physical philosophy of Werner Heisenberg* (The Hague: Nijhoff, 1965).

48 Cited in Popper, *Quantum Theory* (op. cit.), p. 8.

49 Cited in Cushing, *Quantum Mechanics* (op. cit.), p. 108.

50 Ibid, p. 108.

51 Ibid, p. 108.

52 Ibid, p. 52.

53 Einstein, cited in Bell, *Speakable and Unspeakable in Quantum Mechanics* (op. cit.), p. 83.

54 Cushing, op. cit., p. 108.

55 Ibid, p. 204.

56 See McCormick (ed.), *Starmaking* (Note 8, above).

57 Israel Scheffler, 'The Wonderful Worlds of Goodman', in *Starmaking* (op. cit.), pp. 133–42.

58 Nelson Goodman, 'On Starmaking', ibid, pp. 143–50; also *Ways of Worldmaking* and other entries under Note 8, above.

59 Ibid, p. 145.

60 Goodman, 'Notes on the Well-Made World', in *Starmaking* (op. cit.), pp. 151–60; p. 155.

61 See for instance Rorty's various references to Goodman in *Objectivity, Relativism, and Truth* (op. cit.); also – for similar attempts to draw a line between Goodman's and other, more respectable versions of the relativist argument – Joseph Margolis, *Texts Without Referents: reconciling science and narrative* (Oxford: Blackwell, 1989) and *The Truth About Relativism* (Blackwell, 1991); Hilary Putnam, 'Reflections on Goodman's *Ways of Worldmaking*', in *Philosophical Papers*, Vol. 3 (Cambridge: Cambridge University Press, 1983); W.V. Quine, 'Goodman's *Ways of Worldmaking*', in *Theories and Things* (Cambridge, MA: Harvard University Press, 1981).

62 See Notes 8 and 61, above; also Rorty, *Contingency, Irony, and Solidarity* (Cambridge: Cambridge University Press, 1989).

63 See entries under Note 30, above; also Putnam's various contributions to McCormick (ed.), *Starmaking* (op. cit.).

64 See Note 30, above; also Putnam, 'How to Think Quantum-Logically', *Synthèse*, Vol. 29 (1974), pp. 55–61 and the various relevant papers collected in *Mathematics, Matter and Method*, 2nd edn (Cambridge: Cambridge University Press, 1979).

65 See entries under Note 42, above.

66 J.A. Wheeler, 'Delayed Choice Experiments and the Bohr–Einstein Dialogue'. Paper presented at the joint meeting of the American Philosophical Society and the Royal Society, London, 5 June, 1980. My discussion here relies heavily on F. Selleri, 'Wave–Particle Duality: recent proposals for the detection of empty waves', in W. Schommers (ed.), *Quantum Theory and Pictures of Reality: foundations, interpretations, and new aspects* (Berlin: Springer Verlag, 1989), pp. 279–332. See also J.A. Wheeler and W.H. Zurek, *Quantum Theory and Measurement* (Princeton, NJ: Princeton University Press, 1983).

67 See Note 42, above.
68 For the origins of this debate about quantum nonlocality and its possible implications, see Notes 5 and 42, above; also Don Howard, 'Einstein on Locality and Separability', *Studies in the History and Philosophy of Science*, Vol. 16 (1985), pp. 171–201; Tim Maudlin, *Quantum Non-Locality and Relativity: metaphysical intimations of modern science* (Oxford: Blackwell, 1993); Redhead, *Incompleteness, Nonlocality and Realism* (op. cit.).
69 Wheeler, cited in Selleri, 'Wave–Particle Duality' (op. cit.), p. 298.
70 Ibid, p. 298.
71 Cited by Selleri, 'Wave–Particle Duality' (op. cit.), p. 297.
72 Goodman, 'Notes on the Well-Made World' (op. cit.), p. 153.
73 Ibid, p. 152
74 Goodman, *Ways of Worldmaking* (op. cit.), p. 138.
75 See especially Wiggins, *Sameness and Substance* (op. cit.).
76 Goodman, 'On Starmaking' (op. cit.), p. 144.
77 Ibid, p. 145.
78 Goodman, 'On Some Worldly Worries', in McCormick (ed.), *Starmaking* (op. cit.), pp. 165–70; p. 167.
79 Goodman, 'On Starmaking' (op. cit.), p. 145.
80 Scheffler, 'The Wonderful Worlds of Goodman' (op. cit.), p. 138.
81 See entries under Note 61, above.
82 Michael Dummett, 'Can an Effect Precede its Cause?', 'Bringing About the Past', and 'The Reality of the Past' in *Truth and Other Enigmas* (London: Duckworth, 1978), pp. 319–32, 333–50 and 358–74; also *Frege: philosophy of language* (London: Duckworth, 1973).
83 See especially Dummett, 'Truth' and 'Realism', in *Truth and Other Enigmas* (op. cit.), pp. 1–24 and 145–65; also 'What Is a Theory of Meaning?', in S.D. Guttenplan (ed.), *Mind and Language* (Oxford: Oxford University Press, 1975), pp. 97–138 and 'What is a Theory of Meaning? II', in Gareth Evans and John McDowell (eds.), *Truth and Meaning* (Oxford: Oxford University Press, 1976), pp. 67–137.
84 See Dummett, 'The Philosophical Significance of Gödel's Theorem', 'Platonism', and 'The Philosophical Basis of Intuitionist Logic', in *Truth and Other Enigmas* (op. cit.), pp. 186–201, 202–14 and 215–47.
85 For further discussion, see Richard L. Kirkham, 'What Dummett Says about Truth and Linguistic Competence', *Mind*, Vol. 98 (1989), pp. 207–24; Michael Luntley, *Language, Logic and Experience: the case for anti-realism* (London: Duckworth, 1988); N. Tennant, *Anti-Realism and Logic* (Oxford: Clarendon Press, 1987); Alan Weir, 'Dummett on Meaning and Classical Logic', *Mind*, Vol. 95 (1986), pp. 465–77; Timothy Williamson, 'Knowability and Constructivism: the logic of anti-realism', *Philosophical Quarterly*, Vol. 38 (1988), pp. 422–32; Kenneth P. Winkler, 'Scepticism and Anti-Realism', *Mind*, Vol. 94 (1985), pp. 46–52; Crispin Wright, *Realism, Meaning and Truth* (Oxford: Blackwell, 1987).
86 See for instance William P. Alston, *A Realist Conception of Truth* (Ithaca, NY: Cornell University Press, 1996); Roy Bhaskar, *Scientific Realism and Human Emancipation* (London: Verso, 1986); Michael Devitt, *Realism and Truth*, 2nd edn (Oxford: Blackwell, 1986); Penelope Maddy, *Realism in Mathematics* (Oxford: Oxford University Press, 1990); Mark Platts (ed.), *Reference, Truth and Reality: essays on the philosophy of language* (London: Routledge & Kegan Paul, 1980).
87 See especially C.J. Misak, *Verificationism: its history and prospects* (London: Routledge, 1995); also Oswald Hanfling (ed.), *Essential Readings in Logical Positivism* (Oxford: Blackwell, 1981) and Nicholas Rescher (ed.), *The Heritage of Logical Positivism* (Lanham: University Press of America, 1985).
88 Dummett, 'What Is a Theory of Meaning?' (op. cit.), p. 100.

89 A.J. Ayer, *The Foundations of Empirical Knowledge* (London: Macmillan, 1955); William Empson, review of Ayer, in *Argufying: essays on literature and culture*, ed. John Haffenden (London: Chatto & Windus, 1987), pp. 583–4; p. 583.

90 Ibid, p. 584.

91 Popper, *Quantum Theory and the Schism in Physics* (op. cit.).

92 Empson, review of Ayer (op. cit.), p. 584.

93 See Note 86, above; also D.M. Armstrong, *What is a Law of Nature?* (Cambridge: Cambridge University Press, 1983); J. Aronson, R. Harré and E. Way, *Realism Rescued* (op. cit.); Roy Bhaskar, *A Realist Theory of Science* (Leeds: Leeds Books, 1975); Rom Harré and E.H. Madden, *Causal Powers* (Oxford: Blackwell, 1975); Wesley C. Salmon, *Scientific Explanation and the Causal Structure of the World* (Princeton, NJ: Princeton University Press, 1984); Peter J. Smith, *Realism and the Progress of Science* (op. cit.); Michael Tooley, *Causation: a realist approach* (Blackwell, 1988).

94 Dummett, 'The Metaphysics of Verificationism', in L.E. Hahn (ed.), *The Philosophy of A.J. Ayer* (La Salle, IL: Open Court, 1992), p. 146.

95 Dummett, 'The Reality of the Past' (op. cit.), p. 364.

96 Dummett, 'Realism' (op. cit.), p. 146.

97 Alston, *A Realist Conception of Truth* (op. cit.), p. 109.

98 Dummett, *Frege: philosophy of language* (op. cit.), p. 467.

99 Ibid, p. 514.

100 Alston, *A Realist Conception of Truth* (op. cit.), pp. 118–9.

101 Dummett, *Frege: philosophy of language* (op. cit.), p. 467; also Gottlob Frege, 'On Sense and Reference', in P.T. Geach and M. Black (eds.), *Translations from the Philosophical Writings of Gottlob Frege* (Oxford: Blackwell, 1952), pp. 56–78; Ludwig Wittgenstein, *Philosophical Investigations*, trans. G.E.M. Anscombe (Blackwell, 1958).

102 Dummett, 'What is a Theory of Meaning? II' (op. cit.), p. 99.

103 Dummett, 'Is Logic Empirical?', in *Truth and Other Enigmas* (op. cit.), pp. 269–89.

104 Putnam, 'Is Logic Empirical?', *Boston Studies for the Philosophy of Science*, Vol. 5 (1969), eds. R.S. Cohen and M. Wartofsky, pp. 216–41; also entries under Note 64, above.

105 Quine, 'Two Dogmas of Empiricism' (op. cit.), p. 43.

106 See Quine, *Philosophy of Logic* (Englewood Cliffs, NJ: Prentice-Hall, 1970); 'Carnap and Logical Truth', in *The Philosophy of Rudolf Carnap*, ed. P.A. Schilpp (La Salle, IL: Open Court, 1963); reprinted in Quine, *The Ways of Paradox* (New York: Random House, 1966). See also Enrico J. Beltrametti and Bas C. van Fraassen, *Current Issues in Quantum Logic* (New York: Plenum, 1981); Rachel Wallace Garden, *Modern Logic and Quantum Mechanics* (Bristol: A. Hilger, 1983); M. Gardner, 'Is Quantum Logic Really Logic?', *Philosophy of Science*, Vol. 38 (1971), pp. 508–29; Peter Gibbins, *Particles and Paradoxes: the limits of quantum logic* (Cambridge: Cambridge University Press, 1987); Susan Hack, *Deviant Logic: some philosophical issues* (Cambridge University Press, 1974); Peter Mittelstaedt, *Quantum Logic* (Princeton, NJ: Princeton University Press, 1994).

107 Dummett, 'Is Logic Empirical?' (op. cit.), p. 281. All further references given by 'ILE' and page number in the text.

108 Dummett, *Frege: philosophy of language* (op. cit.), p. 467.

109 Dummett, 'What Is a Theory of Meaning? II' (op. cit.), p. 100.

110 See Notes 1, 4, 5, 12, 19, 42 and 47, above.

111 Quine, 'Two Dogmas of Empiricism' (op. cit.); also *Pursuit of Truth*, revised edn. (Cambridge, MA: Harvard University Press, 1990).

112 Gibbins, *Particles and Paradoxes* (op. cit.); p. 166.

113 Ibid, p. 113.

114 Ibid, p. 158

115 See also Dummett, *The Logical Basis of Metaphysics* (Cambridge, MA: Harvard University Press, 1991).
116 Dummett, 'Can Analytical Philosophy be Systematic, and Ought it to Be?', in *Truth and Other Enigmas* (op. cit.), pp. 437–58; p. 454.
117 Ibid, p. 458.
118 See entries under Note 19, above.
119 See Note 22, above.
120 See Notes 4, 5, 6, 12, 18 and 19, above.
121 Popper, *Quantum Theory and the Schism in Physics* (op. cit.).

8 From Copenhagen to the stars

Some ways of quantum worldmaking

I

In this book I have argued that many of the problems with present-day interpretations of quantum mechanics can be traced back to the orthodox ('Copenhagen') theory as propounded in the writings of its earliest and most influential advocate, Niels Bohr.[1] According to Bohr – in a much-quoted statement – 'it would be reasonable to say that no man who is called a philosopher really understands what is meant by complementary descriptions'.[2] His remark seems to me distinctly double-edged in that anyone who is 'called a philosopher' might have good reasons for taking issue with Bohr's doctrine of complementarity and the arguments he offers in support of it. That is to say, they might point to various passages where Bohr advances some large claims with regard to QM and its (supposed) implications for the scope and limits of human knowledge while conspicuously failing to clarify his own philosophical terms and distinctions.

In this final chapter I shall therefore focus on several such passages – mostly arising from his famous series of debates with Einstein on the 'completeness' or otherwise of orthodox QM theory – and ask how far those claims stand up to more detailed and careful analysis.[3] My aim is not only to criticize Bohr's interpretation of quantum mechanics but also to question the widespread belief that the arguments went decisively in Bohr's favour, at least as regards their final and climactic exchange over the so-called 'EPR paradox'. On the orthodox account this is taken to have shown that realism was henceforth not an option for anyone who accepted the validity of quantum mechanics as a matter of overwhelmingly strong observational and predictive warrant.[4] I shall argue, on the contrary, that Bohr's statements are often so vague or conceptually imprecise as to leave the issue unresolved either way but with a strong presumption in favour of realism as an inference to the best (most 'complete' and rational) explanation. However, there are passages in Einstein's argument also where the realist case goes by default through the slippage from an alethic (truth-based or objective) to an epistemic (knowledge-based or verificationist) view of what 'realism' properly entails.[5] Hence – I suggest – a great deal of the confusion that has surrounded this debate during the past six decades and which continues to generate misunderstanding among physicists and philosophers alike. Towards the end of this chapter I shall therefore return to John Wheeler's wild extrapolation from the evidence of QM delayed-choice experiments to the idea of observer-induced retroactive causal effects over vast (astrophysical) distances of space–time separation. For nothing could more clearly illustrate the kinds of conceptual problem that arise when theorists base such far-reaching speculative arguments on the existence of certain deeply disputed quantum-physical phenomena.

II

One fairly obvious sign of that confusion is Bohr's habit of constantly veering about – often within a few sentences – between different senses of the term 'phenomenon' as applied to quantum mechanics. Thus sometimes it bears the usual philosophical (especially Kantian) sense of 'that which can be known or which presents itself to us through the understanding's power of bringing sensuous intuitions under adequate concepts'.[6] On this account QM 'phenomena' are always by very definition subject to the scope and limits of human knowledge, and must therefore crucially *not* be confused with whatever may constitute the ultimate (noumenal) nature of quantum 'reality'. This is why some commentators – Honner among them – have argued that Bohr's thinking is Kantian in certain basic respects, not least its combination of 'empirical realism' with 'transcendental idealism', or its attempt to maintain a viable distinction between the concepts and categories of human knowledge and that which (quite possibly) eludes their utmost powers of adequate comprehension.[7] Elsewhere, however, Bohr can be found speaking of quantum 'phenomena' as if they belonged very firmly on the other (i.e. noumenal, objective, or observer-independent) side of that same Kantian dichotomy. Take for instance the following passage where Bohr reflects on the consequences of Planck's inaugural (1900) discovery of the quantum of action.

> This postulate implies a renunciation as regards the causal space–time co-ordination of atomic processes. Indeed, our usual description of physical phenomena is based entirely on the idea that the phenomena concerned may be observed without disturbing them appreciably .... Now, the quantum postulate implies that any observation of atomic phenomena will involve an interaction with the agency of observation not to be neglected. Accordingly, an independent reality in the ordinary physical sense can neither be ascribed to the phenomena nor to the agencies of observation.[8]

One could devote a large amount of detailed exegesis to the various senses of the word 'phenomena' as required or implied by its contexts of occurrence in this and other passages of Bohr's writing. Sufficient to say, for present purposes, that it shifts across from an apparently objective usage ('the phenomena concerned may be observed without disturbing them appreciably') to a usage – as in the last sentence – that explicitly denies those phenomena any claim to 'independent reality' in 'the ordinary physical sense'. Elsewhere the meaning is ambiguous, as with 'our usual description of physical phenomena' or 'any observation of atomic phenomena', where the grammar strongly suggests *something there* to be described or observed, but where the gist of Bohr's argument clearly goes against that objectivist construal.

I think that these are not just problems of translation or localized instances of vague or opaque phraseology. Rather, they are evidence of the deep confusion that is often to be found in Bohr's statements of the orthodox QM case and which even his best-willed commentators are hard put to explicate in terms that make any kind of logical sense. Again, take this passage from his late essay *The Unity of Human Knowledge* where Bohr is clearly striving to present his argument in terms that so far as possible resist a subjectivist reading. 'From our present standpoint', he writes:

> physics is to be regarded not so much as the study of something *a priori* given, but rather as the development of methods for ordering and surveying human experience. In this respect our task must be to account for such experience in a manner independent of individual subjective judgement and therefore objective in the sense that it can be unambiguously communicated in ordinary human language.[9]

Clearly the phrase '*a priori*' has nothing like its usual (Kantian) import but must rather be taken in the sense: 'whatever is given to human knowledge or experience through our various modes of perceptual-cognitive enquiry'. Bohr seems to emphasize this aspect of givenness – of that which stands prior to any methods we develop for imposing order upon it – with a view to stressing the 'objective' (as opposed to phenomenal or experiential) character of quantum reality. Thus any adequate theory of quantum mechanics cannot be a matter of 'individual subjective judgement', whatever the tendency of some commentators to read Bohr's pronouncements in just that way. Yet it is hard to see how he squares this claim with the idea of 'objective' truth as that which can be 'unambiguously communicated in ordinary human language'. For of course the main point about objectivist arguments – in quantum physics and elsewhere – is that they hold out against any such move to equate truth with our present-best powers of descriptive or causal-explanatory thought. Einstein was the first to protest that a doctrine framed in such ambiguous terms could not be thought of as meeting the criteria for an adequate, let alone – as Bohr so insistently claimed – a 'complete' physical theory.[10] The same view was taken by Schrödinger and also by those later dissident quantum theorists, Bohm and Bell among them, who regarded orthodox QM as *necessarily* incomplete on account of its failure to provide an intelligible picture of physical reality and its resort to such elusive modes of description.[11]

Bell puts the case with typical forthright vigour in a statement which contrasts sharply with the passage from Bohr cited above. It has to do with Bohr's famous theory of complementarity, a doctrine (he remarks) which gets a dutiful mention in most textbook treatments of the topic, but usually 'only in a few lines', perhaps because 'the authors do not understand the Bohr philosophy sufficiently to find it helpful'. Thus:

> Consider for example the elephant. From the front she is head, trunk, and two legs. From the back she is bottom, tail, and two legs. From the sides she is otherwise, and from top and bottom different again. The various views are complementary in the usual sense of the word. They supplement one another, they are consistent with one another and they are all entailed by the unifying concept of 'elephant'. It is my impression that to suppose Bohr used the word 'complementary' in this ordinary way would have been regarded by him as missing his point and trivializing his thought. He seems to insist rather that we must use in our analysis concepts which *contradict* one another, which do not add up to, or derive from, a whole. By 'complementarity' he meant, it seems to me, the reverse: contradictoriness. Bohr seemed to like aphorisms such as: 'the opposite of a deep truth is also a deep truth': 'truth and clarity are complementary'. Perhaps he took a subtle satisfaction in the use of a familiar word with the reverse of its familiar meaning.[12]

I have cited this passage at length because it is one of the few commentaries on Bohr to suggest that maybe the emperor has no clothes – that the orthodox theory and extrapolations from it are perhaps simply incoherent – rather than assuming that any problems encountered must have to do with the depth of Bohr's thinking and the challenge it poses to conventional ideas about physical reality.

Of course this did not prevent Bell from devising his more sophisticated version of the EPR thought experiment which ironically turned out – or so it seemed – to strengthen the orthodox case. That is, he provided an elegant statistical proof to the effect that any hidden-variables theory (i.e. any realist interpretation along the lines proposed by Einstein and Bohm) could match the well-established QM results and predictions only at the cost of admitting nonlocality in a particularly flagrant form.[13] Since Bell's famous argument is more often alluded to than explained or even clearly spelled out let me cite a passage from Clauser and Shimony's 1978 survey of the field which states the main issue with admirable brevity and force.

> Because of the evidence in favour of quantum mechanics from the experiments based upon Bell's theorem, we are forced either to abandon the strong version of EPR's criterion of reality – which is tantamount to abandoning a realistic view of the physical world (perhaps an unheard tree falling in a forest makes no sound after all) – or else to accept some

kind of action-at-a-distance. Either option is radical, and a comprehensive study of their philosophical consequences remains to be made.[14]

In fact there has been a good deal of work since their article appeared that has focused not only on the details of those difficult and complex experiments but also on the more 'philosophical' issues – especially the challenge to relativity theory and Einsteinian local realism – that Clauser and Shimony indicate here.[15] However, there are two points worth noting in our present context of argument. One is the distinctly equivocal use of that familiar topos concerning the tree that falls unheard in the forest, an image with strong phenomenalist (or Berkeleian-idealist) associations. As so often, its deployment can be seen to exploit the ambiguity between 'sound' construed in objective-physical terms, i.e. sound waves caused by an impact and propagated through a fluid medium, and 'sound' as perceived by a sentient creature through the excitation of its auditory system. It is the same ambiguity – as Popper sharply points out – which has characterized a range of phenomenalist doctrines from Berkeley to the logical positivists, and which also typifies the orthodox (Copenhagen) version of QM theory.[16] As regards local realism, the issue stands very much as Clauser and Shimony present it in the above-cited passage. That is to say, Bell's results have been strongly borne out by experiment and leave no choice but to accept nonlocality as a strict entailment of that theory when conjoined with the empirical (i.e. observational-predictive) evidence. However, as we have seen, Bohm was able to accept this fact without prejudice to his realist case just so long as it was shown not to contravene the 'no-first-signal' rule, that is to say, the special-relativity requirement that no messages could be passed between distant observers by means of measurements carried out upon remotely interacting particles.[17] Thus it is far from evident – whatever the orthodox QM wisdom – *either* that Bohr must be judged to have gained the upper hand in his debates with Einstein, *or* that Bell's results must likewise be taken (despite his firmly held realist view) as offering yet further, decisive support for the standard Bohr-derived interpretation.

At this point it is worth going back briefly over the EPR argument and Bohr's objections to it since they constitute the single most important crux in the entire history of quantum-theoretical debate. Moreover, they are closely connected with other recent versions of the dispute between realism and anti-realism, among them (as we have seen) Dummett's programmatic restatement of the anti-realist case in terms of a logico-semantic approach that denies the existence of objective or 'verification-transcendent' truths.[18] Einstein and his colleagues very clearly adopted the contrary (realist) position when they declared as a matter of principle, in the opening sentence of their paper, that '[a]ny serious consideration of a physical theory must take into account the distinction between the objective reality, which is independent of any theory, and the physical concepts with which the theory operates'.[19] This was their basis for asserting the paper's single most widely known and disputed claim: that '[i]f, without in any way disturbing a system, we can predict with certainty (i.e. with probability equal to unity) the value of a physical quantity, then there exists an element of physical reality corresponding to this physical quantity'.[20] Thus for instance – the nub of their case against orthodox QM – one could thought-experimentally disprove Heisenberg's uncertainty thesis by conceiving a case in which separate measurements of non-commuting conjugate variables were carried out on two particles that had previously interacted and could thus be known to yield inverse (anti-correlated) values for any given parameter or orientation. No doubt it was impossible – for reasons explained by Heisenberg – to conduct two such measurements simultaneously on a single particle.[21] But by following the above procedure one could obtain precise values for each particle *with respect to every such property* simply by virtue of the well-established QM principle which guaranteed that any results thus achieved would exhibit perfect anti-correlation for whatever chosen measurement parameter. In which case, according to the EPR authors, realism stood vindicated and there was hence no need – as supposed by advocates of the orthodox theory – to abandon that all-important distinction between 'objective reality' and the 'physical concepts with which the theory operates'.

Bohr's response to EPR took the form of a resolute refusal to concede the very possibility of upholding any such distinction. What he rejected, more specifically, was the EPR claim that the measurements in question might be carried out 'without in any way disturbing the system'. For this phrase made room for a crucial ambiguity, or so Bohr argued in what might be thought a striking case of the pot calling the kettle black. Perhaps there was no question, with the set-up as described, of a 'mechanical disturbance of the system under investigation during the last critical stage of the measuring procedure'. Even so, he went on, 'there is essentially the question of an influence on the very conditions which define the possible types of predictions regarding the future behaviour of the system'.[22] And since those conditions must indeed be supposed – on the orthodox QM account – to 'constitute an inherent element of the description of any phenomenon to which the term "physical reality" can be properly attached', therefore the EPR authors can be shown to have missed a strictly unavoidable complicating factor which undermines both their realist case and their derivative case for the 'incompleteness' of the orthodox QM theory.

However, it is not hard to see that Bohr's argument involves a very pure example of *petitio principii*, that is to say, the logical fallacy of circular reasoning from questionable premises to equally questionable conclusions. For only *if* one takes for granted the correctness of the orthodox theory will it follow that 'there is essentially a question of an influence on the very conditions which define the possible types of predictions regarding the future behaviour of the system'.[23] At very least this assumes a radical revision – more like a downright rejection – of those realist principles enounced by the EPR authors in the opening sentence of their paper. So it is that Bohr can then quickly proceed, by a similar twist of argument, to take up their reference to 'physical reality' and redefine it in orthodox QM terms which assign it a meaning wholly at odds with their intended (realist) purport. Thus the 'conditions' that Bohr lays down – by stipulative warrant – as 'constitut[ing] an inherent element of the description of any phenomenon to which the term "physical reality" can be properly attached' are conditions that derive from his own understanding of the radically holistic nature of quantum phenomena and the hopelessness of striving to maintain those distinctions set forth in the EPR paper. One can sympathize with Einstein – and also with Bell – in their strong suspicion that the 'depth' of Bohr's thought concealed certain basic philosophical confusions which would perhaps lose much of their seeming profundity if exposed to more careful analysis.

Still it may be thought that we can get a firmer hold on the logic of Bohr's argument and thus do it more justice by turning to one of his well-disposed commentators who attempts to clarify what Bohr very often left vaguely or confusedly expressed. Dugald Murdoch belongs very much in that company so I shall now cite a pertinent passage from his book where he sets out the main points at issue between Bohr and Einstein–Podolsky–Rosen.

> Although a measurement on [the particle] S_1, Bohr admits, could not physically disturb S_2, nevertheless the experimental arrangement required for the measurement determines both for S_1 and for S_2 the necessary conditions for the meaningful ascribability of the physical property concerned. For the properties of exact position and exact momentum, moreover, these conditions mutually exclude each other. If the momentum of S_1 were measured, then the necessary conditions for the meaningful ascribability of an exact position to S_1 and S_2 (or even of meaningful talk of such a property) would not be satisfied. *Mutatis mutandis* the same holds for a measurement of the position of S_1. Hence, Bohr holds, Einstein is mistaken in assuming that S_2 must have both an exact, though unknown, simultaneous position and momentum. An object cannot meaningfully be said to have certain properties in the absence of the conditions which make such talk meaningful.[24]

What comes out clearly in this passage – more so than in Bohr's own discussions of the topic – is the extent to which orthodox QM theory rests on a verificationist approach whereby ascriptions of truth or reality must be deemed strictly inadmissible (even meaningless) unless justified on empirical or observational grounds.[25] This is no doubt one of the main reasons why that theory achieved its maximum hold during just the period when mainstream philosophy of science was dominated by similar doctrines such as Machian instrumentalism and Vienna-school logical positivism.[26] But there is also a strong case to be made that the influence worked in both directions and that quantum mechanics on the orthodox (Copenhagen) view has continued to exert a powerful effect upon thinkers like Dummett who espouse an anti-realist position according to which no statement can be meaningful unless it satisfies the standard verificationist requirements.[27] Thus Bohr's chief objection to the EPR paper is precisely that it ascribes 'real' values of position or momentum to particles, and moreover takes them to possess such values *continuously or between measurements*. It is for just this reason that upholders of the orthodox (Copenhagen) QM theory make a point of insisting on that theory's 'completeness' and thus rule out the very possibility of an alternative Bohm-type realist account. For on Bohm's construal – like Einstein's – those values are thought of as pertaining to an objective (observer-independent) physical reality quite aside from whether or not they are subject to some localized act of measurement.[28]

One can therefore understand why philosophical anti-realists such as Dummett should assume that orthodox QM sets the terms for debate and that any viable solution to the quantum paradoxes must accept those terms or otherwise be counted a non-starter.[29] What this amounts to, again, is a verificationist approach that in Dummett takes the form of an argument from the epistemic conditions for ascribing verification criteria or standards of warranted assertability. But it is just that kind of epistemic criterion that the EPR authors explicitly reject in the closing paragraph of their paper where they deny that orthodox QM theory has anything like the evidential weight that could count decisively against the argument for a scientific realism premised on the existence of objective (i.e. verification-transcendent) truths. Thus:

> [o]ne could object to this conclusion [i.e. the 'incompleteness' of orthodox QM] on the grounds that our criterion of reality is not sufficiently restrictive. Indeed, one would not arrive at our conclusion if one insisted that two or more physical quantities can be regarded as simultaneous elements of reality *only when they can be simultaneously measured or predicted*. On this point of view, since either one or the other, but not both simultaneously, of the quantities P [position] and Q [momentum] can be predicted, they are not simultaneously real. This makes the reality of P and Q depend upon the process of measurement carried out on the first system, which does not disturb the second system in any way. No reasonable definition of reality could be expected to permit this.[30]

In other words the orthodox QM case derives its seeming plausibility from a wholly 'unreasonable' requirement, one that could not possibly be satisfied by any practicable system of measurement, whether in the quantum or the non-quantum (macrophysical) domain.

Very often this closing statement is taken – by exponents of the orthodox view – as a poignant admission, on Einstein's part, of the sheer irreducible strangeness of quantum phenomena and his own inability to make sense of them from a standpoint that combined 'classical' realism with the dictates of relativity theory.[31] However, such a reading will impose itself only if one accepts that Bohr's case is sufficiently proven and hence that the EPR authors are labouring against insuperable odds when they strive to reconcile the QM evidence with their preferred realist ontology. Otherwise the passage must surely be construed as a firm re-statement of the realist position along with their associated grounds for holding that the orthodox theory is both 'incomplete' – as shown by its failure (or downright refusal) to assign any physically intelligible content to the QM predictive–observational data – and based on a doctrinaire misunderstanding of what 'realism' properly entails. That is to say, Bohr insists that the impossibility of conducting simultaneous measurements of position and momentum on any given particle is reason enough to reject the idea that the particle none the less possesses such values as a matter of objective (verification-transcendent) truth. For him this follows also – as a matter of necessity – from the QM 'quantum of action', i.e. the fact that any measurement obtained will in some degree affect or 'disturb' the system under observation.[32] Thus, according to Bohr, the EPR authors have based their argument for realism in the quantum domain on a set of naive objectivist assumptions which are simply ruled out by factors intrinsic to any such experimental arrangement as that described in their paper. From the EPR standpoint, conversely, Bohr has himself introduced a whole range of orthodox quantum-theoretical beliefs which are open to challenge first on the grounds that they don't (as claimed) follow necessarily from the QM predictive–observational data, and second that they cannot be assigned any rational content or adequate physical representation.

Bohr's response is simply to repeat that quantum mechanics requires the abandonment of all such classical concepts, chief among them the idea of an 'objective', observer-independent 'reality' assumed to possess certain properties and attributes quite aside from any limits on the knowledge attainable through measurement and observation. More precisely, we have no choice but to carry on using those same concepts (since they are the only ones available to us), but must use them henceforth without the illusion that they 'represent' anything in the nature of quantum reality. What the EPR thought experiment actually reveals, according to Bohr, is 'an essential inadequacy of the customary viewpoint of natural philosophy for a rational account of physical phenomena of the type with which we are concerned'.[33] Thus:

> the finite interaction between object and measuring agencies conditioned by the very existence of the quantum of action entails – because of the impossibility of controlling the reaction of the object on the measuring instruments if these are to serve their purpose – the necessity of a final renunciation of the classical ideal of causality and a radical revision of our attitude towards the problems of physical reality.[34]

However, this assumes, once again, that the EPR thesis involves a commitment to some version of *epistemic* realism (i.e. the doctrine that measurement values must be known or at any rate knowable) in order for the physical elements over which they range to possess any claim to reality. In that case there would clearly be a difficulty with the EPR argument not only as concerns quantum physics – where the measurement problem is especially acute – but also with respect to macrophysical systems where it is likewise impossible, strictly speaking, to obtain precise simultaneous values for variables such as location and momentum. Bohr sees this as yet further evidence for his claim that these problems extend beyond the realm of subatomic phenomena to the entirety of physics and thence to every field of enquiry, whether in the natural or the human sciences.[35] From the EPR standpoint it should rather be viewed as a *reductio ad absurdum* of the orthodox QM doctrine, entailing as it does – most famously in the case of Schrödinger's 'superposed' cat – the impossibility of fixing a definite (non-arbitrary) point of transition from the quantum to the macrophysical domain.[36]

III

Thus it might well seem that the parties to this dispute are working with two such sharply opposed concepts of quantum-physical 'reality' that there is no prospect of resolving the question or even of achieving some minimal consensus on the chief points at issue between them. However, this is an overly defeatist conclusion and one that can best be avoided by looking again at what 'realism' means for the EPR authors, Bohm, and Bell on the one hand and for Bohr and orthodox QM theorists on the other. It will then become apparent that the latter take for granted – *both in their own approach to the issue and in that which they attribute to EPR* – an epistemic conception of realism which entails that all ascriptions of physical reality must meet the verificationist requirement of adequate empirical evidence or observational warrant. Yet this is not at all the version of realism propounded by the EPR authors, at least in their more careful and precise (which is also to say, less pressured or concessive) statements of the realist case. Rather, they are committed to an *alethic* and not an epistemic conception of realism, one that takes the truth-value of any statement to be determined by the way things stand in reality (quantum-physical reality included), and not by the scope or limits of our knowledge with respect to that same reality.

This distinction is brought out very clearly by William Alston when he surveys various kinds of anti-realist argument – among them those of Dummett and the later Putnam – and finds them to rest on a regular confusion between the two kinds of realist case. Thus he 'flatly denies' Dummett's verificationist claim that 'we could hardly grasp what it is for a statement to be true if we had no conception whatever of how it might be known to be true'.[37] After all, as Alston remarks, '[e]ven if we have such a conception it is not at all necessary for this to figure in our understanding of what it is for a statement to be true'.[38] Moreover, we can grasp the truth-conditions of numerous unverified sentences – such as 'theoretical statements in science, lawlike open-ended universal generalizations, and subjunctive conditionals' – without possessing the knowledge required by a Dummett-type verificationist approach. Alston argues a convincing case for this alethic or truth-based concept of realism as one that can successfully meet the whole range of sceptical counter-arguments since these latter always get a purchase through imposing some different (typically epistemic) criterion for ascriptions of reality and truth.

There is a passage in his book that I should like to quote at some length since it relates most closely to issues in the quantum-theoretical domain. Thus:

> Consider the status of speculative suggestions prior to the time at which someone figures out a way of empirically testing them, for example, the first glimmer of atomism in ancient Greece. When the idea first occurred to one that matter is composed of tiny, invisible, and indivisible particles with much empty space in between, no one had any idea of how to put the suggestion to the test because no one had developed ways of embedding this suggestion in a context that made possible the derivation of observationally testable consequences. In subsequent millennia, many such connections were hypothesized, of ever-increasing sophistication and complexity, and this brought the thesis within the range of empirical inquiry. But even at the beginning the sentence, 'All matter is composed of invisible, indivisible particles' had a meaning that was grasped by some people. It must have; otherwise those who worked on ways to connect it up with other suppositions so as to put it to the test would have had nothing to work on. It had a meaning and was understood before there were any 'verification conditions' associated with it.[39]

This seems to me a decisive argument against Dummett's anti-realist position and against any kindred version of the case for denying that truth-values can apply to sentences or statements for which there exists no adequate means of verification. Moreover, the same objection applies when that case is extended – as it is by Dummett – to historical statements, to counterfactual-supporting causal explanations in the natural sciences, and likewise to mathematical or logical theorems which as yet remain strictly 'undecidable' since lacking an adequate proof procedure.[40] In all such instances – according to alethic realism – we can still understand what it would mean for a statement to count as decidably true or false, even though we cannot say (as required by epistemic realism) just which items of more advanced knowledge or improvements of conceptual grasp might be involved in putting that statement to the test. It is a basic premise of alethic realism that truth-values are 'verification-transcendent' in the sense that they obtain in virtue of what is *actually the case* with respect to some given ontological domain, and hence irrespective of our present (or even our future best-possible) state of knowledge concerning them. For the epistemic realist, on the other hand, there will always be a problem in countering arguments – like those raised by Dummett – which push that thesis to its 'logical' (anti-realist) conclusion by denying that we could ever conceivably have knowledge of truths that transcend our utmost powers of empirical observation, factual ascertainment, or intellectual grasp. What gives such arguments their seeming plausibility is precisely the confusion between epistemic warrant and truth as a matter of objective validity conditions. For we can indeed *know* (in the alethic realist's sense) that these latter apply to all well-formed statements that are genuine candidates for truth – including statements that belong to Dummett's 'disputed class' – whether or not we presently possess the kinds of evidence or proof procedure that would suffice to settle the issue.

Now it might well be said that this argument has no bearing on the issue of realism versus anti-realism as concerns quantum mechanics. At least it is more than likely that this would be the position adopted by anyone who followed Bohr in accepting orthodox QM as a 'complete' theory of the given (empirically observable) phenomena and who therefore rejected the EPR argument along with all subsequent, Bohm-type versions of the realist case. That is, they would view such realist commitments as going beyond the empirical evidence and thereby involving a meaningless appeal to 'metaphysical' (verification-transcendent) notions of reality and truth. At which point the alethic realist might counter by adducing Alston's above-quoted passage on the example of ancient atomism. For it is just his point that a verificationist approach along these lines cannot explain the advancement of scientific knowledge from speculative hypotheses – of which it can at least be known that they are *either* true or false with respect to some specified aspect of physical reality – to the stage where those hypotheses achieve the status of conceptually adequate and empirically testable propositions.[41] Thus quantum mechanics would figure as a striking instance of the kinds of epistemic uncertainty that typically attend the early (speculative) phase of development in some new scientific theory but which carry no decisive implications with regard either to the ultimate nature of physical 'reality' or even to the ultimate scope and limits of human understanding. After all, as Alston remarks, '[i]f Newton knew what Einstein knew, he would not be justified in supposing his theory of gravitation to be correct'.[42] By the same token, if present-day orthodox QM theorists knew what might be known to some future (better informed) generation of quantum physicists, then they could just as well find themselves obliged to renounce the completeness doctrine and the concomitant idea that QM entails an irreversible break with every last precept of (so-called) classical realism. However, it is not this appeal to future possible states of knowledge that counts most importantly against the anti-realist ban on any appeal to objective (verification-transcendent) truth-values. Rather it is the fact – in Alston's words – that 'what confers a truth value on a statement is something independent of the cognitive-linguistic goings on that issued in that statement, including any epistemic status of those goings on'.[43] For there will otherwise always come a point in this debate where epistemic realism falls prey to just the kind of anti-realist logic that Dummett and the proponents of orthodox QM bring up in support of their own sceptical position.

One major problem with the EPR paper is that the authors tend to swing back and forth between an alethic and an epistemic concept of realism according to the localized context of argument. In some passages it is clear enough that they are adopting an alethic standpoint and doing so, moreover, for just the sorts of reason that Alston regards as counting strongly against any version of the epistemic case. Thus there is nothing in the least ambiguous about their statement with regard to the 'completeness' issue: that 'every element of the physical reality must have a counterpart in the physical theory'.[44] By framing the issue in just these terms, Einstein and his colleagues are careful to exclude the construal standardly placed upon it by Bohr, Heisenberg, and orthodox QM theorists. That is, they rule out the epistemic interpretation which holds that quantum mechanics must *necessarily* be considered 'complete' since it specifies the degree of uncertainty pertaining to all observations or measurements, and thereby establishes the absolute limits of our knowledge concerning certain physical phenomena. Dugald Murdoch again provides the clearest statement of a doctrine that is central to Bohr's thinking but often very murkily expressed. Thus:

> Mutual exclusiveness [i.e. of wave–particle ontologies or simultaneous measurements of position and momentum] is frequently thought to be the sole condition of Bohr's notion of complementarity .... [However] this is a mistaken view: the notions of mutual exclusiveness and of joint completion are equally necessary, indeed complementary, ingredients in the meaning of Bohr's conception. In the genesis of the conception the notion of joint completion came first (in the acceptance of wave–particle duality); the notion of mutual exclusiveness came later (in the acceptance of the uncertainty principle).[45]

This is why Bohr insisted – against Einstein – that the statistical aspect of orthodox QM was a strictly irreducible component of the theory and not (as in classical mechanics) a result of contingent limitations on our knowledge entailed by the extreme complexity of certain physical phenomena, thus requiring that those phenomena be treated in probabilistic terms. For this was exactly Einstein's point: that the orthodox theory was incomplete since it represented just such a state of limited knowledge, yet one that Bohr had raised to the status of a full-scale *Naturphilosophie* with distinctly irrationalist leanings. In which case there was no reason to deny that quantum mechanics, like its classical counterpart, might at length give way to a better, more complete interpretation which revealed the existence of hidden variables (or underlying causal mechanisms) and thereby removed its 'irreducibly' statistical character.

This is why the two main aspects of Bohr's doctrine – completeness and complementarity – can best be seen from an alethic-realist standpoint *not* as a final vindication of that doctrine but rather as a kind of blocking strategy adopted in order to deflect any challenge along the lines laid out in the EPR paper and then taken up by theorists like Bell and Bohm. However, as I said above, the EPR argument is not so clear when it comes to distinguishing the strong (alethic) from the weaker or more vulnerable (epistemic) conception of what 'realism' means in this context. One passage in particular is worth citing here since it shifts across from the first to the second position within the space of just a few sentences. Thus the authors begin by stating very firmly – in the alethic-realist mode – that any 'serious' evaluation of these issues 'must take into account the distinction between the objective reality, which is independent of any theory, and the physical concepts with which the theory operates'. These concepts, they continue, 'are intended to correspond with the objective reality, and by means of [them] we picture this reality to ourselves'.[46] Here the authors would seem to be steering a careful path between the twin perils of an epistemic realism that makes 'reality' dependent on our knowledge of it as achieved through that picturing relation and, on the other hand, a sceptical outlook – such as Bohr's – which insists that the 'reality' of quantum phenomena is forever beyond reach of any 'picture' (or indeed any concept, theory, representation, etc.) that we could possibly frame concerning it. Nevertheless, in requiring that those concepts should be 'intended to correspond with objective reality', the authors make it plain that the truth (or completeness) of any such theory must be reckoned in terms of the physical elements over which its values range, and not in terms of epistemic criteria relating to the scope or limits of humanly attainable knowledge.

Thus they are still – up to this point – defending an alethic conception of realism in the quantum domain even when they say that it is 'by means of these concepts that we picture this reality to ourselves'. For of course this allows that such a 'picture' may be *objectively* false or inadequate, and hence that – in quantum mechanics as elsewhere – truth is 'verification-transcendent' in the sense of not depending (as Dummett would have it) on our possession of the requisite evidence, knowledge, or means of epistemic access. However, the EPR authors then proceed to reformulate the issue in terms which leave room for a different interpretation, one that would fasten on their shift of emphasis from alethic to epistemic realism. I shall cite the relevant passage at length so as to bring out the tension that exists between these discrepant claims. Thus:

> In attempting to judge the success of a physical theory, we may ask ourselves two questions: (1) 'Is the theory correct?' and (2) 'Is the description given by the theory complete?'. It is only in the case where positive answers may be given to both of these questions, that the concepts of the theory may be said to be satisfactory. The correctness of the theory is judged by the degree of agreement between the conclusions of the

theory and human experience. This experience, which alone enables us to make inferences about reality, in physics takes the form of experiment and measurement.[47]

The first sentence here is perfectly consistent with what had gone before, i.e. with an alethic (truth-based) conception of realism which defines the correctness and completeness of scientific theories as a matter of objective correspondence with the way things stand in physical reality rather than invoking epistemic criteria like those adopted by orthodox QM theorists. In the second sentence also there is a straightforward appeal to objective (verification-transcendent) truth as precisely that condition which a theory has to 'satisfy' in order to count as correct and complete quite aside from any epistemic limits imposed by the available means of observation, measurement, or conceptual-explanatory grasp. However, the EPR authors then suddenly change tack and introduce a quite different criterion of 'correctness', namely 'the degree of agreement between the conclusions of the theory and human experience'. Moreover, it is this experience – which 'in physics takes the form of experiment and measurement' – that 'alone enables us to make inferences about reality'. One could find no more striking example of the slide from alethic to epistemic realism, or from an argument premised on the existence of objective (observer-independent) truths to an argument which lays itself open to just those liabilities exploited by orthodox QM theorists. For it is precisely by asserting this order of dependence – where truth is a matter of epistemic warrant and such warrant is defined in terms of 'human experience' – that anti-realists of various philosophical persuasion (Dummett and Bohr among them) are able to press their case.

IV

It seems to me that this passage was a weak point in the EPR paper which gave Bohr a counter-strategic opening and which has since led many commentators to assume that Bohr undoubtedly gained the upper hand in his discussions with Einstein. What it allowed him to do, in short, was shift the terms of debate onto ground of his own choosing where any claims that Einstein or his colleagues might make with regard to physical 'reality' would have to be justified according to epistemic criteria, or with reference to the kinds of knowledge presently attainable through 'experience', observation, experiment, or measurement. At which point, of course, the way was open for Bohr to raise all manner of objections to the EPR argument, in particular its crucial claim that 'if, without in any way disturbing a system, we can predict with certainty (i.e. with probability equal to unity) the value of a physical quantity, then there exists an element of physical reality corresponding to this physical quantity'.[48] The statement is crucial because – as I have said – it enunciates the chief premise of alethic realism as applied to issues in the quantum-theoretical domain or in any area of research where there is a question concerning the status of truth-claims that go beyond the presently available evidence. However, it can easily be made to appear just a throwback to old (pre-quantum) notions of 'objective' reality and truth once the EPR authors have entered their appeal to 'experience' as the ultimate grounds for deciding what shall count as a verifiable statement or a well-framed (truth-evaluable) hypothesis. Thus, in Bohr's words, '[w]e now see that the wording of the above-mentioned criterion of physical reality proposed by Einstein, Podolsky and Rosen contains an ambiguity as regards the meaning of the expression "without in any way disturbing the system"'.[49] That ambiguity results – so he claims – from their not taking fully into account the uncertainty relations, the quantum of action, and the consequent impossibility of achieving any precise simultaneous measurement of conjugate variables such as position and momentum.

Let me quote Bohr's (supposedly) clinching statement once again since it shows just how much argumentative ground the EPR authors conceded when they made that sudden mid-paragraph switch from an alethic to an epistemic definition of 'physical reality'. What they ignore, Bohr writes,

> is essentially the question of an influence on the very conditions which define the possible types of predictions regarding the future behaviour of the system. Since these conditions constitute an inherent element of the description of any phenomenon to which the term 'physical reality' can be properly attached, we see that the argumentation of the mentioned authors does not justify their conclusion that quantum-mechanical description is essentially incomplete.[50]

Now there is nothing whatsoever in this line of counter-argument that would force any retraction or substantive re-thinking on the EPR authors' part had they but remained unambiguously committed to an alethic-realist standpoint. That is to say, Bohr's objection would have no purchase – would show up as a plain misconstrual of their clearly articulated case – except for that sentence where they yield a hostage to orthodox QM fortune by suggesting (*contra* their statements elsewhere) that '[t]he correctness of the theory is judged by the degree of agreement between the conclusions of the theory and human experience'. For otherwise it can be seen that every claim put forward in the above passage by Bohr presupposes the truth (the correctness and completeness) of orthodox QM and pre-emptively discounts any notion of 'physical reality' as existing quite apart from the various epistemic restrictions that happen to be placed on our knowledge of it or our powers of observation and measurement.

So there is indeed, as Bohr remarks, a crucial 'ambiguity' in the EPR paper but not necessarily where he wishes to locate it or pointing to the kind of resolution he hopes to achieve. That resolution is one that will finally secure the twin foundational postulates of orthodox QM, i.e. the theory's completeness in defining the ultimate scope and limits of our knowledge with respect to quantum phenomena, and also its requirement of complementarity as a means of negotiating just those limits while perforce continuing to use certain 'classical' concepts and terminology. One can see most clearly how this argument works by citing some remarks that follow straight on from the passage quoted above. So far from being 'incomplete', Bohr argues,

> this description ... may be characterized as a rational utilization of all possibilities of *unambiguous* interpretation of measurements, compatible with the finite and uncontrollable interaction between the objects and the measuring instruments in the field of quantum theory. In fact, it is only the mutual exclusion of any two experimental procedures, permitting the *unambiguous* definition of complementary physical quantities, which provides room for new physical laws, the coexistence of which might at first sight appear irreconcilable with the basic principles of science. It is just this entirely new situation as regards the description of physical phenomena, that the notion of complementarity aims at characterizing [my italics].[51]

What is most striking about this passage is the extent to which Bohr's argument relies on a foregone assumption as to the correctness and completeness of orthodox QM, and – following from that – the requirement that all debate should henceforth be conducted in terms that acknowledge the 'entirely new situation' which has thus come about. So if certain QM-derived principles prove to be at odds with certain 'basic principles of science', then these latter need not be abandoned outright but should rather be treated – in 'complementary' fashion – as alternative, incompatible, yet somehow non-conflicting descriptions of the 'same' phenomena. And if this very notion of the 'coexistence' of rival descriptions or theories should still come into conflict with a basic principle of science (that either one or other, or perhaps neither, but at any rate *not both* can be thought of as providing the correct description), then again the issue can only be resolved by way of complementarity. For it is an additional merit of this approach – as Bohr sees it – that rationality is also redefined so as to allow for a suspension of choice between rival descriptions, or the acceptance of both (in any given case) as describing different, complementary aspects of the physical 'reality' in question. Thus the 'mutual exclusion of any two experimental procedures' is an epistemic lesson in the limits of observation–measurement which should lead us to adopt complementarity as a generalized principle which applies at every level of interpretation, or whenever two theories apparently come into conflict. That it nevertheless seems not to apply, strangely enough, in the case of Bohr versus EPR is an anomaly passed over in silence by Bohr and his orthodox followers.

In short, the whole drift of Bohr's argument is to rule out heterodox (realist) construals of quantum mechanics – along with the 'classical' framework of concepts to which they subscribe – while purporting to accept the widest possible range of candidates for treatment as alternative (complementary) modes of description. This emerges most clearly in the way that Bohr pursues his point about the 'ambiguity' of certain formulations in the EPR paper and turns it around so as to confirm the 'unambiguous' correctness and completeness of orthodox QM. (I have italicized the two occurrences of this word in the above-cited passage so as to bring out the emphasis they carry as a key point of strategic purchase in Bohr's handling of the issue.) For if the standard theory may indeed be characterized 'as a rational utilization of all possibilities of unambiguous interpretation of measurements', then of course, according to the orthodox theory, any other interpretation (such that proposed by EPR) will inevitably lapse into ambiguity when it comes to addressing the measurement problem as defined by Bohr. The point is pushed home in the sequel clause where he stipulates that those same possibilities of measurement must be 'compatible with the finite and uncontrollable interaction between the objects and the measuring instruments in the field of quantum theory'. This not only takes for granted the correctness/completeness of that theory but also introduces its own ambiguity as regards the closing phrase. Thus it is far from clear – since the grammar pulls both ways – whether the 'field' in question is that of quantum-physical 'objects', 'measuring instruments', etc., or whether it is the field of 'quantum theory' wherein these issues are subject to debate. But again, on a more charitable reading, perhaps this apparent ambiguity (or downright confusion) should rather be taken as making Bohr's point that no such distinction any longer obtains in the realm of quantum mechanics.

The second occurrence of the word 'unambiguous' signals even more plainly what a weight of interpretive presupposition goes along with Bohr's avowed appeal to the straightforward (uninterpreted) 'evidence' of quantum phenomena. 'In fact', he writes, 'it is only the mutual exclusion of any two experimental procedures, permitting the unambiguous definition of complementary physical quantities, which provides room for new physical laws, the coexistence of which might at first sight appear irreconcilable with the basic principles of science'.[52] I have just commented at length on this passage so will here remark only that it pre-empts the issue with regard to the EPR paper by decreeing in effect that no statement shall count as 'unambiguous' except in so far as it accepts the completeness of orthodox QM theory and hence the idea that it cannot do more than represent one alternative (complementary) description of a quantum 'reality' that must otherwise elude any means of description whatsoever. So it is that Bohr can advance his claim for the necessity of revising even those 'basic principles of science' which fail to find room for complementarity and his other philosophical extrapolations from the standard (Copenhagen) account. For in his view it is simply indefensible – a product of fixed 'metaphysical' prejudice or failure to acknowledge 'this entirely new situation as regards the description of physical phenomena' – that anyone should question the evident need for a wholesale revision of realist principles in line with the requirements of orthodox quantum theory.

Thus what counts as 'unambiguous', on Bohr's understanding, is a principle of generalized complementarity applied to all statements in the quantum-physical domain which can only strike the realist as a massive evasion of the issue, or as a way of seeking refuge in vague pronouncements incapable of any more precise or adequate expression. Even Murdoch – one of Bohr's more sympathetic commentators – feels compelled to admit certain misgivings when it comes to the complementarity doctrine. As he puts it:

> Someone who holds a more realist theory of models would find Bohr's practice very unsatisfactory, for the complementary use of the dual models provides no intelligible conception of the real nature of the electron at all; indeed it makes its nature highly mysterious. Adopting a pragmatist approach, however, Bohr is content, since the complementary use of disparate models is logically consistent and allows *all* empirical data to be subsumed.[53]

In other words QM theory is 'complete' and 'correct' according to just one possible set of criteria, namely those of the logical-empiricist programme, which achieved its widest acceptance among philosophers of science at the time when that theory (on its orthodox construal) was likewise taking command of the field.[54] Moreover – *as we have seen – there is a strong and continuing two-way connection between this interpretation of quantum mechanics and the kinds of anti-realist (or verificationist) approach developed by philosophers, such as Michael Dummett, who stand squarely in the line of descent from 'old-style' logical empiricism. Thus Dummett's proposals display a striking affinity with various aspects of the orthodox QM position, among them his denial of objective or verification-transcendent truths, his rejection of bivalence as applied to statements in the so-called 'disputed class', and his appeal to criteria of epistemic access – or 'warranted assertability' – as a means of effectively ruling out any form of alethic (truth-based) realist argument.[55] Whence Dummett's steadfast refusal to grant that there might be statements for which as yet we possess no adequate evidence or proof procedure, but of which we can none the less assert that they *must* be either true or false in virtue of their making some definite claim with respect to some determinate (though to us unknown or epistemically inaccessible) fact of the matter. It was on this point precisely that Einstein took issue with Bohr when he stated – without the least trace of ambiguity – that the criterion of truth (of correctness and completeness), in quantum mechanics as elsewhere, was that 'every element of the physical reality must have a counterpart in the physical theory'. And it was on this point also that Bohr came back with his arguments against the possibility of maintaining any such realist standpoint while taking account of the quantum of action, the measurement problem, and other such (in his view) strictly inescapable entailments of orthodox QM theory.

V

It should be clear by now that I regard these problems as resulting from a number of deep-laid philosophic confusions on Bohr's part rather than as posing an ultimate obstacle to any statement of the realist case. I doubt whether the orthodox theory would have gained such a hold had the EPR authors been more consistent in adopting an alethic (objective and truth-based) conception of scientific realism, and thus avoiding the all too easy slide from the weaker (epistemic) version of their argument to a verificationist stance which allowed Bohr to claim victory on his own terms. Most likely it was Einstein's recognition of this fact that led him to renounce his earlier (Mach-inspired) positivist belief that scientific theories should 'save the phenomena' and had no business advancing realist or causal-explanatory theories which claimed to go beyond the empirical evidence.[56] At any rate the EPR argument – as I have construed it here – is one that requires both a correspondence theory of truth and a conception of scientific realism that indeed goes beyond such evidence in supposing that there exist features of 'physical reality' (subatomic structures, charges on particles, causal dispositions, and so forth) which determine the truth or otherwise of any theory we may hold concerning them.

Moreover, this case is no way threatened by the standard range of orthodox QM counter-arguments from observer 'interference' or the limits of precise measurement as applied to the quantum domain. For it is another chief point of the realist (EPR or Bohm-type) interpretation that such effects are themselves to be understood as a component part of the given physical 'situation', and hence that any uncertainties encountered have to do with the limits of our present-best knowledge, rather than pertaining to the very nature of quantum-physical 'reality'. Einstein made this point most explicitly in a passage from his 1949 'Reply to Criticisms':

> I am, in fact, firmly convinced that the essentially statistical character of contemporary quantum theory is solely to be ascribed to the fact that it operates with an incomplete description of physical systems .... Assuming the success of efforts to accomplish a complete physical description, the statistical quantum theory would, within the framework of future physics, take an approximately analogous position to statistical mechanics within the framework of classical mechanics.[57]

It is a crucial passage for the obvious reason that orthodox QM staked its claim to 'completeness' very largely on the contrary assumption, i.e. that the statistical character of quantum mechanics was an intrinsic property or feature which could not be resolved by any future advance toward a better (more complete) understanding. What the Einstein passage brings out with particular force is the close link which he takes to exist between the correspondence theory of truth as applied to statements in the quantum-physical domain and the realist requirement that these should be construed as possessing a more than merely empirical or observational content. That is to say, the physicist is justified – according to Einstein – in treating the data as evidence for a theory that is verification-transcendent in the sense that it postulates 'elements of physical reality' which cannot be observed (for all the well-known QM reasons) but which *can*, none the less, be known to exist in virtue of certain 'basic principles of science'. Of course it is those same principles that Bohr considers to require drastic revision in consequence of 'this entirely new situation as regards the description of physical phenomena'. But here one may note the saving ambiguity about that term 'phenomena' which on the one hand enables Bohr to preserve some idea of a quantum-physical domain 'out there' to be described and theorized, while on the other it leaves him free to suggest that no such distinction any longer obtains between 'objective' reality and whatever we can know of it through experiment or observation. This is what Popper called the 'great quantum muddle' – referring specifically to the orthodox QM position as regards statistical evidence – and it does seem to me that Bohr's statements invite such a charge.[58]

Of course there are problems – too well-known to require any detailed rehearsal here – with the correspondence theory of truth, especially when stated (as by Einstein) in terms of a direct relation between 'elements of physical reality' and 'elements of physical theory'.[59] Chief among them is that of explaining just what this correspondence relation is supposed to entail since most of the standard answers can be shown to involve some form of apparently tautologous or circular reasoning. Thus it might be said that the truth or falsehood of a given proposition depends quite simply on whether or not it corresponds to the facts of the case. But this invites the rejoinder that 'facts' are themselves discursive constructs – or can only take the form of statements, propositions, sentences, etc. – rather than objectively existent entities belonging on the other (real-world) side of the supposed correspondence relation. Then again, it may be argued that truth ascriptions (such as 'the charge on every electron is negative') are valid just so long as it is *actually the case* that this claim holds good of physical reality as concerns the charge on electrons. However, at this point the practised anti-realist will once again object that such talk merely disguises the circular equivalence between 'truth' and what is 'actually the case', both of which can only be construed in linguistic, discursive, or representational terms.

This circularity thesis has played a central role in a great many recent rebuttals of the realist position, from Richard Rorty's all-purpose version of the argument to Dummett's more logically refined variations on the theme.[60] All the same, it is a highly questionable thesis, as Alston remarks in a clear-headed passage that merits extended quotation. '"Is true" and "is the case that" are importantly different', he remarks.

> 'Is true' applies to the likes of propositions, statements, and beliefs, items on the *thought* side of the thought-world relationship, whereas 'is the case

that' belongs on the other side. A proposition is *true* when it is related in the right kind of way (identity of content) to something that *is the case*. A state of affairs' being the case is the worldly realization that renders the proposition *true*. What is the case is the *truth maker*. What is true is the *truth bearer*. And this is precisely the realist conception of truth. To bring this out is to bring out the content of that concept.[61]

It seems to me that this distinction has been blurred or completely lost from sight in much of the post-EPR debate about quantum mechanics and its supposed implications for our knowledge of physical reality. In Bohr's case it is assumed as a matter of orthodox QM principle that no such distinction can possibly hold for anyone who has grasped the elements of quantum theory. Hence his famous assertion that 'nobody who is called a philosopher' can be thought to have properly understood quantum mechanics and the challenge it offers to our normal conceptions of reality, logic, and truth.[62] On the contrary, I have argued: this challenge will appear insurmountable *only if* one accepts that orthodox QM is the sole interpretation compatible with the evidence, and must therefore set the terms for all further debate as regards its putative bearing on epistemological and ontological issues. Otherwise there is nothing – orthodox prejudice aside – to prevent one from following Einstein, Bell, and Bohm in attributing the measurement problem and its various related quandaries to the limits of our present-best knowledge, rather than treating them as somehow intrinsic to the nature of quantum 'reality'.

One sure sign that there is something amiss with the orthodox Copenhagen view – that, in Popper's caustic words, there is 'something rotten in the state of Denmark' – is the support that it has offered for a range of extravagantly counter-intuitive or downright implausible conjectures. Chief among them is the 'many-worlds' QM interpretation, put forward by its advocates – including David Deutsch – as a strictly unavoidable consequence of the theory in its basic 'uninterpreted' form, yet involving a baroque multiplicity of coexistent worlds or 'universes' which harks back to Leibniz and the period of high pre-Kantian rationalist metaphysics.[63] (See Chapters 4 and 5 of this book for a detailed discussion of the many-worlds theory and what I take to be its various philosophical confusions.) Then again, there is John Wheeler's equally fantastic quantum-astrophysical deduction from the evidence of nonlocal interaction phenomena[64] (construed in orthodox QM terms) to what he takes as the entailed necessity that we can somehow actually *affect or decide* what appears to have occurred billions of light-years 'ago' in some remote corner of the radiotelescopically observable cosmos.[65] Thus, in Wheeler's words, 'the observing device in the here and now, according to its last-minute setting one way or the other, has an irretrievable consequence for what one has the right to say about a photon that was given out long before there was any life in the universe'.[66] And again, lest the point should not register with sufficient impact, 'what we have the right to say of past spacetime, and past events, is decided by choices – of what measurements to carry out – made in the near past and now. The phenomena called into being by these decisions reach backward in time ... to the earliest days of the universe'.[67]

I am grateful to one of my anonymous readers at typescript stage for clarification of the thinking that lies behind Wheeler's extraordinary claim. In brief, it has to do with the fact (in their words) that 'photon propagation is instantaneous from the point of view of a frame travelling with the speed of light', and moreover that 'the distance travelled also tends to zero', in which case the velocity will remain at c (the speed of light) for all frame-internal measurement purposes. Hence the apparent plausibility of Wheeler's argument, premised as it is on a combination of orthodox QM theory and theses derived from special relativity. However, this yoking together of the two theories ignores all the salient points of conflict between them, in particular – as we have seen – the problem with conceiving how effects of time-reversal at the subatomic (quantum) level can possibly be thought of as carrying across into the macrophysical domain. No doubt there is a sense in which quantum mechanics and relativity theory both involve renouncing any notion of absolute time, or any classical (Newtonian) conception of time intervals as possessing a determinate value irrespective of the kind of measurement that is carried out or the reference frame in question. But this does not licence a conflation of the two theories so as to bring them out jointly in support of a claim like Wheeler's, one that would entail such wildly paradoxical or downright nonsensical implications. After all, as Eberhard remarks:

> [c]ausal effects backward in time may create a problem because of causal loops. Tomorrow, we can set up a parameter that influences an observation made today and, if the theory allows information understandable to humans to flow backward in time, we can send ourselves a message from the future to the present. Then the theory would allow us to inform ourselves of the occurrence of some events tomorrow and we could take an action today to prevent their occurrence. Equations have to be carefully set up to avoid such nonsense.[68]

These imaginary scenarios – and the conceptual problems they engender – will of course be familiar enough to devotees of science-fiction and films like *Back to the Future*. What is so striking about Wheeler's hypothesis (such as Deutsch's multiverse conjecture) is the way that it jumps straight to the most extravagantly far-fetched conclusions on the basis of a speculative theory which by its very nature could not be borne out by empirical testing on the relevant physical scale. That is to say, it extrapolates from the evidence of delayed-choice experiments at the quantum level to the idea of retroactive causal influence across vast distances of space–time separation as if this involved no conflict either with relativity-theory or with the actual, real-world operative constraints upon our knowledge of remote astrophysical events.

Nothing could be further from Eberhard's statement of the basic realist premise, i.e. that if a theory fails to respect those constraints or produces results that contravene them then 'equations have to be carefully set up to avoid such nonsense'. Or again – following Bohm – some way must be found to acknowledge the existence of remote superluminal correlation effects while also explaining why these do not (and cannot) have the kinds of strictly unthinkable consequence that Wheeler is happy to take on board.[69] One could scarcely wish for a plainer, more striking example of the confusion that results from a failure to distinguish epistemic (or empirical) from alethic (objective or ontological) orders of truth claim. Thus, in the passage cited two paragraphs above, Wheeler is able to advance his case for retroactive causation and yet avoid manifest absurdity only by couching that claim in epistemic terms ('what we have the right to say'), while none the less appearing to assert a far more radical thesis with respect to the influence of present 'choices' on events in the remote astrophysical past. Here again, as with Bohr, the argument is carried very largely on certain ambiguous terms – 'phenomenon' chief among them – which allow Wheeler to equivocate between the weak (philosophically innocuous) and the strong (more dramatic and indeed quite mind-boggling) version of his argument. On the former, such puzzles can most plausibly be put down to observational anomalies, artefacts of measurement, or the limits of our current conceptual grasp as concerns those same 'phenomena'. On the latter, such caution is thrown to the wind and one has to interpret Wheeler as claiming that momentary choices of measurement parameter taken in the 'here-and-now' of observation can literally decide what *shall have occurred* all those billions of light-years 'away' or 'ago'. This follows – it is argued – by the strictest order of necessity if one accepts the empirical QM data (especially the results of Bell-type delayed-choice experiments[70]) along with the orthodox interpretation as regards superposed quantum states and their 'collapse' through the act of observation–measurement so as to yield determinate values of position or momentum.

Thus, as Wheeler tautologically puts it, '[n]o elementary phenomenon is a phenomenon until it is a registered (observed) phenomenon'. And again:

> In the delayed-choice version of the split-beam experiment … we have no right to say what the photon is doing in all its long course from point of entry to point of detection. Until the act of detection the phenomenon-to-be is not yet a phenomenon. We could have intervened at some point along the way with a new measuring device; but then regardless whether it is the new measuring device or the previous one that happens to be triggered we have a new phenomenon. We have come no closer than before to penetrating to the untouchable interior of the phenomenon. For a process of creation that can and does operate everywhere, that reveals itself and yet hides itself, what could one have dreamed up out of pure imagination more magic – and more fitting – than this?[71]

Moreover, there is no reason why this experiment should not be extended 'from the laboratory level to the cosmological scale' by way of the analogy with gravitational lensing and its effect upon our knowledge of (or influence upon) remote and long-past astrophysical events.[72] Here again, like Bohr, Wheeler equivocates between a weak (epistemic) and a strong (ontological) version of the claim. That is, it has to do *either* with the extent of our observational capacities, means of measurement, powers of conceptual grasp, etc., *or* with whatever must have been the case in respect of those past events and the (supposed) retrocausal influence upon them of measurements conducted in the 'here-and-now'. Wheeler may appear to disavow this latter interpretation when he denies us any 'right to say' what the photon did – or which path it travelled – from source to point of detection. Of course this follows as a matter of principle from the orthodox QM theory according to which no 'phenomenon' exists until it is measured or observed. In that case his argument, though highly counter-intuitive, may still be seen as making an essentially epistemic point and hence as carrying no further, more dramatic implications as concerns retroactive causal influence or the effect of present choices on past events. Yet clearly Wheeler *does* want to make a much stronger claim since his whole purpose in projecting the delayed-choice experiment onto a cosmological scale is to assert that such effects are not just conceivable but must rather be thought of as *actually having occurred* in the kind of complex physical system (quasar + intervening galaxy + suitably disposed measuring apparatus + sentient observer) described in his paper. Thus Wheeler's more reserved statements – and his talk of 'phenomena' as a usefully ambiguous means of evading the issue – should not be taken as retracting or significantly qualifying his claim that present causes can indeed have past effects over vast spacetime intervals.

It is worth trying to sort out just what is involved in this claim since (like the case of Schrödinger's superposed cat) it may be thought to constitute a striking *reductio ad absurdum* of orthodox QM theory.[73] In the original delayed-choice experiment, as Wheeler describes it,

> one decides whether to put in the half-silvered mirror or take it out at the very last minute. Thus one decides whether the photon 'shall have come by one route, or by both routes', after it has '*already done* its travel' .... In this sense, we have a strange inversion of the normal order of time. We, now, by moving the mirror in or out have an unavoidable effect on what we have a right to say about the *already* past history of that photon.[74]

Here again there is a hint that room is being left for the weaker (epistemic or phenomenalist) version of the claim through the inclusion of that telltale phrase 'what we have a right to say', suggesting as it does that these puzzles have to do with the limits of measurement (or maybe the limits of human conceptual grasp) rather than the ultimately puzzling nature of quantum-astrophysical reality. Yet of course the whole gist of Wheeler's argument is to cancel this concessionary clause and assert that we *do* quite literally decide by which path the photon shall have come 'after it has already done its travel'; hence that we really *do* have to reckon with 'a strange reversal of time'. In which case, even though the 'phenomenon-yet-to-be' is 'not yet a phenomenon', still it must be thought of as becoming a phenomenon (along with one or other of its retro-causally actualized alternative histories) at the moment of detection or insertion/non-insertion of the half-silvered mirror. Moreover, the very term 'phenomenon' at this point assumes a very different meaning, since it is no longer effectively confined to its orthodox QM usage, i.e. 'whatever shows up through the act of measurement quite aside from any putative "reality" behind appearances'. Rather it has to be construed in ontological terms as applying to an element of physical reality (the photon) which has undergone certain physically specifiable events, actions, influences, etc., and which has thus – strange as this may seem – had its 'previous' history eventually decided by a 'future' choice of measurement parameter. One can hardly blame Wheeler for inserting concessionary clauses which, unlike the half-silvered mirror, tend very often to moderate or soften the paradoxical import of his argument. Still it is well to be clear about just how far that argument conflicts with any realist conception of space and time or any physical theory (special and general relativity included) that does not accept orthodox QM as setting the very terms and conditions for all subsequent debate.

This is yet more evident when Wheeler goes on to extrapolate from the small-scale (laboratory) version of the delayed-choice experiment to its supposed cosmological implications. 'We get up in the morning', he writes,

> and spend the day in meditation whether to observe 'by which route' or to observe interference between 'both routes'. When night comes and the telescope is at last usable we leave the half-silvered mirror out or put it in, according to our choice. The monochromatizing filter placed over the telescope makes the counting rate low. We may have to wait an hour for the first photon. When it triggers a counter, we discover 'by which route' it came with the one arrangement; or, by the other, what the relative phase is of the waves associated with the passage of the photon from source to receptor 'by both routes' – perhaps 50,000 light years apart as they pass the lensing galaxy G-1. But the photon has already *passed* that galaxy billions of years before we made our decision. This is the sense in which, in a loose way of speaking, we decide what the photon *shall have done* after it has *already* done it. In actuality it is wrong to talk of the 'route' of the photon. For a proper way of speaking we recall once more that it makes no sense to talk of the phenomenon until it has been brought to a close by an irreversible act of amplification.[75]

I have cited this passage at length because it seems to me a very striking example of the various philosophical confusions brought about – here on a massively 'amplified' scale – by adherence to orthodox QM doctrine. Once again, there is the routine scattering of scare-quotes ('by which route', 'by both routes', etc.) and the equally routine reminder that such phrases are not to be taken literally or assigned a definite physical reference, any more than the 'loose way of speaking' which has us 'decide what the photon shall have done after it has actually done it'. For Wheeler is quite aware – like Dummett – that one has to tread warily in arguing the case for time reversal or retroactive causation lest one be confronted with a whole range of intractable problems and paradoxes.[76]

Yet there is no getting around Wheeler's claim that our present choice as to which kind of measurement to make must itself decide which of the various possibilities shall have been actualized, and therefore which astrophysical sequence of events – from source, via lensing galaxy, to point of detection – explains the observational results as measured. Any doubt on this score is soon dispelled when he reverts to 'strong' retrocausalist form and declares that 'the phenomena called into being by these decisions reach back in time in their consequences ... back even to the earliest days of the universe'.[77] This passage effectively forces the issue since its claim is either trivial ('phenomena' = 'what we are entitled to make of these appearances and their prehistory according to orthodox QM doctrine') or committed to the full-scale ontological version of the thesis ('what those appearances tell us concerning astrophysical events maximally remote from us in space and time'). There is a similar decision to be made about Wheeler's next sentence where he declares that '[r]egistering equipment operating in the here and now has an undeniable part in bringing about that which appears to have happened'.[78] Again, this is either a trivial truth – one that self-evidently applies to every act of observation on whatever scale and in whatever physical context – or (if one distributes the emphasis differently) another full-scale statement of the case for retroactive causal influence. However, such doubts are pretty much resolved by the two sentences which follow. Thus, '[useful as it is under everyday circumstances to say that the world exists "out there" independent of us, that view can no longer be upheld. There is a strange sense in which this is a "participatory universe"'.[79]

VI

Physicists very often – and sometimes with good reason – get impatient with philosophers who presume to tell them how science ought to be done or how best to make sense of their own observations, methods, theories, etc. However, as I have argued, quantum mechanics is a field so rife with philosophical confusions that they (the philosophers) have an interest and perhaps even a competence in sorting such matters out. The above-cited passages from Wheeler are a fair sample of what goes wrong when a phenomenalist doctrine such as that of orthodox QM becomes mixed up with far-reaching though obscure ontological claims that lead on to quasi-mystical talk about the 'untouchable interior of the phenomenon'. Or again, '[f]or a process of creation that can and does operate everywhere, that reveals itself and yet hides itself, what could one have dreamed up out of pure imagination more magic – and more fitting – than this?'[80] My point is not (of course) that such statements are typical of Wheeler's thinking, nor that they represent the kinds of conjecture typically drawn from orthodox QM premises by theorists of a speculative bent. Rather, it has been my purpose here to examine those other, less extravagant but more pervasive sorts of confusion that have dogged the debate around quantum mechanics from its inception to the present day.

This is not to deny that philosophers have both contributed to the prevailing anti-realist consensus in QM theory and used it on occasion to derive support for their own kindred views. Nor is it to claim (absurdly) that philosophers are always better placed to resist those kinds of doctrinaire reasoning from dubious premises that have characterised so much of that debate in the orthodox QM line of descent. Thus Nelson Goodman, for one, finds welcome support from Wheeler for his extreme constructivist idea that we 'make' stars – just as we 'make' constellations – through a process of radical 'worldmaking' which extends not only to the furthest reaches of present observation but also right back to the origins of the universe since space and time cannot be conceived as existing independently of human knowledge or experience.[81] Kant has a lot to answer for here, along with those later philosophers, Putnam among them, who have enlisted Kant on the side of their own – sometimes QM-influenced – modes of anti-realist (or framework-relativist) thinking.[82] On the other hand these *are* philosophical issues in the sense that they involve matters of interpretation which cannot be settled – *pace* theorists like Deutsch and Wheeler – by a straightforward appeal to the quantum-physical 'evidence' which is then taken to justify all manner of far-flown speculative claims. At the very least it seems wise to suspend judgement with regard to the kinds of metaphysical extravagance proposed by advocates of the many-worlds theory or the stance adopted by those of a more sceptic disposition who would count reality a world well lost in response to the challenge of quantum mechanics. For there can scarcely be rational warrant for drawing such extreme or paradoxical conclusions from a field of thought where problems with the 'orthodox' model have not been resolved – rather, if anything, deepened and multiplied – despite almost a century of intensive debate.

Endnotes

1 See Niels Bohr, *Atomic Theory and the Description of Nature* (Cambridge: Cambridge University Press, 1961); *Atomic Physics and Human Knowledge* (New York: Wiley, 1958); *The Philosophical Writings of Niels Bohr*, 3 vols. (Woodbridge, CT: Oxbow Press, 1987); *Collected Works*, eds. Léon Rosenfield and Erik Rüdinger, 6 vols. (Amsterdam: North-Holland, 1962–85); also Henry J. Folse, *The Philosophy of Niels Bohr: the framework of complementarity* (North-Holland, 1985); John Honner, *The Description of Nature: Niels Bohr and the philosophy of quantum physics* (Oxford:

Clarendon Press, 1987); Dugald Murdoch, *Niels Bohr's Philosophy of Physics* (Cambridge University Press, 1987).

2 Cited in James T. Cushing, *Quantum Mechanics: historical contingency and the Copenhagen hegemony* (Chicago: University of Chicago Press, 1994), p. 32.

3 See Albert Einstein, B. Podolsky and N. Rosen, 'Can Quantum-Mechanical Description of Reality be Considered Complete?', *Physical Review*, series 2, Vol. 47 (1935), pp. 777–80, reprinted in J.A. Wheeler and W.H. Zurek (eds.), *Quantum Theory and Measurement* (Princeton: Princeton University Press, 1983); all further references will be to this edition and indicated by 'EPR' with page number. See also Niels Bohr, article under the same title, *Physical Review*, Vol. 48 (1935), pp. 696–702; Bohr, 'Conversation with Einstein on Epistemological Problems in Atomic Physics', in P.A. Schilpp (ed.), *Albert Einstein: philosopher–scientist* (La Salle, IL: Open Court, 1969), pp. 199–241; Arthur Fine, *The Shaky Game: Einstein, realism, and quantum theory* (Chicago: University of Chicago Press, 1986); Tim Maudlin, *Quantum Non-Locality and Relativity: metaphysical intimations of modern science* (Oxford: Blackwell, 1993); Michael Redhead, *Incompleteness, Nonlocality and Realism: a prolegomenon to the philosophy of quantum mechanics* (Oxford: Clarendon Press, 1987).

4 See entries under Note 3, above; also – from a range of interpretative standpoints – Evadro Agazzi (ed.), *Realism and Quantum Physics* (Amsterdam: Rodopi, 1997); David Z. Albert, *Quantum Mechanics and Experience* (Cambridge, MA: Harvard University Press, 1993); Bernard D'Espagnat, *Veiled Reality: an analysis of present-day quantum-mechanical concepts* (Reading, MA: Addison-Wesley, 1995); Peter Holland, *The Quantum Theory of Motion* (Cambridge: Cambridge University Press, 1993); Dipankar Home, *Conceptual Foundations of Quantum Mechanics: an overview from modern perspectives* (New York: Plenum, 1998); Karl Popper, *Quantum Theory and the Schism in Physics* (London: Routledge, 1992); Alastair I.M. Rae, *Quantum Mechanics: illusion or reality?* (Cambridge University Press, 1986); Euan Squires, *The Mystery of the Quantum World*, 2nd edn. (Bristol & Philadelphia: Institute of Physics Publishing, 1994); A. Sudbury, *Quantum Mechanics and the Particles of Nature* (Cambridge University Press, 1986).

5 On this distinction, see especially William P. Alston, *A Realist Conception of Truth* (Ithaca, NY: Cornell University Press, 1996); also *Epistemic Justification* (Cornell University Press, 1979).

6 Immanuel Kant, *Critique of Pure Reason*, trans. N. Kemp Smith (London: Macmillan, 1964).

7 Honner, *The Description of Nature* (op. cit.) and other entries under Note 1, above.

8 *The Philosophical Writings of Niels Bohr* (op. cit.), Vol. 1, pp. 53–4.

9 Bohr, *Essays 1958–1962 on Atomic Physics and Human Knowledge* (London: Wiley, 1963), p. 10.

10 See Note 3, above.

11 J.S. Bell, *Speakable and Unspeakable in Quantum Mechanics: collected papers on quantum philosophy* (Cambridge: Cambridge University Press, 1987); David Bohm, *Causality and Chance in Modern Physics* (London: Routledge & Kegan Paul, 1957); Bohm and B.J. Hiley, *The Undivided Universe: an ontological interpretation of quantum theory* (London: Routledge, 1993).

12 Bell, *Speakable and Unspeakable in Quantum Mechanics* (op. cit.), p. 190.

13 See entries under Note 3, above; also James T. Cushing and Ernan McMullin (eds.), *Philosophical Consequences of Quantum Theory: reflections on Bell's Theorem* (Notre Dame, IN: University of Notre Dame Press, 1989).

14 J.F. Clauser and A. Shimony, 'Bell's Theorem: experimental tests and implications', *Reports on Progress in Physics*, Vol. 41 (1978), pp. 1881–1927; p. 1921.

15 See entries under Notes 3, 4, 11, 13 and 14, above.

16 Karl Popper, *Quantum Theory and the Schism in Physics* (op. cit.).

17 For further discussion of this and related issues see Evadro Agazzi (ed.), *Realism and Quantum Mechanics* (Amsterdam and Atlanta: Rodopi, 1997); David Z. Albert, 'Bohm's Alternative to Quantum Mechanics', *Scientific American*, No. 270 (May 1994), pp. 58–63; F.J. Belinfante, *A Survey of Hidden Variable Theories* (Oxford: Pergamon Press, 1973); S.V. Bhave, 'Separable Hidden Variables Theory to Explain the Einstein-Podolsky-Rosen Paradox', *British Journal for the Philosophy of Science*, Vol. 37 (1986), pp. 467–75; Peter Holland, *The Quantum Theory of Motion* (Cambridge: Cambridge University Press, 1993); J.R. Lucas and P.E. Hodgson, *Spacetime and Electromagnetism* (Oxford: Oxford University Press, 1990).

18 See Michael Dummett, *Truth and Other Enigmas* (London: Duckworth, 1978); also Michael Luntley, *Language, Logic and Experience: the case for anti-realism* (London: Duckworth, 1988); N. Tennant, *Anti-Realism and Logic* (Oxford: Clarendon Press, 1987); Crispin Wright, *Realism, Meaning and Truth* (Oxford: Blackwell, 1987) and *Truth and Objectivity* (Cambridge, MA: Harvard University Press, 1992).

19 EPR (note 3, above), p. 138.

20 Ibid, p. 138.

21 See Werner Heisenberg, *The Physical Principles of the Quantum Theory* (New York: Dover, 1949) and *Physics and Philosophy* (New York: Harper & Row, 1958); also Patrick A. Heelan, *Quantum Mechanics and Objectivity: a study of the physical philosophy of Werner Heisenberg* (The Hague: Nijhoff, 1965).

22 Wheeler and Zurek (eds.), *The Quantum Theory of Measurement* (op. cit.), p. 148.

23 Ibid, p. 148.

24 Murdoch, *Niels Bohr's Philosophy of Physics* (op. cit.), p. 170.

25 See Dummett, *Truth and Other Enigmas* (op. cit.); also A.J. Ayer (ed.), *Logical Positivism* (New York: Free Press, 1959); Oswald Hanfling (ed.), *Essential Readings in Logical Positivism* (Oxford: Blackwell, 1981); C.J. Misak, *Verificationism: its history and prospects* (London: Routledge, 1995); Bas C. van Fraassen, *The Scientific Image* (Oxford: Clarendon Press, 1980) and *Quantum Mechanics: an empiricist view* (Clarendon Press, 1992).

26 See Note 25, above; also Ernst Mach, *The Science of Mechanics: a critical and historical account of its development*, trans. T.J. McCormack (La Salle, IL: Open Court, 1960).

27 See especially Dummett, 'Is Logic Empirical?', in *Truth and Other Enigmas* (op. cit.), pp. 269–89.

28 See Notes 11, 13 and 14, above.

29 See Note 27, above.

30 EPR, p. 141.

31 See for instance Fine, *The Shaky Game* (op. cit.); also Arkady Plotnitsky, *Complementarity: anti-epistemology after Bohr and Derrida* (Durham, N.C.: Duke University Press, 1994).

32 See Notes 1, 21 and 22, above.

33 Bohr, in Wheeler and Zurek (eds.), op. cit., p. 148.

34 Ibid, p. 146.

35 See entries under Note 1, above.

36 See Erwin Schrödinger, *Letters on Wave Mechanics* (New York: Philosophical Library, 1967).

37 Dummett, 'What Is a Theory of Meaning? II', in Gareth Evans and John McDowell (eds.), *Truth and Meaning* (Oxford: Oxford University Press, 1976), p. 100.

38 Alston, *A Realist Conception of Truth* (op. cit.), p. 120.

39 Ibid, p. 113.

40 Dummett, *Truth and Other Enigmas* (op. cit.); *Frege: philosophy of language* (London: Duckworth, 1973); *Elements of Intuitionism* (Oxford: Oxford University Press, 1977); *The Interpretation of Frege's Philosophy* (Duckworth, 1981). See also entries under Note 18 (above) and Richard L. Kirkham, 'What Dummett Says about Truth and Linguistic Competence', *Mind*, Vol. 98 (1989), pp. 207–24; Alan Weir,

'Dummett on Meaning and Classical Logic', *Mind*, Vol. 95 (1986), pp. 465–77; Timothy Williamson, 'Knowability and Constructivism: the logic of anti-realism', *Philosophical Quarterly*, Vol. 38 (1988), pp. 422–32; Kenneth P. Winkler, 'Scepticism and Anti-Realism', *Mind*, Vol. 94 (1985), pp. 46–52.

41 See especially M. Gardner, 'Realism and Instrumentalism in Nineteenth-Century Atomism', *Philosophy of Science*, Vol. 46 (1979), pp. 1–34; also J. Perrin, *Atoms*, trans D.L. Hammick (New York: Van Nostrand, 1923) and Mary Jo Nye, *Molecular Reality* (London: MacDonald, 1972).

42 Alston, *A Realist Theory of Truth* (op. cit.), p. 193.

43 Ibid, p. 84.

44 EPR, p. 138.

45 Murdoch, *Niels Bohr's Philosophy of Physics* (op. cit.), p. 61.

46 EPR, p. 138.

47 EPR, p. 138.

48 EPR, p. 138.

49 In Wheeler and Zurek (eds.), *The Quantum Theory of Measurement* (op. cit.), p. 148.

50 Ibid, p. 148.

51 Ibid, p. 148.

52 Ibid, p. 148.

53 Murdoch, *Niels Bohr's Philosophy of Physics* (op. cit.), p. 232.

54 For more in this connection, see Cushing, *Quantum Mechanics* (op. cit.); Misak, *Verificationism: its history and prospects* (op. cit.); Hans Reichenbach, *Experience and Prediction* (Chicago: University of Chicago Press, 1938) and *Philosophic Foundations of Quantum Mechanics* (Berkeley & Los Angeles: University of California Press, 1944); Wesley C. Salmon, *Hans Reichenbach: logical empiricist* (Dordrecht: D. Reidel, 1979); van Fraassen, *Quantum Mechanics: an empiricist view* (op. cit.).

55 See Notes 18, 25, 27 and 40 (above).

56 See Einstein, 'Autobiographical Notes', in P.A. Schilpp (ed.), *Albert Einstein: philosopher–scientist* (op. cit.), pp. 3–105; also Fine, *The Shaky Game* (op. cit.).

57 Einstein, 'Reply to Criticisms', in Schilpp (ed.), *Albert Einstein* (op. cit.), pp. 665–88.

58 Popper, *Quantum Theory and the Schism in Physics* (op. cit.).

59 See for instance – from a range of philosophical standpoints – Donald Davidson, 'The Structure and Content of Truth', *Journal of Philosophy*, Vol. 87 (1990), pp. 279–328; Michael Devitt, *Realism and Truth*, 2nd edn. (Oxford: Blackwell, 1986); Brian Ellis, *Truth and Objectivity* (Blackwell, 1990); A.C. Ewing, *Non-Linguistic Philosophy* (London: Allen & Unwin, 1968); Paul Horwich, *Truth* (Blackwell, 1990); Lawrence E. Johnson, *Focusing on Truth* (London: Routledge, 1992); Richard L. Kirkham, *Theories of Truth* (Cambridge, MA: M.I.T. Press, 1992); D.J. O'Connor, *The Correspondence Theory of Truth* (London: Allen & Unwin, 1975); H. Price, *Facts and the Function of Truth* (Blackwell, 1988); Crispin Wright, *Realism, Meaning and Truth* (op. cit.).

60 See Dummett, *Truth and Other Enigmas* (op. cit.) and entries under Notes 25, 27, 37 and 40, above; also Richard Rorty, *Consequences of Pragmatism* (Brighton: Harvester, 1982); *Objectivity, Relativism, and Truth* (Cambridge: Cambridge University Press, 1991); *Truth and Progress* (Cambridge University Press, 1998).

61 Alston, *A Realist Conception of Truth* (op. cit.), p. 52.

62 See Note 2, above.

63 David Deutsch, *The Fabric of Reality* (Harmondsworth: Penguin, 1997); also Bryce S. DeWitt and Neill Graham (eds.), *The Many-Worlds Interpretation of Quantum Mechanics* (Princeton, NJ: Princeton University Press, 1973).

64 A. Aspect, P. Grangier and C. Roger, 'Experimental Realization of the E–P–R Paradox', *Physical Review*, Vol. 48 (1982), pp. 91–4; J.S Bell, *Speakable and*

Unspeakable in Quantum Mechanics (op. cit.); Cushing and McMullin (eds.), *Philosophical Consequences of Quantum Theory* (op. cit.).

65 See for instance J.A. Wheeler, 'Delayed Choice Experiments and the Bohr-Einstein Dialogue', paper presented at the joint meeting of the American Philosophical Society and the Royal Society, London, June 5th, 1980; also F. Selleri, 'Wave–Particle Duality: recent proposals for the detection of empty waves', in W. Schommers (ed.), *Quantum Theory and Pictures of Reality: foundations, interpretations, and new aspects* (Berlin: Springer Verlag, 1989), pp. 279–32; Wheeler and Zurek (eds.), *Quantum Theory and Measurement* (op. cit.).

66 Wheeler and Zurek (eds.), *Quantum Theory and Measurement*, p. 191.

67 Ibid, p. 194.

68 P.H. Eberhard, 'The EPR Paradox: roots and ramifications', in Schommers (ed.), *Quantum Theory and Pictures of Reality* (op. cit.), pp. 48–88; p. 77.

69 See Note 11, above.

70 See Notes 3, 64 and 65, above.

71 Wheeler and Zurek (eds.), *Quantum Theory and Measurement* (op. cit.), p. 189.

72 See Note 65, above.

73 See Erwin Schrödinger, *Letters on Wave Mechanics* (op. cit.); also John Gribbin, *In Search of Schrödinger's Cat: quantum physics and reality* (New York: Bantam Books, 1984).

74 Wheeler, in Wheeler and Zurek (eds.), *Quantum Theory and Measurement* (op. cit.), pp. 183–4.

75 Ibid, p. 192.

76 Dummett, 'Can an Effect Precede its Cause', 'Bringing About the Past', 'A Defence of McTaggart's Proof of the Unreality of Time', and 'The Reality of the Past', in *Truth and Other Enigmas* (op. cit.), pp. 319–32, 333–50, 351–7, and 358–74.

77 Wheeler, in Wheeler and Zurek (eds.), *Quantum Theory and Measurement* (op. cit.), p. 194.

78 Ibid, p. 194.

79 Ibid, p. 194.

80 Ibid, p. 189.

81 See the various contributions by Nelson Goodman – along with the various responses by his critics – in Peter J. McCormick (ed.), *Starmaking: realism, anti-realism, and irrealism* (Cambridge, MA: M.I.T. Press, 1996); also Goodman, *Ways of Worldmaking* (Indianapolis: Bobbs-Merrill, 1978) and *Problems and Projects* (Bobbs-Merrill, 1972).

82 See for instance Hilary Putnam, *Representation and Reality* (Cambridge: Cambridge University Press, 1988); *The Many Faces of Realism* (La Salle, IL: Open Court, 1987); *Realism With a Human Face* (Harvard University Press, 1990); *Renewing Philosophy* (Harvard University Press, 1992).

Index of names

Adorno, T.W. 144, 162n
Agazzi, E. 37n
Albert, D.Z. 36–7n, 68n, 161n, 223n
Alston, W.P. 38n, 40–1, 67n, 69n, 214, 215, 228n, 240–3, 251–2, 259n
Angel, R.B. 133n
Anscombe, G.E.M. 145–6, 162n
Aristotle 33, 62, 141, 147, 155, 156
Armstrong, D.M. 38n, 70n, 190n, 229n
Aspect, A. 22–3, 37n, 64, 88, 190n, 206, 227n, 261n
Audi, M. 131n
Augustine, St. 141
Ayer, A.J. 101n, 212–13, 229n
Ayers, M.R. 191n, 226n

Bacon, F. 155
Baker, G.P. 180, 192n
Barnes, B. 69n, 103n
Barrow, J. 68n
Beedle, V. 133n, 226n
Belinfante, F.J. 37n
Bell, J.S. 3, 5, 14, 16–24, 29, 36n, 49, 64–65, 68n, 71n, 74, 80, 83, 85–93, 95, 102n, 106, 165, 182, 185–6, 200–1, 205, 206, 219, 227n, 233–35, 237, 240, 244, 252, 254, 259n
Bellarmine, Cardinal 183
Beltrametti, E.G. 104n
Berkeley, G. 24, 31, 45, 113, 116, 117, 160, 212, 235
Bernstein, R.J. 225n
Bhave, S.H. 37n
Bhaskar, R. 38n, 67n, 70n, 228n
Birkhoff, G. 192n
Bloor, D. 69n, 70n, 103n, 191n
Bohm, D. 2–5, 14–16, 20, 26–7, 30, 32–3, 35, 36–7n, 45, 61, 62, 68n, 74–81, 83, 86–93, 95–6, 98, 100–1, 102n, 108–9, 113–14, 120–5, 129, 132n,

134–5, 142–4, 146, 159, 161n, 165, 182, 185, 188n, 198–201, 203, 204, 222, 225n, 233–5, 238, 240, 244, 250, 252, 254, 259n
Bohr, N. 2–5, 8–18, 20, 21, 24, 25, 30–2, 36n, 41, 44, 48, 50–1, 60, 63, 66, 67n, 70n, 73–4, 77, 79, 81, 82, 83–87, 89, 91–2, 97–100, 102n, 105, 111–12, 125, 128, 131, 132n, 134, 152, 160n, 165, 176, 183, 185, 186, 194, 196, 198–204, 207, 222, 224n, 233–40, 242–52, 254, 255, 258n, 259n
Boltzmann, L. 41
Borges, J.-L. 135
Born, M. 93
Boswell, J. 116
Boyd, R. 182, 193n
Braithwaite, R.B. 101n
Brentano, F. 138
Brown, J. 143
Brown, J.R. 37n, 104n, 164n
Bruner, J.S. 161n, 190n
Bruno, G. 205
Burtt, E.A. 131, 133n

Callinicos, A. 161n
Calvino, I. 135
Carl, W. 163n
Carnap, R. 101n, 133n, 162n, 225n
Chisholm, R.M. 190n
Churchland, P. 29, 38n
Clarke, S. 152, 163n
Clauser, J.F. 234–5, 259n
Coffa, J.A. 36n, 71n, 104n, 133n
Copernicus, N. 6, 48, 160, 183
Cushing, J. 37n, 67n, 74–81, 87, 93–6, 99–100, 102n, 144, 161n, 163n, 200, 201, 203, 204, 225n, 259n

Dalton, J. 34, 49, 154, 170, 178, 179